Andrew Reid, Pierre Joseph Macquer

Elements of the Theory and Practice of Chymistry

Vol. 1

Andrew Reid, Pierre Joseph Macquer

Elements of the Theory and Practice of Chymistry
Vol. 1

ISBN/EAN: 9783337105815

Printed in Europe, USA, Canada, Australia, Japan

Cover: Foto ©berggeist007 / pixelio.de

More available books at **www.hansebooks.com**

ELEMENTS

OF THE

Theory and Practice

OF

CHYMISTRY.

TRANSLATED

From the French of M. Macquer.

Member of the Royal Academy of Sciences, and Pro-
feſſor of Medicine in the Univerſity of Paris.

IN TWO VOLUMES.

VOL. I.

THE THIRD EDITION.

LONDON,
Printed for J. Nourse, W. Strahan,
J. and F. Rivington, T. Longman,
T. Cadell, and E. Johnston.
MDCCLXXV.

RIGHT HONOURABLE

THE

Earl of BUTE.

My LORD,

THIS Tranſlation of M. MACQUER's celebrated ELEMENTS of CHYMISTRY was undertaken with the ſole View of rendering ſome ſmall Service to my Country; as I hoped it might contribute both to recommend and to facilitate the Study of a SCIENCE, which, though more entertaining, inſtructive, and extenſively uſeful than any other, hath of late been too much neglected in this Iſland.

I think

DEDICATION.

I think myself very happy in seeing this
Design approved of, and its Success in a
manner assured, by the Honour which
Your Lordship hath been pleased to do me,
in condescending to take it under your Pro-

THE

AUTHOR'S

PREFACE.

AN hundred and fifty years are scarce elapsed since the clouds of prejudice, which had long overspread the world, began to clear up, and men were convinced, by cultivating the Sciences and attending to Nature, that no fanciful hypotheses would ever lead them to the true causes of those various phenomena that inceffantly and every where meet the obferver's eye; but that the narrow limits of the human underftanding confine the courfe of our refearches to one fingle path; namely, that of Experiment, or the Ufe of our Senfes. Yet, in this fhort period, Natural Philofophy hath rifen to a high pitch of improvement, and may with truth be faid to have made much greater advances towards perfection, fince the experimental method was introduced, than in the many ages before.

This is true with regard to every branch of Natural Philofophy; but more particularly with regard to Chymiftry. Though this Science cannot be faid to have ever exifted without experiments, yet it laboured under the fame difadvantages with the reft; becaufe thofe who ftudied it made all their experiments with a view to confirm their own Hypothefes, and in confequence of principles which had no foundation whatever, but in their wild imaginations.

Hence

Hence arofe that enormous heap, that incongru-
ous jumble of facts, which fome time ago conftituted all the knowledge of Chymifts. Moft of
them, and efpecially thofe who affumed the pom-
pous title of Alchymifts, were perfuaded that all
the Metals were no other than Nature's rude unfi-
nifhed effays towards making Gold; which, by
means of due coction in the bowels of the earth,
advanced gradually towards maturity, till at laft
they were perfectly converted into that beautiful
and precious Metal.

On this principle, which, if not demonftrably
falfe, is at leaft utterly deftitute of proof, and un-
fupported by a fingle obfervation, they attempted
to finifh what Nature had begun, by procuring to
the imperfect Metals this much defired coction.
To attain it they made an infinite number of experiments and trials; which all confpired to detect
the falfity of their fyftem, and to fatisfy men of
fenfe, that the methods they employed were very
far from anfwering the purpofe.

However, as facts always promote the knowledge of Nature, it happened that thofe experiments, though quite ufelefs with regard to the end
for which they were originally made, proved the
occafion of feveral curious difcoveries.

Thefe lucky confequences of their miftaken labours raifed the courage of the Chymifts, or rather Alchymifts, who looked upon every fuch inftance of fuccefs as a new ftep towards the Grand
Work, and greatly increafed the fond opinion they
entertained of themfelves, and of their Art, which
on that account they fet up very high above all
other Sciences. Nay, they carried this notion of
fuperiority fo far, as to hold the reft of mankind
unworthy, or incapable, of rifing to fuch fublime
knowledge. In confequence thereof Chymiftry
became an occult and myfterious Science; its ex-

preffions

preffions were all tropes and figures, its phrafes metaphorical, and its axioms fo many enigmas : in fhort, an obfcure unintelligible jargon is the jufteft character of the Alchymiftic language.

Thus, by endeavouring to conceal their fecrets, thofe gentlemen rendered their art ufelefs to man-kind, and brought it into deferved contempt. But at length the genius of true Philofophy prevailed in Chymiftry, as well as in the other fciences. Some great men arofe, who had generofity enough to think their knowledge no otherways valuable than as it proved of fervice to Society. They did their utmoft to introduce both the knowledge and the practice of many important fecrets, till then of no ufe; they drew afide the veil which hid the charms of Chymiftry; and that Science emerging from the profound obfcurity, in which it had for many ages lain concealed, gained the admiration of the world as foon as it appeared in open day. Several focie-ties of ingenious men were formed in the moft learned countries of Europe, who vied with one another in their labours to execute the noble fcheme, and affifted each other by mutually com-municating their difcoveries. Chymiftry made the moft rapid progrefs, enriching and perfecting the Arts derived from, or depending on it. In a word, it put on a new face, and became truly wor-thy of the title of Science; founding its principles and its proceffes on folid experiments, and on juft confequences deduced from them.

Since that time the Art is become fo extenfive, by the numerous difcoveries which Chymifts have already made, and are daily making, that large volumes are required to contain a complete Treatife on the fubject. In fhort, Chymiftry may now, in fome degree, be compared to Geometry : each of thefe Sciences takes in a moft ample field

of

of enquiry, which every day enlarges very confiderably ; from each are derived feveral Arts, not only ufeful but even neceffary to Society ; each hath its Axioms and its undeniable principles, either demonftrated from internal evidence, or founded on conftant experience ; fo that the one, as well as the other, may be reduced to certain fundamental truths, on which all the reft are built. Thefe fundamental truths connected together, and laid down with order and precifion, form what we call the Elements of a Science. It is well known that there are many fuch works relating to Geometry, but it is not fo with regard to Chymiftry ; there being very few books which treat of this Science in an Elementary manner.

Yet it muft be owned that performances of this kind are exceedingly ufeful. Many who have a relifh for the Sciences, but have not leifure to read elaborate Works which treat of them minutely, are glad to meet with a book from which, without facrificing too much of their time, or neglecting their ordinary bufinefs, they may obtain a tafte or juft notion of a Science that is not their principal ftudy. Thofe who incline to go further, and learn more, may, by reading an Elementary tract, be enabled to underftand Authors who, as they commonly write only for proficients in the Art, are obfcure and hardly intelligible to mere beginners. Nay, I prefume to fay that an Elementary Treatife of Chymiftry may prove a very ufeful book, even to thofe who have made fome progrefs in the Science : for as it contains only the fundamental propofitions, and indeed is an abftract of the whole Art, it may help them to recollect the moft important parts of what they have read in many different works, and fix in their memories the moft effential truths, which might elfe be either confounded with others, or entirely forgot. And thefe

are

are the motives which determined me to compose the Work which I now offer to the Publick.

The general Plan on which I proceed is to suppose my Reader an absolute Novice in Chymistry; to lead him from the most simple truths, and such as imply the lowest degree of knowledge, to such as are more complex, and require a greater acquaintance with Nature. This order, which I have laid down for my rule, hath obliged me to begin with examining the most simple substances that we know, and which we consider as the elements whereof others are composed; as, by knowing the properties of these elementary parts, we are naturally led to those of their several combinations; and on the other hand, in order to know the properties of compound bodies, it is necessary we should be first acquainted with the properties of their principles. The same reason induced me, when enquiring into the properties of one substance, to take no notice of those which relate to any other substance not treated of before. For example: as I treat of Acids before Metals, I say nothing under the head of those Acids concerning their power of dissolving Metals: that I defer till I come to the subject of Metals: and thus I avoid speaking prematurely of a substance with which I suppose my Reader wholly unacquainted. And this method I was so much the more easily induced to follow, that I know of no Chymical book written on the same Plan.

After discoursing of Elements in general, I treat next of such substances as are immediately composed of them, and are, next to them, the most simple: such are all saline substances. This head comprehends mineral Acids, fixed Alkalis, and their several combinations: the volatile sulphureous spirit, sulphur, phosphorus, and the Neutral salts which have an earth or fixed Alkali for their basis:

6

those

thofe which have for their bafis either a volatile
Alkali, or fome metallic fubftance, are referred,
according to my general Plan, to the heads under
which I treat of thofe fubftances.

Metallic fubftances are fcarcely more compounded than the faline; which induces me to confider
them next. I begin with thofe which are the moft
fimple, or at leaft feem to be fo; becaufe their principles, being very ftrongly connected together, are
feparated with the greateft difficulty : fuch are the
Metals properly fo called ; namely Gold, Silver,
Copper, Iron, Tin, and Lead. After thefe come
the Semi-metals in order; to wit, Regulus of
Antimony, Zinc, Bifmuth, and Regulus of Arfenic. Mercury being a doubtful fubftance, which
fome Chymifts rank with the Metals, and others
with the Semi-metals, becaufe it actually poffeffes
certain properties in common with each, I have
treated of it in a feparate Chapter, which ftands
between the Metals and Semi-metals.

I next proceed to examine the feveral forts of
Oils, whether Vegetable, which are divided into
fat, effential, and empyreumatic ; or Animal, and
Mineral Oils.

By examining thefe fubftances we obtain ideas of
all the principles which enter into the compofition
of Vegetable and Animal bodies ; that is, of thofe
fubftances that are capable of fermentation : this
enables me to treat of fermentation in general ; of
its three different degrees or kinds, the fpirituous,
acetous, and putrid ; and of the products of thofe
fermentations, ardent fpirits, acids analogous to
thofe of vegetables and animals, and volatile alkalis.

The order in which I treat of all thofe fubftances
being different from that in which they are obtained
from compound bodies, I give, in a diftinct chapter, a general Idea of Chymical Decompofition,
with a view to fhew the order in which they are
feparated

feparated, from the feveral bodies in the compofition whereof they are found. This brings them
a fecond time under review, and gives me an opportunity of diftinguifhing thofe which exift naturally in compound bodies, from thofe which are
only the refult of a new combination of fome of
their principles produced by the fire.

The fucceeding Chapter explains the late Mr.
Geoffroy's Table of Affinities; which I take to be
of great ufe at the end of an Elementary Tract like
this, as it collects into one point of view the moft
effential and fundamental doctrines which are difperfed through the work.

I conclude with an account of the Conftruction
of fuch Veffels and Furnaces as are ufually employed in Chymiftry.

In this Part I fay nothing of any manual operations, or the feveral ways of performing Chymical
proceffes; referving thefe particulars for my Treatife of Practical Chymiftry, to which this muft be
confidered as an Introduction.

Elements of the THEORY of Chymiſtry.

Elements of the PRACTICE of Chymistry.

2. To

CONTENTS.

THEORY of CHYMISTRY.

CHAP. I.

Of the PRINCIPLES *of Bodies.*

THE Object and chief End of Chymiftry is to feparate the different fubftances that enter into the compofition of bodies; to examine each of them apart; to difcover their properties and relations; to decompofe thofe very fubftances, if poffible; to compare them together, and combine them with others; to reunite them again into one body, fo as to reproduce the original compound with all its properties; or even to produce new compounds that never exifted among the works of nature, from mixtures of other matters differently combined.

But this Analyfis, or Decompofition, of Bodies is finite; for we are unable to carry it beyond a certain limit. In whatever way we attempt to go

VOL. I.   B   further;

further, we are always flopped by fubftances in
which we can produce no change, which are incapable of being refolved into others, and which
ftand as fo many firm barriers obftrucfing our progrefs.

To thefe fubftances we may, in my opinion,
give the title of Principles or Elements : at 'leaft
they are really fuch with regard to us. Of this
kind the principal are Earth, Water, Air, and .
Fire. For though there be reafon to think that thefe
are not the firft component parts, or the moft fimple
elements, of matter ; yet, as we know by experience, that our fenfes cannot poffibly difcover
the principles of which they are compofed, it feems
more reafonable to fix upon them, and confider
them as fimple homogeneous bodies, and the
principles of the reft, than to tire our minds with
vain conjectures about the parts or elements of
which they may confift ; feeing there is no criterion
by which we can know whether we have hit upon
the truth, or whether the notions we have formed
are mere fancies. We fhall therefore confider thefe
four fubftances as the principles or elements of all
the various compounds which nature prefents to our
enquiries : becaufe of all thofe we know they are in
fact the moft fimple ; and becaufe all our decompofitions, all our experiments on other bodies, plainly prove that they are at laft refolvable into thefe
primary parts.

These principles do not enter in the fame proportion into all bodies : there are even fome mixts in
the compofition of which this or that particular
principle is not to be found. Thus Air and Water feem to be wholly excluded from the texture of
Metals : at leaft all the experiments hitherto made
on them feem to eftablifh this opinion.

The fubftances compofed immediately of thefe
Firft Elements we fhall call *Secondary* Principles ;
becaufe

becaufe in reality their feveral combinations with each other, the interchangeable coalitions that take place between them, conftitute the different natures of all other bodies ; which, as they refult from the union both of primary and fecondary principles, are properly entitled to the name of Compounds or Mixts.

Before we enter upon the examination of Compound Subftances, it is neceffary to confider the moft Simple ones, or our four Firft Principles, with fome attention, in order to difcover their chief properties.

§. I. *Of* Air.

Air is that Fluid which we conftantly breathe, and which encompaffes the whole furface of the terreftrial globe. Being heavy, like all other bodies, it penetrates into all places that are not either abfolutely inacceffible, or filled with fome other body heavier than itfelf. Its principal property is to be fufceptible of condenfation and rarefaction ; fo that the very fame quantity of Air may occupy a much greater, or a much fmaller fpace, according to the different ftate it is in. Heat and cold, or, if you will, the prefence and the abfence of the particles of Fire, are the moft ufual caufes, and indeed the meafures, of its condenfation and rarefaction : for if a certain quantity of Air be heated, its bu'k enlarges in proportion to the degree of heat applied to it ; the confequence whereof is, that the fame fpace now contains fewer particles of Air than it did before. Cold again produces juft the contrary effect.

On this property which Air has, of being condenfed and dilated by heat, its elafticity chiefly depends. For if Air were forced by condenfation into a lefs compafs than it took up before, and then expofed to a very confiderable degree of cold, it would remain quite inactive, without exert-

ing fuch an effort as it ufually makes againft the
compreffing body. On the other hand, the elafticity
of heated Air arifes only from hence, that being
rarefied by the action of fire, it requires much
more room than it occupied before.

Air enters into the compofition of many fubftances, efpecially vegetable and animal bodies : for
by analyfing moft of them fuch a confiderable quantity thereof is extricated, that fome naturalifts have
fufpected it to be altogether deftitute of elafticity
when thus combined with the other principles in the
compofition of bodies. According to them the efficacy of the elaftic power of the Air is fo prodigious, and its force when compreffed fo exceffive, that
it is not poffible the other component parts of bodies fhould be able to confine fo much of it, in that
ftate of compreffion which it muft needs undergo, if
retaining its elafticity it were pent up among them.

However that be, this elaftic property of the Air
produces the moft fingular and important phenomena, obfervable in the refolution and compofition of bodies.

§. II. *Of* W A T E R.

W A T E R is a thing fo well known, that it is
almoft needlefs to attempt giving a general idea of
it here. Every one knows that it is a tranfparent,
infipid fubftance, and ufually fluid. I fay it is
ufually fo ; for being expofed to a certain degree of
cold it becomes folid : folidity therefore feems to
be its moft natural ftate.

Water expofed to the fire grows hot ; but only
to a limited degree, beyond which its heat never
rifes, be the force of fire applied to it ever fo violent : it is known to have acquired this degree of
heat by its boiling up with great tumult. Water
cannot be made hotter, becaufe it is volatile, and
incapable of enduring the heat, without being evaporated and entirely diffipated.

If

If fuch a violent and fudden heat be applied to
Water, as will not allow it time to exhale gently in
vapours, as when, for inftance, a fmall quantity
thereof is thrown upon a metal in fufion, it is diffipated at once with vaft impetuofity, producing a
moft terrible and dangerous explofion. This furprifing effect may be deduced from the inftantaneous dilatation of the parts of the Water itfelf, or rather of the Air contained in it. Moreover, Water
enters into the texture of many bodies, both compounds and fecondary principles; but, like Air, it
feems to be excluded from the compofition of all
metals and moft minerals. For although an im-
menfe quantity of Water exifts in the bowels of the
Earth, moiftening all its contents, it does not therefore follow that it is one of the principles of minerals. It is only interpofed between their parts; for
they may be entirely robbed of it, without any fort
of decompofition : indeed it is not capable of an
intimate connection with them.

§. III. *Of* EARTH.

We obferved that the two principles above treated of are volatile; that is, the action of Fire feparates them from the bodies they help to compofe,
carrying them quite off, and diffipating them.
That of which we are now to fpeak, namely Earth,
is fixed, and, when it is abfolutely pure, refifts the
utmoft force of Fire. So that, whatever remains of
a body, after it hath been expofed to the power of
the fierceft Fire, muft be confidered as containing
nearly all its earthy principle, and confifting chiefly
thereof. I qualify my expreffion thus for two reafons : the firft is, becaufe it often happens, that this
remainder does not actually contain all the Earth
which exifted originally in the mixt body decompofed by Fire ; fince it will afterwards appear that
Earth, though in its own nature fixed, may be rendered

dered volatile by being intimately united with other
fubftances which are fo; and that, in fact, it is
common enough for part of the Earth of a body to
be thus volatilized by its other principles: the fecond is, that what remains after the calcination of
a body is not generally its Earth in perfect purity,
but combined with fome of its other principles,
which, though volatile in their own natures, have
been fixed by the union contracted between it and
them.   We fhall, in the fequel, produce fome ex-
amples to illuftrate this theory.

Earth therefore, properly fo called, is a fixed
principle, which is permanent in the Fire. There is
reafon to think it very difficult, if not impoffible, to
obtain the terrene principle wholly free from every
other fubftance : for after our utmoft endeavours to
purify them, the Earths we obtain from different
compounds are found to have different properties,
according to the different bodies from which they
are procured ; or elfe, if thofe Earths be pure, we
muft allow them to be effentially different, feeing
they have different properties.

Earth, in general, with regard to its properties,
may be diftributed into *fufible*, and *unfufible* ; that
is, into Earth that is capable of melting or becoming fluid in the Fire, and Earth that conftantly
remains in a folid form, never melting in the ftrongeft degree of heat to which we can expofe it.

The former is alfo called *vitrifiable*, and the fecond *unvitrifiable* Earth ; becaufe, when Earth is
melted by the force of fire, it becomes what we
call *glafs*, which is nothing but the parts of Earth
brought into nearer contact, and more clofely united
by the means of fufion.   Perhaps the Earth, which
we look upon as uncapable of vitrification, might
be fufed if we could apply to it a fufficient degree
of heat.   It is at leaft certain that fome Earths, or
ftones, which feparately refift the force of Fire fo
that

that they cannot be melted, become fufible when mixed together. Experience convinced Mr. du Hamel that lime-ftone and flate are of this kind. It is however undoubtedly true that one Earth differs from another in its degree of fufibility : and this gives ground to believe, that there may be a fpecies of Earth abfolutely unvitrifiable in its nature, which, being mixed in different proportions with fufible Earths, renders them difficult to melt.

Whatever may be in this, as there are Earths which we are abfolutely unable to vitrify, that is a fufficient reafon for our divifion of them. Unvitrifiable Earths feem to be porous, for they imbibe Water ; whence they have alfo got the name of *Abforbent Earths,*

§. IV. *Of* FIRE.

THE Matter of the Sun, or of Light, the Phlogifton, Fire, the Sulphureous Principle, the Inflammable Matter, are all of them names by which the Element of Fire is ufually denoted. But it fhould feem, that an accurate diftinction hath not yet been made between the different ftates in which it exifts ; that is, between the phenomena of Fire actually exifting as a principle in the compofition of bodies, and thofe which it exhibits when exifting feparately and in its natural ftate : nor have proper diftinct appellations been affigned to it in thofe different circumftances. In the latter ftate we may properly give it the names of Fire, Matter of the Sun, of Light, and of Heat ; and may confider it as a fubftance compofed of infinitely fmall particles, continually agitated by a moft rapid motion, and of confequence, effentially fluid.

This fubftance, of which the fun may be called the general refervoir, feems to flow inceffantly from that fource, diffufing itfelf over the world, and through all the bodies we know ; but not as a principle,

ciple,

ciple, or effential part of them, fince they may be
deprived thereof, at leaft in a great meafure, without fuffering any decompofition. The greateft
change produced on them, by its prefence or its abfence, is the rendering them fluid or folid : fo that
all other bodies may be deemed naturally folid; Fire
alone effentially fluid, and the principle of fluidity
in others. This being prefuppofed, Air itfelf might
become folid, if it could be entirely deprived of the
Fire it contains; as bodies of moft difficult fufion
become fluid, when penetrated by a fufficient quantity of the particles of Fire.

One of the chief properties of this pure Fire is to
penetrate eafily into all bodies, and to diffufe itfelf
among them with a fort of uniformity and equality :
for if a heated body be contiguous to a cold one, the
former communicates to the latter all its excefs of
heat, cooling in exact proportion as the other warms,
till both come to have the very fame degree of
heat. Heat, however, is naturally communicable
foonest to the upper parts of a body ; and confequently, when a body cools, the under parts become
foonest cold. It hath been obferved, for inftance,
that the lower extremity of a heated body, freely
fufpended in the air, grows cold fooner than the
upper; and that when a bar of iron is red-hot at one
end, and cold at the other, the cold end is much
fooner heated by placing the bar fo that the hot
end may be undermoft, than when that end is
turned uppermoft. The levity of the matter of
Fire, and the vicinity of the Earth, may poffibly be
the caufes of this phenomenon.

Another property of Fire is to dilate all bodies
into which it penetrates. This hath already been
fhewn with regard to Air and Water ; and it produces the fame effect on Earth.

Fire is the moft powerful agent we can employ
to decompofe bodies ; and the greateft degree of
heat

heat producible by man, is that excited by the rays of the sun collected in the focus of a large burning-glass.

§. V. *Of the* PHLOGISTON.

FROM what hath been said concerning the nature of Fire, it is evidently impossible for us to fix and confine it in any body. Yet the phenomena attending the combustion of inflammable bodies shew, that they really contain the matter of Fire as a constituent principle. By what mechanism then is this fluid, which is so subtle, so active, so difficult to confine, so capable of penetrating into every other substance in nature; how comes it, I say, to be so fixed as to make a component part of the most solid bodies? It is no easy matter to give a satisfactory answer to this question. But, without pretending to guess the cause of the phenomenon, let us rest contented with the certainty of the fact, the knowledge of which will undoubtedly procure us considerable advantages. Let us therefore examine the properties of Fire thus fixed and become a principle of bodies. To this substance, in order to distinguish it from pure and unfixed Fire, the Chymists have assigned the peculiar title of the *Phlogiston*, which is indeed no other than a Greek word for the Inflammable Matter; by which latter name, as well as by that of the Sulphureous Principle, it is also sometimes called. It differs from elementary Fire in the following particulars. 1. When united to a body, it communicates to it neither heat nor light. 2. It produces no change in its state, whether of solidity or fluidity; so that a solid body does not become fluid by the accession of the Phlogiston, and *vice versa*; the solid bodies to which it is joined being only rendered thereby more apt to be fused by the force of the culinary fire. 3. We can convey it from the body, with which it is joined, into another body,

body, fo that it fhall enter into the compofition thereof, and remain fixed in it.

On this occafion both thefe bodies, that which is deprived of the Phlogifton and that which receives it, undergo very confiderable alterations; and it is this laft circumftance, in particular, that obliges us to diftinguifh the Phlogifton from pure Fire, and to confider it as the element of Fire combined with fome other fubftance, which ferves it as a bafis for conftituting a kind of fecondary principle. For if there were no difference between them, we fhould be able to introduce and fix pure Fire itfelf, where-ever we can introduce and fix the Phlogifton: yet this is what we can by no means do, as will appear from experiments to be afterwards produced.

Hitherto Chymifts have never been able to obtain the Phlogifton quite pure, and free from every other fubftance: for there are but two ways of feparating it from a body of which it makes a part; to wit, either by applying fome other body with which it may unite the moment it quits the former; or elfe by calcining and burning the compound from which you defire to fever it. In the former cafe it is evi-dent that we do not get the Phlogifton by itfelf, becaufe it only paffes from one combination into another; and in the latter, it is entirely diffipated in the decompofition, fo that no part of it can pof-fibly be fecured.

The inflammability of a body is an infallible fign that it contains a Phlogifton; but from a body's not being inflammable, it cannot be inferred that it contains none: for experiments have demonftra-ted that certain metals abound with it, which yet are by no means inflammable.

We have now delivered what is moft neceffary to be known concerning the principles of bodies in general. They have many other qualities befides thofe above-mentioned; but we cannot properly

take

take notice of them here, becaufe they prefuppofe
an acquaintance with fome other things relating to
bodies, of which we have hitherto faid nothing; in-
tending to treat of them in the fequel as occafion
fhall offer. We fhall only obferve in this place,
that when animal and vegetable matters are burnt,
in fuch a manner as to hinder them from flaming,
fome part of the Phlogifton contained in them
unites intimately with their moft fixed earthy parts,
and with them forms a compound, that can be confumed only by making it red-hot in the open air,
where it fparkles and waftes away, without emitting
any flame. This compound is called a *Coal*. We
fhall enquire into the properties of this Coal under
the head of Oils : at prefent it fuffices that we know
in general what it is, and that it readily communicates to other bodies the Phlogifton it contains.

CHAP. II.

A general View of the Relations or Affinities between Bodies.

BEFORE we can reduce compound Bodies
to the firft principles above pointed out, we
obtain, by analyfing them, certain fubftances which
are indeed more fimple than the bodies they helped
to compofe, yet are themfelves compofed of our primary principles. They are therefore at one and the
fame time both principles and compounds; for
which reafon we fhall, as was before faid, call them
by the name of Secondary Principles. Saline and
oily matters chiefly conftitute this clafs. But before
we enter upon an examination of their properties,
it is fit we lay before the Reader a general view of
what Chymifts underftand by the Relations or Affinities

ties of Bodies ; becaufe it is neceffary to know thefe, in order to a diftinct conception of the different combinations we are to treat of.

All the experiments hitherto made concur with daily obfervation to prove that different bodies, whether principles or compounds, have fuch a mutual Conformity, Relation, Affinity, or Attraction, if you will call it fo, as difpofes fome of them to join and unite together, while they are incapable of contracting any union with others. This effect, whatever be its caufe, will enable us to account for, and connect together, all the phenomena that Chymiftry produces. The nature of this univerfal affection of matter is diftinctly laid down in the following propofitions.

Firft, If one fubftance hath any Affinity or conformity with another, the two will unite together, and form one compound.

Secondly, It may be laid down as a general rule, that all fimilar fubftances have an Affinity with each other, and are confequently difpofed to unite ; as water with water, earth with earth, &c.

Thirdly, Subftances that unite together lofe fome of their feparate properties ; and the compounds refulting from their union partake of the properties of thofe fubftances which ferve as their principles.

Fourthly, The fimpler any fubftances are, the more perceptible and confiderable are their Affinities : whence it follows, that the lefs bodies are compounded, the more difficult it is to analyfe them ; that is, to feparate from each other the principles of which they confift.

Fifthly, If a body confift of two fubftances, and to this compound be prefented a third fubftance that has no Affinity at all with one of the two primary fubftances aforefaid, but has a greater Affinity with the other than thofe two fubftances have with each other, there will enfue a decompofition, and a
new

new union; that is, the third substance will separate the two compounding substances from each other, coalesce with that which has an Affinity with it, form therewith a new combination, and disengage the other, which will then be left at liberty, and such as it was before it had contracted any union.

Sixthly, It happens sometimes that when a third substance is presented to a body consisting of two substances, no decomposition follows; but the two compounding substances, without quitting each other, unite with the substance presented to them, and form a combination of three principles: and this comes to pass when that third substance has an equal, or nearly equal, Affinity with each of the compounding substances. The same thing may also happen even when the third substance hath no Affinity but with one of the compounding substances only. To produce such an effect, it is sufficient that one of the two compounding substances have to the third body a Relation equal, or nearly equal, to that which it has to the other compounding substance with which it is already combined. Thence it follows, that two substances, which, when apart from all others, are incapable of contracting any union, may be rendered capable of incorporating together in some measure, and becoming parts of the same compound, by combining with a third substance with which each of them has an equal Affinity.

Seventhly, A body, which of itself cannot decompose a compound consisting of two substances, because, as we just now said, they have a greater Affinity with each other than it has with either of them, becomes neverthelefs capable of separating the two by uniting with one of them, when it is itself combined with another body, having a degree of Affinity with that one, sufficient to compensate its own want thereof. In that case there are two Affinities, and thence ensues a double decomposition and a double combination.

Thefe

These fundamental truths, from which we shall deduce an explanation of all the phenomena in Chymistry, will be confirmed and illustrated by applying them, as we shall do, to the several cases, of which our design in this Treatise obliges us to give a circumstantial account.

CHAP. III.

Of Saline Substances in general.

IF a particle of water be intimately united with a particle of earth, the result will be a new compound, which, according to our third proposition of Affinities, will partake of the properties of earth and of water; and this combination principally forms what is called a *Saline Substance.* Consequently every saline Substance must have an affinity with earth and with water, and be capable of uniting with both or either of them, whether they be separate or mixed together: and accordingly this property characterises all Salts, or Saline Substances, in general.

Water being volatile and Earth fixed, Salts in general are less volatile than the former, and less fixed than the latter; that is, fire, which cannot volatilize and carry off pure earth, is capable of rarefying and volatilizing a Saline Substance; but then this requires a greater degree of heat than is necessary for producing the same effects on pure water.

There are several sorts of Salts, differing from one another, in respect either of the quantity, or the quality of the earth in their composition; or lastly, they differ on account of some additional principles, which not being combined with them in sufficient quantity to hinder their Saline properties from appearing,

permit

permit them to retain the name of Salts, though
they render them very different from the simplest
Saline Substances.

It is easy to infer, from what has been said of Salts
in general, that some of them must be more, some
less, fixed or volatile than others, and some more,
some less, disposed to unite with water, with earth,
or with particular sorts of earth, according to the
nature or the proportion of their principles.

Before we proceed further, it is proper just to
mention the principal reasons, which induce us to
think that every Saline Substance is actually a combination of earth and water, as we supposed at our
entering on this subject. The first is, the conformity
Salts have with earth and water, or the properties
they possess in common with both. Of these properties we shall treat fully, as occasion offers to consider them, in examining the several sorts of Salts.
The second is, that all Salts may be actually resolved into earth and water by sundry processes;
particularly by repeated dissolution in water, evaporation, desiccation, and calcination. Indeed the
Chymists have not yet been able to procure a Saline
Substance, by combining earth and water together.
This favours a suspicion, that, besides these two,
there is some other principle in the composition of
Salts, which escapes our researches, because we
cannot preserve it when we decompose them: but
it is sufficient to our purpose, that water and earth
are demonstrably amongst the real principles of Saline Substances, and that no experiment hath ever
shewn us any other.

§. I. *Of* ACIDS.

OF all Saline Substances, the simplest is that called
an *Acid*, on account of its taste; which is like that of
verjuice, sorrel, vinegar, and other sour things, which
for the same reason are also called Acids. By this

peculiar

peculiar tafte are Acids chiefly known. They have moreover the property of turning all the blue and violet colours of vegetables red, which diftinguifhes them from all other falts.

The form, under which Acids moft commonly appear, is that of a tranfparent liquor; though folidity is rather their natural ftate. This is owing to their affinity with water; which is fo great that, when they contain but juft as much of it as is neceffary to conftitute them Salts, and confequently have a folid form, they rapidly unite therewith the moment they come into contact with it: and as the air is always loaded with moifture and aqueous vapours, its contact alone is fufficient to liquify them; becaufe they unite with its humidity, imbibe it greedily, and by that means become fluid. We therefore fay, they attract the moifture of the air. This change of a falt from a folid to a fluid ftate, by the fole contact of the air, is alfo called *Deliquium*; fo that when a falt changes in this manner from a folid into a fluid form, it is faid to run *per deliquium*. Acids being the fimpleft fpecies of Saline bodies, their affinities with different fubftances are ftronger than thofe of any other fort of falt with the fame fubftances; which is agreeable to our fourth propofition concerning Affinities.

Acids in general have a great affinity with earths: that with which they moft readily unite is the unvitrifiable earth to which we gave the name of abforbent earth. They feem not to act at all upon vitrifiable earths, fuch as fand; nor yet upon fome other kinds of earths, at leaft while they are in their natural ftate. Yet the nature of thefe earths may be in fome meafure changed, by making them red hot in the fire, and then quenching them fuddenly in cold water: for by repeating this often they are brought nearer to the nature of abforbent earths, and rendered capable of uniting with Acids.

3   When

When an acid liquor is mixed with an abforbent earth, for inftance with chalk, thefe two fubftances inftantly rufh into union, with fo much impetuofity, efpecially if the acid liquor be as much dephlegmated, or contain as little water, as may be, that a great ebullition is immediately produced, attended with confiderable hiffing, heat, and vapours, which rife the very inftant of their conjunction.

From the combination of an Acid with an abforbent earth there arifes a new compound, which fome Chymifts have called *Sal Salfum*; becaufe the Acid by uniting with the earth lofes its four tafte, and acquires another not unlike that of the common fea falt ufed in our kitchens; yet varying according to the different forts of Acids and earths combined together. The Acid at the fame time lofes its property of turning vegetable blues and violet colours red.

If we enquire what is become of its propenfity to unite with water, we fhall find that the earth, which of itfelf is not foluble in water, hath by its union with the Acid acquired a facility of diffolving therein; fo that our *Sal Salfum* is foluble in water. But, on the other hand, the Acid hath, by its union with the earth, loft part of the affinity it had with water; fo that if a *Sal Salfum* be dried, and freed of all fuperfluous humidity, it will remain in that dry folid form, inftead of attracting the moifture of the air, and running *per deliquium*, as the Acid would do if it were pure and unmixed with earth. However, this general rule admits of fome exceptions; and we fhall have occafion in another place to take notice of certain Combinations of Acids with earths, which ftill continue to attract the moifture of the air, though not fo ftrongly as a pure Acid.

Acids have likewife a great affinity with the Phlogifton. When we come to treat of each Acid in particular, we fhall examine the combinations of

each with the Phlogifton: they differ fo widely
from one another, and many of them are fo little
known, that we cannot at prefent give any general
idea of them.

§ II. *Of* ALKALIS.

ALKALIS are Saline combinations in which there
is a greater proportion of earth than in Acids. The
principal arguments that may be adduced to prove
this fact are thefe: Firft; if they be treated in the
manner propofed above for analyzing Saline Subftances, we obtain from them a much greater
quantity of earth than we do from Acids. Secondly; by combining certain Acids with certain
earths we can produce Alkalis; or at leaft fuch
faline compounds as greatly refemble them. Our
third and laft argument is drawn from the properties of thofe Alkalis which, when pure and unadulterated with any other principle, have lefs affinity with water than Acids have, and are alfo more
fixed, refifting the utmoft force of fire. On this
account it is that they have obtained the title of
Fixed, as well as to diftinguifh them from another
fpecies of Alkali, to be confidered hereafter, which
is impure and volatile.

Though fixed Alkalis, when dry, fuftain the utmoft violence of fire without flying off in vapours,
it is remarkable that, being boiled with water in an
open veffel, confiderable quantities of them rife
with the fteam: an effect which muft be attributed
to the great affinity between thefe two fubftances,
by means whereof water communicates fome part
or its volatility to the fixed falt.

Alkalis freed of their fuperfluous humidity by
calcination attract the moifture of the air, but not
fo ftrongly as Acids: fo that it is eafier to procure
and preferve them in a folid form.

They

They flow in the fire, and are then capable of uniting with vitrifiable earths, and of forming therewith true glafs, which, however, will partake of their properties, if they be ufed in fufficient quantity.

As they melt more readily than vitrifiable earth, they facilitate its fufion; fo that a weaker fire will reduce it to glafs, when a fixed Alkali is joined with it, than will melt it without that addition.

Alkalis are known by their tafte, which is acrid and fiery; and by the properties they poffefs of turning certain vegetable blues and violet colours green; particularly fyrop of violets.

Their affinity with Acids is greater than that of abforbent earths; and hence it comes to pafs that if an Alkali be prefented to a combination of an Acid with an abforbent earth, the earth will be feparated from the Acid by the Alkali, and a new union between the Acid and the Alkali will take place. This is both an inftance and a proof of our fifth propofition concerning Affinities.

If a pure Alkali be prefented to a pure Acid, they rufh together with violence, and produce the fame phenomena as were obferved in the union of an abforbent earth with an Acid; but in a greater and more remarkable degree.

Fixed Alkalis may in general be divided into two forts: one of thefe hath all the above recited properties; but the other poffeffes fome that are peculiar to itfelf. We fhall confider this latter fort more particularly under the head of Sea-Salt.

§ III. *Of* NEUTRAL SALTS.

THE Acid and the Alkali thus uniting mutually rob each other of their characteriftic properties; fo that the compound refulting from their union produces no change in the blue colours of vegetables, and has a tafte which is neither four nor

acrid,

acrid, but faltifh. A faline combination of this kind is for that reafon named *Sal Salfum*, *Sal Medium*, or a *Neutral Salt*. Such combinations are alfo called by the plain general name of *Salts*.

It muft be obferved that, in order to make thefe Salts perfectly Neutral, it is neceffary that neither of the two faline principles of which they are compounded be predominant over the other; for in that cafe they will have the properties of the prevailing principle. The reafon is this: neither of thefe faline fubftances can unite with the other but in a limited proportion, beyond which there can be no further coalition between them. The action by which this perfect union is accomplifhed is termed *Saturation*; and the inftant when fuch proportions of the two faline fubftances are mixed together, that the one is incorporated with as much of the other as it can poffibly·take up, is called the *Point of Saturation*. All this is equally applicable to the combination of an Acid with an abforbent earth.

The combination is known to be perfect, that is, the Point of Saturation is known to be obtained, when after repeated affufions of an Acid in fmall quantities to an Alkali, or an abforbent earth, we find thofe phenomena ceafe, which in fuch cafes conftantly attend the conflict of union, as we faid above, namely, ebullition, hiffing, &c. and we may be affured the Saturation is complete when the new compound hath neither an acid nor an acrid tafte, nor in the leaft changes the blue colours of vegetables.

Neutral Salts have not fo great an affinity with water as either Acids or Alkalis have; becaufe they are more compounded: for we obferved before, that the affinities of the moft compounded bodies are generally weaker than thofe of the moft fimple. In confequence hereof few Neutral Salts, when dried, attract the moifture of the air; and thofe that do,

attract

attract it more flowly, and in lefs quantity, than either Acids or Alkalis do.

All Neutral Salts are foluble in water ; but more or lefs readily, and in a greater or fmaller quantity, according to the nature of their component principles.

Water made boiling hot diffolves a greater quantity of thofe falts which do not attract the moifture of the air, than when it is cold ; and indeed it muft be boiling hot to take up as much of them as it is capable of diffolving : but as for thofe which run in the air, the difference, if there be any, is imperceptible.

Some Neutral Salts have the property of fhooting into Cryftals, and others have it not.

The nature of Cryftallization is this : Water cannot diffolve, nor keep in folution, more than a determinate quantity of any particular Salt : when therefore fuch a quantity of water is evaporated from the folution of a Salt capable of cryftallization, that the remainder contains juft as much Salt as it can diffolve, then by continuing the evaporation the Salt gradually recovers its folid form, and concretes into feveral little tranfparent maffes called Cryftals. Thefe cryftals have regular figures, all differing from one another according to the fpecies of Salt of which they are formed. Different methods of evaporating faline folutions have different effects on the figure and regularity of the cryftals ; and each particular fort of Salt requires a peculiar method of evaporation to make its cryftals perfectly regular.

A folution of Salt defigned for cryftallization is ufually evaporated by means of fire to a pellicle ; that is, till the Salt begin to concrete ; which is perceived by a kind of thin dark fkin that gathers on the furface of the liquor, and is formed of the cryftallized particles of Salt. When this pellicle appears the folution is fuffered to cool, and the cry-

ftals

ftals form therein fafter or flower according to the fort of Salt in hand. If the evaporation be carried on brifkly to perfect drynefs, no cryftals will be formed, and only an irregular mafs of Salt will be obtained..

The reafons why no cryftals appear when the evaporation is haftily performed, and carried on to drynefs, are, firft, that the particles of Salt, being always in motion while the folution is hot, have not time to exert their mutual affinities, and to unite together as cryftallization requires: fecondly, that a certain quantity of water enters into the very compofition of cryftals; which is therefore abfolutely neceffary to their formation, and in a greater or fmaller proportion according to the nature of the Salt *.

If thefe cryftallized Salts be expofed to the fire, they firft part with that moifture which is not neceffary to a faline concretion, and which they retained only by means of their cryftallization: afterwards they begin to flow, but with different degrees of fufibility.

It muft be obferved that certain Salts melt as foon as they are expofed to the fire; namely, thofe which retain a great deal of water in cryftallizing. But this fluor which they fo readily acquire muft be carefully diftinguifhed from actual fufion: for it is owing only to their fuperfluous humidity, which heat renders capable of diffolving and liquifying them; fo that when it is evaporated the Salt ceafes to be fluid, and requires a much greater degree of fire to bring it into real fufion.

The Neutral Salts that do not cryftallize may, indeed, be dried by evaporating the water which

* Thofe who have the curiofity to fee a more particular account of the Cryftallization of Neutral Salts, may read Mr. *Rouelle's* excellent Memoir on that Subject, among thofe of the Academy of Sciences for 1744.

keeps

keeps them fluid; but by becoming solid they acquire no regular form; they again attract the moisture of the air, and are thereby melted into a liquor. These may be called *Liquefcent Salts.*

Moft of the Neutral Salts, that confift of an Acid joined with a fixed Alkali, or with an abforbent earth, are themfelves fixed, and refift the force of fire; yet feveral of them, if they be diffolved in water and the folution boiled and evaporated, fly off along with the fteams.

CHAP. IV.

Of the feveral Sorts of Saline Subftances.

§. I. *Of the* UNIVERSAL ACID.

THE Univerfal Acid is fo called, becaufe it is in fact the Acid which is moft univerfally diffufed through all nature, in waters, in the atmofphere, and in the bowels of the earth. But it is feldom pure; being almoft always combined with fome other fubftance. That from which we obtain it with moft eafe and in the greateft quantity is Vitriol, a mineral which we fhall confider afterwards: and this is the reafon why it is called the *Vitriolic Acid*; the name by which it is beft known.

When the Vitriolic Acid contains but little phlegm, yet enough to give it a fluid form, it is called *Oil of Vitriol*; on account of a certain unctuofity belonging to it. In truth this name is very improperly beftowed on it; for we fhall afterwards fee that, bating this unctuoufnefs, it has none of the properties of oils. But this is not the only impropriety in names that we fhall have occafion to cenfure.

C 4   If

If the Vitriolic Acid contain much water it is then called *Spirit of Vitriol.* When it does not contain enough to render it fluid, and fo is in a folid form, it is named the *Icy Oil of Vitriol.*

When Oil of Vitriol highly concentrated is mixed with water, they rufh into union with fuch impetuofity that, the moment they touch each other, there arifes a hiffing noife, like that of red-hot iron plunged in cold water, together with a very confiderable degree of heat, proportioned to the degree to which the Acid was concentrated.

If inftead of mixing this concentrated Acid with water, you only leave it expofed to the air for fome time, it attracts the moifture thereof, and imbibes it moft greedily. Both its bulk and its weight are increafed by this acceffion; and if it be under an icy form, that is, if it be concreted, the phlegm thus acquired will foon refolve it into a fluid.

The addition of water renders the Vitriolic Acid, and indeed all other Acids, weaker in one fenfe; which is, that when they are very aqueous they leave on the tongue a much fainter tafte of acidity, and are lefs active in the folution of fome particular bodies : but that occafions no change in the ftrength of their affinities, but in fome cafes rather enables them to diffolve feveral fubftances, which, when well dephlegmated, they are not capable of attacking.

The Vitriolic Acid combined to the point of faturation with a particular abforbent earth, the nature of which is not yet well known, forms a Neutral Salt that cryftallizes. This Salt is called *Alum,* and the figure of its cryftals is that of an octahedron, or folid of eight fides. Thefe octahedra are triangular pyramids, the angles of which are fo cut off that four of the furfaces are hexagons, and the other four triangles.

There

There are feveral forts of Alum, which differ according to the earths combined with the Vitriolic Acid. Alum diffolves eafily in water, and in cryftallization retains a confiderable quantity of it; which is the reafon that being expofed to the fire it readily melts, fwelling and puffing up as its fuperfluous moifture exhales. When that is quite evaporated, the remainder is called *Burnt Alum*, and is very difficult to fufe. The Acid of the Alum is partly diffipated by this calcination. Its tafte is faltifh, with a degree of roughnefs and aftringency.

The Vitriolic Acid combined with certain earths forms a kind of Neutral Salt called *Selenites*, which cryftallizes in different forms according to the nature of its earth. There are numberlefs fprings of water infected with diffolved Selenites; but when this Salt is once cryftallized, it is exceeding difficult to diffolve it in water a fecond time. For that purpofe a very great quantity of water is neceffary, and moreover it muft boil; for as it cools moft of the diffolved Selenites takes a folid form, and falls in a powder to the bottom of the veffel.

If an Alkali be prefented to the Selenites, or to Alum, thefe Salts, according to the principles we have laid down, will be thereby decompofed; that is, the Acid will quit the earths, and join the Alkali, with which it hath a greater affinity. And from this conjunction of the Vitriolic Acid with a fixed Alkali there refults another fort of Neutral Salt, which is called *Arcanum duplicatum, Sal de duobus*, and *Vitriolated Tartar*, becaufe one of the fixed Alkalis moft in ufe is called Salt of Tartar.

Vitriolated Tartar is almoft as hard to diffolve in water as the Selenites. It fhoots into eight-fided cryftals, having the apices of the pyramids pretty obtufe. Its tafte is faltifh, inclining to bitter; and it decrepitates on burning coals. It requires a very great degree of fire to make it flow.

The

The Vitriolic Acid is capable of uniting with the Phlogifton, or rather it has a greater affinity with it than with any other body: whence it follows that all compounds, of which it makes a part, may be decompofed by means of the Phlogifton.

From the conjunction of the Vitriolic Acid with the Phlogifton arifes a compound called *Mineral Sulphur*, becaufe it is found perfectly formed in the bowels of the earth. It is alfo called *Sulphur vivum*, or fimply, *Sulphur*.

. Sulphur is abfolutely infoluble in water, and incapable of contracting any fort of union with it. It melts with a very moderate degree of heat, and fublimes in fine light downy tufts called *Flowers of Sulphur*. By being thus fublimed it fuffers no decompofition, let the operation be repeated ever fo often; fo that Sublimed Sulphur, or Flower of Sulphur, hath exactly the fame properties as Sulphur that has never been fublimed.

If Sulphur be expofed to a brifk heat in the open air, it takes fire, burns, and is wholly confumed. This deflagration of Sulphur is the only means we have of decompofing it, in order to obtain its Acid in purity. The Phlogifton is deftroyed by the flame, and the Acid exhales in vapours : thefe vapours collected have all the properties of the Vitriolic Acid, and differ from it only as they ftill retain fome portion of the Phlogifton; which, however, foon quits them of its own accord, if the free accefs of the common air be not precluded.

The portion of Phlogifton retained by the Acid of Sulphur is much more confiderable when that mineral is burnt gradually and flowly : in that cafe the vapours which rife from it have fuch a penetrating odour, that they inftantaneoufly fuffocate any perfon who draws in a certain quantity of them with his breath. Thefe vapours conftitute what is called the *Volatile Spirit of Sulphur*. There is reafon to
think

think this portion of Phlogifton which the Acid retains is combined therewith in a manner different
from that in which thefe two are united in the Sulphur itfelf; for, as has juft been obferved, nothing
but actual burning is capable of feparating the Vitriolic Acid and the Phlogifton, which by their union
form Sulphur; whereas in the Volatile Spirit of
Sulphur they feparate fpontaneoufly when expofed
to the open air; that is, the Phlogifton flies off and
leaves the Acid, which then becomes in every refpect fimilar to the Vitriolic Acid.

That the Volatile Spirit of Sulphur is a compound, as we have afferted it to be, appears evidently
from hence, that whenever the Vitriolic Acid touches
any fubftance containing the Phlogifton, provided
that Phlogifton be difengaged or opened to a certain degree, a Volatile Spirit of Sulphur is infallibly and immediately generated. This Spirit hath
all the properties of Acids, but confiderably weakened, and of courfe lefs perceptible. It unites with
abforbent earths or fixed Alkalis; and with them
forms Neutral Salts: but when combined therewith
it may be feparated from them by the Vitriolic
Acid, and indeed by any of the mineral Acids,
becaufe its affinities are weaker. Sulphur hath the
property of uniting with abforbent earths, but not
near fo intimately as with fixed Alkalis.

If equal parts of Sulphur and an Alkali be melted
together, they incorporate with each other; and
from their conjunction proceeds a compound of a
moft unpleafant fmell, much like that of rotten
eggs, and of a red colour nearly refembling that of
an animal liver, which has occafioned it to bear the
name of *Hepar Sulphuris*, or *Liver of Sulphur*.

In this compofition the fixed Alkali communicates
to the Sulphur the property of diffolving in water:
and hence it comes that Liver of Sulphur may be
made as well when the Alkali is diffolved by water

into

into a fluid, as when it is fufed by the action of fire.

. Sulphur has lefs affinity than any Acid with the fixed Alkalis: and therefore Liver of Sulphur may be decompounded by any Acid whatever; which will unite with the fixed Alkali, form therewith a Neutral Salt, and feparate the Sulphur.

If Liver of Sulphur be diffolved in water, and an Acid poured thereon, the liquor, which was tranfparent before, inftantly turns to an opaque white; becaufe the Sulphur, being forced to quit its union with the Alkali, lofes at the fame time the property of diffolving in water, and appears again in its own opaque form. The liquor thus made white by the Sulphur is called *Milk of Sulphur*.

If this liquor be fuffered to ftand ftill for fome time, the particles of Sulphur, now moft minutely divided, gradually approach each other, unite, and fall infenfibly to the bottom of the veffel; and then the liquor recovers its tranfparency. The Sulphur thus depofited on the bottom of the veffel is called the *Magiftery* or *Precipitate of Sulphur*. The names of Magiftery and Precipitate are alfo given to all fubftances whatever that are feparated from another by this method; which is the reafon that we ufe the expreffion of precipitating one fub-ftance by another, to fignify the feparating one of them by means of the other,

§. II. *Of the* NITROUS ACID.

It is not certainly known what conftitutes the difference between the Nitrous Acid and the Vitrio-lic Acid, with regard to the conftituent principles of each. The moft probable opinion is, that the Nitrous Acid is no other than the Vitriolic Acid combined with a certain quantity of Phlogifton by

the

the means of putrefaction. If it be fo, the Phlogifton muft be united with the Univerfal Acid in another manner than it is in fulphur, and in its volatile fpirit: for the Nitrous Acid differs from them both in its properties. What gives ground for this opinion is, that the Nitrous Acid is never found but in earths and ftones which have been impregnated with matters fubject to putrefaction, and which therefore muft contain the Phlogifton. For it is neceffary juft to obferve here, though it be not yet proper to enter particularly into the fubject, that all fubftances fufceptible of putrefaction really contain the Phlogifton.

The Nitrous Acid combined with certain abforbent earths, fuch as chalk, marle, boles, forms Neutral Salts which do not cryftallize; and which, after being dried, run in the air *per deliquium*.

All thofe Neutral Salts which confift of the Nitrous Acid joined to an earth, may be decompofed by a fixed Alkali, with which the Acid unites, and deferts the earth; and from this union of the Nitrous Acid with a fixed Alkali refults a new Neutral Salt which is called *Nitre*, or *Salt-petre*. This latter name fignifies the *Salt of Stone*; and in fact Nitre is extracted from the ftones and plaifter, in which it forms, by boiling them in water faturated with a fixed Alkali.

Nitre fhoots in long cryftals adhering fideways to each other? it has a faltifh tafte, which produces a fenfation of cold on the tongue.

This Salt eafily diffolves in water; which, when boiling hot, takes up ftill a greater quantity thereof.

It flows with a pretty moderate degree of heat, and continues fixed therein: but being urged by a brifk fire, and in the open air, it lets go fome part of its Acid, and indeed flies off itfelf in part.

The

The moſt remarkable property of Nitre, and that which characterizes it, is its Fulmination or Explo-ſion; the nature of which is as follows:

When Nitre touches any ſubſtance containing a Phlogiſton, and actually ignited, that is, actually on fire, it burſts out into a flame, burns, and is decompounded with much noiſe.

In this deflagration the Acid is diſſipated, and totally ſeparated from the Alkali, which now re-mains by itſelf.

Indeed the Acid, at leaſt the greateſt part of it, is by this means quite deſtroyed. The Alkali which is left when Nitre is decompounded by defla-gration, is called in general *Fixed Nitre*, and, more particularly, Nitre fixed by ſuch and ſuch a ſub-ſtance as was uſed in the operation. But if Nitre be deflagrated with an inflammable ſubſtance con-taining the vitriolic Acid, as ſulphur, for inſtance, the fixed Salt produced by the deflagration is not a pure Alkali, but retains a good deal of the vi-triolic Acid, and, by combining therewith, hath now formed a neutral Salt.

Hitherto Chymiſts have been at a loſs for the reaſon why Nitre flames, and is decompounded in the manner above mentioned, when it comes in contact with a Phlogiſton properly circumſtanced. For my part, I conjecture it to be for the ſame reaſon that vitriolated tartar is alſo decompounded by the addi-tion of a Phlogiſton; *viz.* the Nitrous Acid, having a greater affinity with the Phlogiſton than with the fixed Alkali, naturally quits the latter to join with the former, and ſo produces a kind of ſulphur, dif-fering probably from the common ſulphur, formed by the vitriolic Acid, in that it is combuſtible to ſuch a degree, as to take fire and be conſumed in the very moment of its production; ſo that it is impoſſible to prevent its being thus deſtroyed, and conſequently impoſſible to ſave it. In ſupport of this opinion

let

let it be confidered, that the concurrence of the
Phlogifton is abfolutely neceffary to produce this
deflagration, and that the matter of pure fire is
altogether incapable of effecting it : for though
Nitre be expofed to the moft violent degree of fire,
even that in the focus of the moft powerful burning-glafs, it will not flame; nor will that effect ever
happen till the Nitre be brought into contact with
a Phlogifton properly fo called, that is, the matter
of fire exifting as a principle of fome body; and
it is moreover neceffary that this Phlogifton be
actually on fire, and agitated with the igneous motion, or elfe that the Nitre itfelf be red-hot, and fo
penetrated with fire as to kindle any inflammable
matter that touches it.

This experiment, among others, helps to fhew the
diftinction that ought to be made between pure elementary fire, and fire become a principle of bodies,
to which we have given the name of Phlogifton.

Before we leave this fubject, we fhall obferve that
Nitre deflagrates only with fuch fubftances as contain the Phlogifton in its fimpleft and pureft form;
fuch as charcoal, fulphur, and the metalline fubftances; and that, though it will not deflagrate
without the addition of fome combuftible matter, it
is neverthelefs the only known body that will burn,
and make other combuftibles burn with it, in clofe
veffels, without the admiffion of frefh air.

The Nitrous Acid hath not fo great an affinity
with earths and Alkalis as the vitriolic Acid hath
with the fame fubftances; whence it follows that the
vitriolic Acid decompofes all neutral falts arifing from
a combination of the Nitrous Acid with an earth or
an Alkali. The vitriolic Acid expels the Nitrous
Acid, unites with the fubftance which ferved it for
a bafis, and therewith forms a neutral falt, which is
an Alum, a Selenites, or a vitriolated Tartar, according to the nature of that bafis.

The

The Nitrous Acid, when thus feparated from its bafis by the vitriolic Acid, is named *Spirit of Nitre*, or *Aqua Fortis*. If it be dephlegmated, or contain but little fuperfluous water, it exhales in reddifh vapours; thefe vapours, being condenfed and collected, form a liquor of a brownifh yellow, that inceffantly emits vapours of the fame colour, and of a pungent difagreeable fmell. Thefe characters have procured it the names of *Smoaking Spirit of Nitre*, and *Yellow Aqua Fortis*. This property in the Nitrous Acid, of exhaling in vapours, fhews it to be lefs fixed than the vitriolic Acid; for the latter, though ever fo thoroughly dephlegmated, never yields any vapours, nor has it any fmell.

§. III. *Of the* A c i d   o f   S e a - S a l t.

The Acid of Sea-Salt is fo called becaufe it is in fact obtained from fuch Sea-Salt as is ufed in our kitchens. It is not certainly known in what this Acid differs from the vitriolic and the nitrous, with regard to its conftituent parts. Several of the ableft Chymifts, fuch as Becher and Stahl, are of opinion that the Marine Acid is no other than the Univerfal Acid united to a particular principle which they call a Mercurial Earth. Concerning this earth we fhall have occafion to fay more, when we come to treat of metallic fubftances: But in the mean time it muft be owned, that the truth of this opinion is fo far from being proved by a fufficient number of experiments, that the very exiftence of fuch a mercurial earth is not yet well eftablifhed; and therefore, that we may not exceed the bounds of our knowledge, we fhall content ourfelves with delivering here the properties which characterize the Acid in queftion, and by which it is diftinguifhed from the two others confidered above.

When it is combined with abforbent earths, fuch as lime and chalk, it forms a neutral falt that does

7   not

not cryftallize, and, when dried, attracts the moifture of the air. If the abforbent earth be not fully
faturated with the Marine Acid, the falt thereby
formed has the properties of a fixed Alkali; and
this is what made us fay, when we were on the fubject of thofe falts, that they might be imitated by
combining an earth with an Acid. The Marine
Acid, like the reft, hath not fo great an affinity
with earths as with fixed Alkalis.

When it is combined with the latter, it forms a
neutral falt which fhoots into cubical cryftals. This
falt is inclined to grow moift in the air, and is confequently one of thofe which water diffolves in
equal quantities, at leaft as to fenfe, whether it be
boiling hot or quite cold.

The affinity of this Acid with Alkalis and abforbent Earths is not fo great as that of the vitriolic
and nitrous Acids with the fame fubftances: whence
it follows that, when combined therewith, it may
be feparated from them by either of thofe Acids.

The Acid of Sea-Salt, thus difengaged from the
fubftance which ferved it for a bafis, is called *Spirit*
of Salt. When it contains but little phlegm it is
of a lemon colour, and continually emits many
white, very denfe, and very elaftic vapours; on
which account it is named the *Smoaking Spirit of*
Salt. Its fmell is not difagreeable, nor much unlike that of faffron; but extremely quick and fuffocating when it fmokes.

The Acid of Sea-Salt, like the other two, feems
to have a greater affinity with the Phlogifton, than
with fixed Alkalis. We are led to this opinion
by a very curious operation, which gives ground to
think that Sea-Salt may be decompofed by the proper application of a fubftance containing the Phlogifton.

From the Marine Acid combined with a Phlogifton refults a kind of Sulphur, differing from the
common fort in many refpects; but particularly in
this property, that it takes fire of itfelf upon being
expofed to the open air. This combination is
called *Englifh Phofphorus*, *Phofphorus of Urine*, becaufe it is generally prepared from urine; or, only
Phofphorus.

This combination of the Marine Acid with a
Phlogifton is not eafily effected; becaufe it requires
a difficult operation in appropriated veffels. For
thefe reafons it does not always fucceed; and Phofphorus is fo fcarce and dear, that hitherto Chymifts
have not been able to make on it the experiments
neceffary to difcover all its properties. If Phofphorus be fuffered to burn away in the air, a fmall
quantity of an acid liquor may be obtained from it,
which feems to be fpirit of falt, but either altered,
or combined with fome adventitious matter; for it
has feveral properties that are not to be found in
the pure Marine Acid; fuch as, leaving a fixed
fufible fubftance behind it when expofed to a ftrong
fire, and being eafily combined with the Phlogifton
fo as to reproduce a Phofphorus.

Phofphorus refembles fulphur in feveral of its
properties: it is foluble in oils; it melts with a
gentle heat; it is very combuftible; it burns without producing foot; and its flame is vivid and
bluifh.

From what has been faid of the union of the
Acid of Sea-Salt with a fixed Alkali, and of the
neutral falt refulting therefrom, it may be concluded
that this neutral falt is no other than the common
kitchen-falt. But it muft be obferved that the fixed
Alkali, which is the natural bafis of the common
falt obtained from fea-water, is of a fort fomewhat
differing from fixed Alkalis in general, and hath
certain properties peculiar to itfelf. For,

3   1. The

1. The basis of Sea-Salt differs from other fixed Alkalis in this, that it crystallizes like a neutral salt.

2. It does not grow moist in the air : on the contrary, when exposed to the air, it loses part of the water that united with it in crystallization, by which means its crystals lose their transparency, become, as it were, mealy, and fall into a fine flour.

3. When combined with the vitriolic Acid to the point of saturation, it forms a neutral salt, differing from vitriolated tartar, first, in the figure of its crystals, which are oblong six-sided solids; secondly, in its quantity of water, which in cryftallization unites therewith in a much greater proportion than with vitriolated tartar ; whence it follows, that this salt dissolves in water more readily than vitriolated tartar; thirdly, in that it flows with a very moderate degree of heat, whereas vitriolated tartar requires a very fierce one.

If the Acid of Sea-Salt be separated from its basis by means of the vitriolic Acid, it is easy to see that, when the operation is finished, the salt we have been speaking of must be the result. A famous Chymist, named Glauber, was the first who extracted the Spirit of Salt in this manner, examined the neutral salt resulting from his procefs, and, finding it to have some singular properties, called it his *Sal mirabile*, or wonderful Salt: on this account it is still called Glauber's *Sal mirabile*, or plainly *Glauber's Salt*.

4. When the basis of Sea-Salt is combined with the nitrous Acid to the point of saturation, there results a neutral salt, or a sort of nitre, differing from the common nitre, first, in that it attracts the moisture of the air pretty strongly; and this makes it difficult to cryftallize : secondly, in the figure of its cryftals, which are parallelopipeds; and this has procured it the name of *Quadrangular Nitre.*

Common

Common falt, or the neutral falt formed by combining the Marine Acid with this particular fort of fixed Alkali, has a tafte well known to every body. The figure of its cryftals is exactly cubical. It grows moift in the air, and, when expofed to the fire, it burfts, before it melts, into many little fragments, with a crackling noife; which is called the *Decrepitation* of Sea-Salt.

That neutral falt mentioned above, which is formed by combining the Marine Acid with a common fixed Alkali, and called *Sal febrifugum Sylvii*, hath alfo this property.

India furnifhes us with a faline fubftance, known by the name of *Borax*, which flows very eafily, and then takes the form of glafs. It is of great ufe in facilitating the fufion of metallic fubftances. It poffeffes fome of the properties of fixed Alkalis, which has induced certain Chymifts to reprefent it, through miftake, as a pure fixed Alkali.

By mixing borax with the vitriolic Acid, Mr. Homberg obtained from it a falt, which fublimes in a certain degree of heat, whenever fuch a mixture is made. This falt has very fingular properties; but its nature is not yet thoroughly underftood. It diffolves in water with great difficulty; it is not volatile, though it rifes by fublimation from the borax. According to Mr. Rouelle's obfervation, it rifes then only by means of the water which carries it up: for, when once made, it abides the fierceft fire, flows and vitrifies juft as borax does; provided care be taken to free it previoufly from moifture by drying it properly. Mr. Homberg called it *Sedative Salt*, on account of its medical effects. The fedative falt hath the appearance, and fome of the properties, of a neutral falt; for it fhoots into cryftals, and does not change the colour of violets: but it acts the part of an Acid with regard to Alkalis, uniting with them to the point of faturation, and thereby forming

ing a true neutral falt. It alfo acts like the Acid
of vitriol on all neutral falts; that is, it difcharges
the Acid of fuch as have not the vitriolic Acid in
their compofition.

Since Mr. Homberg's time it hath been difcovered, that a fedative falt may be made either with the
nitrous or with the marine Acid; and that fublimation is not neceffary to extract it from the borax,
but that it may be obtained by cryftallization only.
For this latter difcovery we are indebted to Mr.
Geoffroy, as we are to Mr. Lemery for the former.

Since that time M. Baron d'Henouville, an able
Chymift, hath fhewn that a fedative falt may be
obtained by the means of vegetable Acids; and
hath lately demonftrated, in fome excellent papers
publifhed in the collection of Memoirs written by
the correfpondents of the Academy of Sciences,
that the fedative falt exifts actually and perfectly in
the borax, and that it is not produced by mixing
Acids with that faline fubftance, as it feems all the
Chymifts before him imagined.   This he proves
convincingly from his analyfis of borax, (which
thereby appears to be nothing elfe but the fedative
falt united with that fixed Alkali which is the bafis
of Sea-Salt) and from his regenerating the fame
borax by uniting together that Alkali and the fedative falt: a proof the moft complete that can poffibly be produced in natural philofophy, and equi-
valent to demonftration itfelf.

In order to finifh what remains to be faid upon
the feveral forts of faline fubftances, we fhould now
fpeak of the Acids obtained from vegetables and animals, and alfo of the volatile Alkalis: but, feeing
thefe faline fubftances differ from thofe of which we
have already treated, only as they are varioufly altered by the unions they have contracted with certain
principles of vegetables and animals, of which nothing has been yet faid, it is proper to defer being

D 3   particular

particular concerning them, till we have explained thofe principles.

CHAP. V.
Of LIME.

ANY fubftance whatever, that has been roafted a confiderable time in a ftrong fire without melting, is commonly called a *Calx*. Stones and metals are the principal fubjects that have the pro-perty of being converted into *Calces*. We fhall treat of Metalline *Calces* in a fubfequent chapter, and in this confine ourfelves to the *Calx* of *Stone*, known by the name of *Lime*.

In treating of earths in general we obferved, that they may be divided into two principal kinds; one of which actually and properly flows when ex-pofed to the action of fire, and turns to glafs; whence it is called a *fufible* or *vitrifiable* earth; the other refifts the utmoft force of fire, and is therefore faid to be an *unfufible* or *unvitrifiable* earth. The latter is alfo not uncommonly called *calcinable* earth; though fundry forts of unfufible earths are inca-pable of acquiring by the action of fire all the qualities of *calcined* earth, or *Lime* properly fo called: fuch earths are particularly diftinguifhed by the denomination of *refractory* earths.

As the different forts of ftones are nothing more than compounds of different earths, they have the fame properties with the earths of which they are compofed, and may like them, be divided into fu-fible or vitrifiable, and unfufible or calcinable. The fufible ftones are generally denoted by the name of *Flints*; the calcinable ftones, again, are the feveral forts of marbles, cretaceous ftones, thofe commonly
called

called free-ftones, &c. fome of which, as they make the beft Lime, are, by way of eminence, called *Lime-ftones*. Sea-fhells alfo, and ftones that abound with foffile fhells, are capable of being burnt to Lime.

' All thefe fubftances, being expofed, for a longer or fhorter time, as the nature of each requires, to the violent action of fire, are faid to be *calcined*. By calcination they lofe a confiderable part of their weight, acquire a white colour, and become friable though ever fo folid before.; as, for inftance, the very hardeft marbles. Thefe fubftances, when thus calcined, take the name of *Quick Lime.*

Water penetrates Quick Lime, and rufhes into it with vaft activity. If a lump of newly calcined Lime be thrown into water, it inftantly excites almoft as great a noife, ebullition, and fmoke, as would be produced by a piece of red-hot iron; with fuch a degree of heat too, that, if the Lime be in due proportion to the water, it will fet fire to combuftible bodies; as hath unfortunately happened to veffels laden with Quick Lime, on their fpringing a fmall leak.

As foon as Quick Lime is put into water, it fwells, and falls afunder into an infinite number of minute particles: in a word, it is in a manner diffolved by the water, which forms therewith a fort of white pafte called *Slacked Lime.*

If the quantity of water be confiderable enough for the Lime to form with it a white liquor, this liquor is called *Lac Calcis*; which, being left fome time to fettle, grows clear and tranfparent, the Lime which was fufpended therein, and occafioned its opacity, fubfiding to the bottom of the veffel. Then there forms on the furface of the liquor a cryftalline pellicle, fomewhat opaque and dark coloured, which being fkimmed off is reproduced from time to time. This matter is called *Cremor Calcis.*

Slacked

Slacked Lime gradually grows dry, and takes the form of a folid body, but full of cracks and deftitute of firmnefs. The event is different when you mix it up, while yet a pafte, with a certain quantity of uncalcined ftony matter, fuch as fand for example: then it takes the name of *Mortar*, and gradually acquires, as it grows drier and older, a hardnefs equal to that of the beft ftones. This is a very fingular property of Lime, nor is it eafy to account for it: but it is a beneficial one; for every body knows the ufe of mortar in building.

Quick Lime attracts the moifture of the air, in the fame manner as concentrated acids, and dry fixed alkalis; but not in fuch quantities as to render it fluid: it only falls into extremely fmall particles, takes the form of a fine powder, and the title of *Lime flacked in the air.*

Lime once flacked, however dry it may afterwards appear, always retains a large portion of the water it had imbibed; which cannot be feparated from it again but by means of a violent calcination. Being fo recalcined it returns to be Quick Lime, recovering all its properties.

Befides this great affinity of Quick Lime with water, which difcovers a faline character, it has feveral other faline properties, to be afterwards examined, much refembling thofe of fixed alkalis. In Chymiftry it acts very nearly as thofe falts do, and may be confidered as holding the middle rank between a pure abforbent earth and a fixed alkali: and this hath induced many Chymifts to think that Lime contains a true falt, to which all the properties it poffeffes in common with falts may be attributed.

But as the chymical examination of this fubject hath long been neglected, the exiftence of a faline fubftance in Lime hath been long doubtful. Mr. du Fay, author of fome excellent chymical

mical experiments, was one of the firft who obtained a Salt from Lime, by lixiviating it with a
great deal of water, which he afterwards evaporated.
But the quantity of falt he obtained by that means
was very fmall; nor was it of an alkaline nature,
as one would think it fhould have been, confidering
the properties of Lime. Mr. du Fay did not carry
his experiments on this fubject any further, probably for want of time; nor did he determine of
what nature the falt was.

Mr. Malouin had the curiofity to examine this
falt of Lime, and foon found that it was nothing elfe
but what was above called *Cremor Calcis.* He found
moreover, that, by mixing a fixed alkali with
Lime-water, a vitriolated tartar was formed; that,
by mixing therewith an alkali like the bafis of fea
falt, a Glauber's falt was produced; and laftly, by
combining Lime with a fubftance abounding in
phlogifton he obtained a true fulphur. Thefe very
ingenious experiments prove to a demonftration
that the vitriolic acid conftitutes the falt of Lime:
for, as hath been fhewn, no other acid is capable of
forming fuch combinations. On the other hand,
Mr. Malouin, having forced the vitriolic acid of
this falt to combine with a phlogifton, found its
bafis to be earthy, and analogous to that of the
felenites: whence he concluded that the falt of
Lime is a true neutral falt, of the fame kind as the
felenites. Mr. Malouin tells us he found feveral
other falts in Lime. But as none of them was a
fixed alkali, and as all the faline properties of Lime
have an affinity with thofe of that kind of falt,
there is great reafon to think that all thofe falts are
foreign to Lime, and that their union with it is
merely accidental.

I myfelf have made feveral experiments in order
to get fome infight into the faline nature of Lime,

and

and shall here produce the result with all possible
concisenefs. · I took several stones of different
kinds, some of which produced by calcination a
very strong Lime, and others but a very weak one.
These I impregnated with different saline substances,
acids, alkalis, and neutrals, and then exposed them
all to the same degree of fire, which was a pretty
strong one, and long enough continued to have
made very good Lime of stones the most difficult
to calcine. The consequence was, that, in the first
place, those stones which naturally made but a weak
Lime were not by this process converted into a
stronger Lime; and, moreover, that none of these
stones, even such as would naturally have produced
the most active Lime, had acquired the properties
of Lime. These experiments I varied many ways,
employing different proportions of saline matters,
and almost every possible degree of fire, and constantly observed, after calcination, that all those
stones were so much the further from the nature of
Lime, as they had been combined with larger doses
of salts. Among those which were impregnated with
the greatest proportion of salts, and had suffered the
greatest violence of fire, I observed some that had
begun to flow, and were in a manner vitrified.
Now as the same subject cannot be, at one and the
same time, in the state of glass and of Lime too ;
as a body cannot approach to one of these states but
in proportion as it recedes from the other ; and as
salts in general dispose those bodies to fusion and
vitrification which are in themselves the most averse
to either, I concluded from my experiments, that
the saline substances I used had, by acting as fluxes
upon the stones, prevented their calcination ; that
consequently we may suspect there is no saline
matter in the composition of Lime, as Lime ; and
that Lime does not owe its saline and alkaline pro-

perties

perties to any falt; or at leaft that if it does owe
thofe properties to a falt, fuch falt muft be naturally and originally combined with the matter of
the ftone in fo juft a proportion, that it is impoffible
to increafe the quantity thereof without prejudicing
the Lime, and depriving it in fome meafure of its
virtue. This theory agrees perfectly with the illuftrious Stahl's opinion; for he thinks, as we obferved in difcourfing of Salts in general, that every
faline fubftance is but an earth combined in a certain manner with water. This notion he applies to
Lime, and fays that fire only fubtilizes and attenuates the earthy matter, and thereby renders it
capable of uniting with water in fuch a manner,
that the refult of their combination fhall be a fubftance having faline properties; and that Lime accordingly never acquires thefe properties till it be
combined with water.

. I have dwelt longer on the Salt of Lime than I
fhall on any other particular; becaufe the fubject,
though in itfelf of great importance, has hitherto
been but little attended to, and becaufe the experiments here recited are entirely new.

. Lime unites with all acids, and in conjunction
with them exhibits various phenomena.

The vitriolic acid poured upon Lime diffolves it
with effervefcence and heat. From this mixture
there exhales a great quantity of vapours, in fmell
and colour perfectly like thofe of fea-falt; from
which however they are found to be very different
when collected into a liquor. From this combination of the vitriolic acid with Lime arifes a neutral
falt, which fhoots into cryftals, and is of the fame
kind with the felenitic falt obtained from Lime by
Mr. Malouin.

The nitrous acid poured upon Lime diffolves it
in like manner with effervefcence and heat: but the
folution is tranfparent, and therein differs from the
former,

former, which is opaque. From this mixture there arifes a neutral falt, which does not cryftallize, and has withal the very fingular property of being volatile, and rifing wholly by diftillation in a liquid form. This phenomenon is fo much the more remarkable, as Lime, the bafis of this falt, is one of the moft fixed bodies known in Chymiftry.

With the acid of fea-falt Lime forms alfo a fingular fort of falt, which greedily imbibes the moifture of the air. We fhall have occafion to take further notice of it in another place.

Thefe experiments made on Lime with acids are likewife quite new. We are indebted for them to Mr. Du Hamel of the Academy of Sciences, whofe admirable Memoirs on feveral fubjects fhew his extenfive knowledge in all parts of Natural Philofophy.

Lime applied to fixed alkalis adds confiderably to their cauftic quality, and makes them more penetrating and active. An alkaline lixivium in which Lime hath been boiled, being evaporated to drynefs, forms a very cauftic fubftance, which flows in the fire much more eafily, attracts and retains moifture much more ftrongly, than fixed alkalis that have not been fo treated. An alkali thus acuated by Lime is called the *Cauftic Stone*, or *Potential Cautery*; becaufe it is employed by furgeons to produce efchars on the fkin and cauterize it.

C H A P.

CHAP. VI.

Of Metallic Subſtances in general.

METALLIC Subſtances are heavy, glittering, opaque, fuſible bodies. They conſiſt chiefly of a vitrifiable earth united with the phlogiſton.

Several Chymiſts inſiſt on a third principle in theſe bodies, and have given it the name of *Mercurial earth*; which, according to Becher and Stahl, is the very ſame that being combined with the vitriolic acid forms and characterizes the acid of ſeaſalt. The exiſtence of this principle hath not yet been demonſtrated by any deciſive experiment; but we ſhall ſhew that there are pretty ſtrong reaſons for admitting it.

We ſhall begin with mentioning the experiments which prove Metallic Subſtances to conſiſt of a vitrifiable earth united with the phlogiſton. The firſt is this: if they be calcined in ſuch a manner as to have no communication with any inflammable matter, they will be ſpoiled of all their properties, and reduced to an earth or calx, that has neither the ſplendour nor the ductility of a metal, and in a ſtrong fire turns to an actual glaſs, inſtead of flowing like a metal.

The ſecond is, that the calx or the glaſs reſulting from a metal thus decompoſed, recovers all its metalline properties by being fuſed in immediate contact with an inflammable ſubſtance, capable of reſtoring the phlogiſton of which calcination had deprived it.

On this occaſion we muſt obſerve that Chymiſts have not yet been able, by adding the phlogiſton, to

give

give the properties of metals to all forts of vitrifiable earths indiscriminately; but to such only as originally made a part of some metallic body. For example, a compound cannot be made with the phlogiston and sand that shall have the least resemblance of a metal: and this is what seems to point out the reality of a third principle, as necessary to form the metalline combination. This principle may probably remain united with the vitrifiable earth of a metallic substance, when reduced to a glass; whence it follows, that such vitrified metals require only the addition of a phlogiston to enable them to appear again in their pristine form.

It may be inferred from another experiment, that the calx and the glass of a metal are not its pure vitrifiable earth, properly so called: for by repeated or long continued calcinations, such a calx or glass may be rendered incapable of ever resuming the metalline form, in whatever manner the phlogiston be afterwards applied to it; so that by this means it is brought into the condition of a pure vitrifiable earth, absolutely free from any mixture. Those Chymists who patronise the Mercurial earth, produce many other proofs of the existence of that principle in metallic substances; but they would be misplaced in an elementary treatise like this.

When by adding the phlogiston to a metallic glass we restore it to the form of a metal, we are said to *reduce, resuscitate,* or *revivify* that metal.

Metallic substances are of different kinds, and are divided into *Metals* and *Semi-metals.*

Those are called Metals which, besides their metalline splendour and appearance, are also malleable; that is, have the property of stretching under the hammer, and by that means of being wrought into different forms without breaking.

Those

Thofe which have only the metalline fplendour and appearance, without malleability, are called Semi-metals.

Metals are alfo further fubdivided into two forts; *viz. Perfect* and *Imperfect* Metals.

The Perfect Metals are thofe which fuffer no damage or change whatever by the moft violent and moft lafting action of fire.

The Imperfect Metals are thofe which by the force of fire may be deprived of their phlogifton, and confequently of their metalline form.

When but a moderate degree of fire is employed to deprive a Metal of its phlogifton, the metal is faid to be *calcined*; and then it appears in the form of a powdered earth, which is called a *Calx*: and this metalline calx being expofed to a more violent degree of fire melts and turns to glafs.

Metallic Subftances have an affinity with acids: but not equally with all; that is, every Metallic Subftance is not capable of uniting and joining with every acid.

When an acid unites with a Metallic Subftance there commonly arifes an ebullition, attended with a kind of hiffing noife and fuming exhalations. By degrees, as the union becomes more perfect, the particles of the metal combining with the acid become invifible: this is termed *Diffolution*; and when a metalline mafs thus difappears in an acid, the metal is faid to be *diffolved* by that acid. It is proper to obferve that acids act upon metalline fubftances, in one refpect, juft as they do upon alkalis and abforbent earths: for an acid cannot take up above fuch a certain proportion thereof as is fufficient to faturate it, to deftroy feveral of its properties, and weaken others. For example, when an acid is combined with a metal to the point of faturation, it lofes its tafte, does not turn the blue colour of a vegetable red, and its affinity with water

is

is confiderably impaired. On the other hand, Metalline Subftances, which when pure are incapable
of uniting with water, by being joined with an
acid acquire the property of diffolving in water.
Thefe combinations of Metalline Subftances with
acids form different forts of neutral falts; fome of
which have the property of fhooting into cryftals,
while others have it not: moft of them, when
thoroughly dried, attract the moifture of the air.

The affinity which Metalline fubftances have with
acids is lefs than that which abforbent earths and
fixed alkalis have with the fame acids: fo that all
metalline falts may be decompounded by one of
thefe fubftances, which will unite with the acid,
and precipitate the metal.

Metalline Subftances thus feparated from an
acid folvent are called *Magifteries*, and *Precipitates*,
of metals. None of thefe precipitates, except thofe
of the perfect metals, retain the metalline form:
moft of their phlogifton hath been deftroyed
by the folution and precipitation, and muft be reftored before they can recover their properties. In
fhort, they are nearly in the fame ftate with metalline fubftances deprived of their phlogifton by calcination; and accordingly fuch a precipitate is
called a *Calx*.

A metalline calx prepared in this manner lofes a
greater or a lefs portion of its phlogifton, the
more or lefs effectually and thoroughly the metalline fubftance, of which it made a part, was diffolved by the acid.

Metallic Subftances have affinities with each other
which differ according to their different kinds: but
this is not univerfal; for fome of them are incapable
of any fort of union with fome others.

It muft be obferved that Metallic Subftances will
not unite, except they be both in a fimilar ftate; that
is, both in a Metalline form, or both in the form
of

of a Glafs ;. for a metalline fubftance retaining its phlogifton cannot contraft an union with any metallic glafs, even its own.

C H. A P. VII.

Of M E T A L S.

THERE are fix Metals, of which two are Perfeft and four Imperfeft. The perfeft Metals are Gold and Silver ; the others are Copper, Tin, Lead, and Iron. Some Chymifts admit a feventh Metal, to wit, Quick-filver : but as it is not malleable, it has been generally confidered as a metallic body of a particular kind. We fhall foon have occafion to examine it more minutely.

The ancient Chymifts, or rather the Alchymifts, who fancied a certain relation or analogy between Metals and the Heavenly Bodies, beftowed on the feven Metals, reckoning Quick-filver one of them, the names of the feven Planets of the Ancients, according to the affinity which they imagined they obferved between thofe feveral bodies. Thus Gold was called *Sol*, Silver *Luna*, Copper *Venus*, Tin *Jupiter*, Lead *Saturn*, Iron *Mars*, and Quick-filver *Mercury*. Though thefe names were affigned for reafons merely chimerical, yet they ftill keep their ground ; fo that it is not uncommon to find the Metals called by the names, and denoted by the charafters, of the Planets, in the writings even of the beft Chymifts. Metals are the heavieft bodies known in nature.

§. I. *Of* G O L D.

GOLD is the heavieft of all Metals. The arts of wire-drawing and gold-beating fhew its wonderful

derful ductility. The greateft violence of fire is not able to produce any alteration in it. Indeed Mr. Homberg, a famous Chymift, pretended that he had made this metal fume, and even vitrified it, by ex-pofing it to the focus of one of the beft burning glaffes, known by the name of the Lens of the *Palais Royal:* but there are very good reafons for calling in queftion the experiments he made on this occafion, or rather for thinking that he was quite miftaken. For,

1. No man hath fince been able to vitrify Gold, though feveral good Experimenters have affiduoufly tried to effect it, by expofing it to the focus of the fame lens, and of other burning-glaffes ftill ftronger.

2. It hath been obferved that though Gold, when expofed to the focus of thofe glaffes, did indeed emit fome vapours and decreafe in weight; yet thofe vapours being carefully collected on a piece of pa-per, proved to be true Gold, in no degree vitrified, and which confequently had fuffered no change but that of being carried away by the violence of the heat, its nature not being in the leaft altered.

3. The fmall portion of vitrified matter, which was formed on the arm that fupported the Gold in Mr. Homberg's experiment, may have come either from the arm itfelf, or rather from fome heteroge-neous particles contained in the Gold ; for it is al-moft impoffible to have it perfectly pure.

4. Neither Mr. Homberg, nor any that have re-peated his experiment, ever reduced this pretended glafs of Gold by reftoring its phlogifton, as is done with other metallic glaffes.

5. To render the experiment decifive, the whole mafs of Gold employed ought to have been vi-trified ; which was not the cafe.

Neverthelefs I do not pretend that this metal is in its own nature abfolutely indeftructible, and un-

7   vitrifiable :

vitrifiable: but there is reason to think that nobody hath hitherto found the means of producing those effects on it, probably for want of a sufficient degree of fire; at least the point is very doubtful.

Gold cannot be diffolved by any pure acid: but if the acid of nitre be mixed with the acid of fea-falt, there refults a compound acid liquor, with which it has fo great an affinity that it is capable of being perfectly diffolved thereby. The Chymifts have called this folvent *Aqua Regis*, on account of its being the only acid that can diffolve Gold, which they confider as the King of Metals. The folution of Gold is of a beautiful orange colour.

If Gold diffolved in *aqua regis* be precipitated by an alkali or an abforbent earth, the precipitate gently dried, and then expofed to a certain degree of heat, is inftantly difperfed into the air, with a moft violent explofion and noife: Gold thus precipitated is therefore called *Aurum Fulminans*. But if the precipitated Gold be carefully wafhed in plenty of water, fo as to clear it of all the adhering faline particles, it will not fulminate; but may be melted in a crucible without any additament, and will then appear in its ufual form. The acid of vitriol being poured on *aurum fulminans* likewife deprives it of its fulminating quality.

Gold does not begin to flow till it be red-hot like a live coal. Though it be the moft malleable and moft ductile of all metals, it has the fingular property of lofing its ductility more eafily than any of them: even the fumes of charcoal are fufficient to deprive it thereof, if they come in contact with it while it is in fufion.

The malleability of this metal, and indeed of all the reft, is alfo confiderably diminifhed by expofing it fuddenly to cold when it is red-hot; for

example,

example, by quenching it in water, or even barely exposing it to the cold air.

The way to reftore ductility to Gold, when loft by its coming in contact with the vapour of coals, and in general to any metal rendered lefs malleable by being fuddenly cooled, is to heat it again, to keep it red-hot a confiderable time, and then to let it cool very flowly and gradually; this operation frequently repeated will by degrees much increafe the malleability of a metal.

Pure fulphur hath no effect on Gold; but being combined with an alkali into a *hepar fulphuris*, it unites therewith very readily. Nay, fo intimate is their union, that the Gold by means thereof becomes foluble in water; and this new compound of Gold and liver of fulphur, being diffolved in water, will pafs through the pores of brown paper without fuffering any decompofition; which does not happen, at leaft in fuch a manifeft degree, to other metallic fubftances diffolved by liver of fulphur.

Aurum fulminans mixed and melted with flower of fulphur lofes its fulminating quality : which arifes from hence, that on this occafion the fulphur burns, and its acid, which is the fame with the vitriolic, being thereby fet at liberty becomes capable of acting upon the Gold as a vitriolic acid would; which, as was faid above, deprives the Gold of its fulminating quality.

§. II. *Of* S I L V E R.

NEXT to Gold, Silver is the moft perfect metal. Like Gold it refifts the utmoft violence of fire, even that in the focus of a burning-glafs. However it holds only the fecond place among metals; becaufe it is lighter than Gold by almoft one half; is alfo fomewhat lefs ductile; and laftly, becaufe it is acted upon by a greater number of folvents.

Yet

Yet Silver hath one advantage over Gold, namely that of being a little harder; which makes it alfo more fonorous.

This metal, like Gold, begins to flow when it is fo thoroughly penetrated by the fire as to appear ignited like a live coal.

While this metal is in fufion, the immediate contact of the vapour of burning coals deprives it almoft entirely of its malleability, in the fame manner as we obferved happens to Gold: but both thefe metals eafily recover that property by being melted with nitre.

The nitrous acid is the true folvent of Silver, and being fomewhat dephlegmated will very readily and eafily take up a quantity of Silver equal in weight to itfelf.

Silver thus combined with the nitrous acid forms a metallic falt which fhoots into cryftals, called by the name of *Lunar Cryftals*, or *Cryftals of Silver*.

Thefe cryftals are moft violently cauftic: applied to the fkin they quickly affect it much as a live coal would; they produce a blackifh efchar, corroding and entirely deftroying the parts they touch. Surgeons ufe them to eat away the proud fungous flefh of ulcers. As Silver united with the nitrous acid hath the property of blackening all animal fubftances, a folution of this metallic falt is employed to dye hair, or other animal matters, of a beautiful and durable black.

Thefe cryftals flow with a very moderate heat, and even before they grow red. Being thus melted they form a blackifh mafs; and in this form they are ufed by Surgeons, under the title of *Lapis Infernalis*, *Infernal Stone*, or *Lunar Cauftic*.

Silver is alfo diffolved by the vitriolic acid: but then the acid muft be concentrated, and in quantity double the weight of the Silver; nor will the folution fucceed without a confiderable degree of heat.

Spirit

Spirit of falt and *aqua regis*, as well as the other acids, are incapable of diffolving this metal ; at leaft in the ordinary way.

Though Silver be not foluble in the acid of fea-falt, nor eafily in the acid of vitriol, as hath juft been obferved, it doth not follow that it hath but a weak affinity with the latter, and none at all with the former : on the contrary, it appears from experiment that it hath with thefe two acids a much greater affinity than with the acid of nitre : which is fingular enough, confidering the facility with which this laft acid diffolves it.

. The experiment which proves the fact, is this. To a folution of Silver in the nitrous acid, add the acid either of vitriol or of fea-falt, and the Silver will inftantly quit its nitrous folvent to join with the fuperadded acid.

Silver thus united with the vitriolic or the marine acid is lefs foluble in water than when combined with the nitrous acid : and for this reafon it is, that when either of thefe two acids is added to a folution of Silver, the liquor immediately becomes white, and a precipitate is formed, which is no other than the Silver united with the precipitating acid. If the precipitation be effected by the vitriolic acid, the precipitate will difappear upon adding a fufficient quantity of water, becaufe there will then be water enough to diffolve it. But the cafe is not the fame when the precipitation is made by the marine acid : for Silver combined therewith is fcarce foluble in water.

This Precipitate of Silver procured by means of the marine acid is very eafily fufed, and when fufed changes to a fubftance in fome meafure tranfparent and flexible ; which hath occafioned it to be called by the name of *Luna Cornea*. If it be propofed to decompound this *luna cornea*, that is, to feparate the marine acid from the Silver with which

it

it is united, the *luna cornea* muft be melted along with fatty and abforbent matters, with which the acid will unite, and leave the metal exceeding pure.

It muft be obferved that if, inftead of the marine acid, fea-falt in fubftance be added to a folution of Silver in the nitrous acid, a Precipitate is alfo produced, which by fufion appears to be a true *luna cornea*. The reafon is, that the fea-falt is decompofed by the nitrous acid, which feizes its bafis, as having a greater affinity therewith than its own acid hath; and this acid being confequently difengaged and fet at liberty unites with the Silver, which, as has been fhewn, has a greater affinity with it than with the nitrous acid. This is an inftance of decompofition effected by means of one of thofe double affinities mentioned by us in our feventh propofition concerning Affinities.

From what hath been already faid it is clear, that all thefe combinations of Silver with acids may be decompounded by abforbent earths and by fixed alkalis; it being a general law with regard to all metallic fubftances. We fhall not therefore repeat this obfervation when we come to treat of the other metals; unlefs fome particular occafion require it.

With regard to Silver I muft take notice that, when feparated by thefe means from the acids in which it was diffolved, it requires nothing but fimple fufion to reftore it to its ufual form; becaufe it does not, any more than Gold, lofe its phlogifton by thofe folutions and precipitations.

Silver unites with fulphur in fufion. If this metal be only made red-hot in a crucible, and fulphur be then added, it immediately flows; the fulphur acting as a flux to it. Silver thus united with fulphur forms a mafs that may be cut, is half malleable, and hath nearly the colour and confiftence of Lead. If this fulphurated Silver be kept a long time in fufion, and in a great degree

E 4   of

of heat, the sulphur flies off and leaves the Silver pure. But if the sulphur be evaporated by a violent heat, it carries off with it part of the Silver.

Silver unites and mixes perfectly with Gold in fusion. The two metals thus mixed form a compound with properties partaking of both.

Metallurgists have hitherto sought in vain for a perfectly good and easy method of separating these two metals by the *dry way* only : (this term is used to signify all operations performed by fusion :) but they are conveniently enough parted by the *moist way*, that is, by acid solvents. This method is founded on the above-mentioned properties of Gold and Silver with respect to acids. It hath been shewn that *aqua regis* only will dissolve Gold ; that Silver, on the contrary, is not soluble by *aqua regis*, and that its proper solvent is the acid of nitre : consequently, when Gold and Silver are mixed together, if the compound mass be put into *aqua fortis*, this acid will take up all the Silver, without dissolving a particle of the Gold, which will therefore remain pure ; and by this means the desired separation is effected. This method, which is commonly made use of by Goldsmiths and in Mints, is called the *Parting Assay*.

It is plain that if *aqua regis* were employed instead of *aqua fortis*, the separation would be equally effected ; and that the only difference between this process and the former would consist in this, that now the Gold would be dissolved, and the Silver remain pure. But the operation by *aqua fortis* is preferable ; because *aqua regis* does take up a little Silver, whereas *aqua fortis* hath not the least effect on Gold.

It must be observed that when Gold and Silver are mixed together in equal parts they cannot be parted by the means of *aqua fortis*. To enable the *aqua fortis* to act duly on the Silver, this metal must

muſt be, at leaſt, in a triple proportion to the Gold. If it be in a leſs proportion, you muſt either employ *aqua regis* to make the ſeparation, or, if you prefer the uſe of *aqua fortis,* melt the metalline maſs, and add as much Silver as is neceſſary to make up the proportion above-mentioned : and hence this Procefs is called *Quartation.*

This effect, which is pretty ſingular, probably ariſes from hence, that when the Gold exceeds or even equals the Silver in quantity, the parts of both being intimately united, the former are capable of coating over the latter, and covering them ſo as to defend them from the action of the *aqua fortis;* which is not the caſe when there is thrice as much Silver as Gold.

There is one thing more to be taken notice of with regard to this procefs ; which is, that perfectly pure *aqua fortis* is rarely to be met with, for two reaſons ; firſt, it is difficult in making it wholly to prevent the riſing of the medium employed to diſengage the nitrous acid ; that is, a little of the vitriolic acid will mix with the vapours of the *aqua fortis :* ſecondly, unleſs the ſaltpetre be very well purified it will always hold ſome ſmall portion of ſea-ſalt, the acid of which, we know, is very readily ſet looſe by the vitriolic acid, and conſequently riſes together with the vapours of the *aqua fortis.* It is eaſy to ſee that *aqua fortis* mixed either with the one or the other is not proper for the Parting Procefs; becauſe, as has juſt been ſaid, the vitriolic and the marine acid equally precipitate Silver diſſolved in the nitrous acid ; by which means, when they are united with that acid they weaken its action upon the Silver, and hinder the diſſolution. Add that *aqua fortis* adulterated with a mixture of ſpirit of ſalt becomes an *aqua regis,* and conſequently is rendered capable of
                                                                diſſolving

diffolving Gold, in proportion as its action upon Silver is diminifhed.

In order to remedy this inconvenience, and free *aqua fortis* from the vitriolic or marine acid with which it is tainted, Silver muft be diffolved therein : by degrees as the metal diffolves, thofe heterogeneous acids lay hold of it, and precipitate with it in the form of a white powder, as we obferved before. This precipitate being wholly fallen, the liquor grows clear ; after which, if it be found capable of diffolving more Silver, without turning milky, it may be depended on as a perfectly pure *aqua fortis.* Then filtre it, diffolve more Silver in it, as long as it will take up any, and you will have a folution of Silver in a very pure *aqua fortis.* By means of this folution may other *aqua fortis* be purified : for pour a few drops thereof into a very impure *aqua fortis,* and immediately the vitriolic or marine acid, with which that *aqua fortis* is contaminated, will join the Silver and fall therewith to the bottom. When the folution of Silver prepared as above does not in the leaft affect the tranfparency of the *aqua fortis,* it is then very pure, and fit for the purpofes of Quartation.

This operation of purifying *aqua fortis* by a folution of Silver is called the *Precipitation* of *aqua fortis* ; and *aqua fortis* thus purified is called *Precipitated Aqua Fortis.*

When Silver is diffolved in *aqua fortis* it may be feparated therefrom, as hath been fhewn, by abforbent earths and fixed alkalis.

We fhall fee by and by that there are other means of effecting this : but whatever way it be feparated from its folvent it recovers its metalline form, as Gold does, by being fimply fufed without any additament.

§. III. *Of*

§. III. *Of* Copper.

Of all the imperfect metals Copper comes the neareft to Gold and Silver. Its natural colour is a deep-red yellow. It refifts a very violent degree of fire for a confiderable time; but lofing its phlogifton at laft, it changes its metalline form for that of a calx, or a pure reddifh earth. This calx is hardly, if at all, reducible to glafs, without the addition of fomething to promote its fufion; all that the fierceft heat can do being only to render it foft. Copper, even while it retains its metalline form, and is very pure, requires a confiderable degree of fire to melt it, and does not begin to flow till long after it is red-hot. When in fufion it communicates a greenifh colour to the flame of the coals.

This metal is inferior to Silver in point of gravity; nor is its ductility fo great, though it be pretty confiderable: but, on the other hand, it exceeds that metal in hardnefs. It unites readily with Gold and Silver; nor does it greatly leffen their beauty when added to them in a fmall quantity: nay, it even procures them fome advantages; fuch as making them harder, and lefs fubject to lofe their ductility, of which thofe metals are often liable to be deprived, by the mixture of the fmalleft heterogeneous particle. This may probably arife from hence, that the ductility of Copper has the peculiarity of refifting moft of thofe caufes which rob the perfect metals of theirs.

The property, which other metalline fubftances have in common with Copper, of lofing the phlogifton by calcining and then vitrifying, furnifhes us with a method of feparating them from Gold and Silver, when they are combined therewith. Nothing more is required than to expofe the mafs compounded of the perfect metals and other metalline fubftances to

a de-a degree of heat fufficient to calcine whatever is not either Gold or Silver. It is evident that by this means thefe two metals will be obtained as pure as is poffible; for, as hath already been faid, no metalline calx or glafs is capable of uniting with metals poffeffed of their phlogifton. On this principle is formed the whole bufinefs of refining Gold and Silver.

When the perfect metals have no other alloy but Copper, as this metal is not to be calcined or vitrified without great difficulty, which is increafed by its union with the unvitrifiable metals, it is eafy to fee that it is almoft impoffible to feparate them without adding fomething to facilitate the vitrification of the Copper. Such metals as have the property of turning eafily to glafs are very fit for this purpofe; and it is neceffary to add a certain quantity thereof, when Gold or Silver is to be purified from the alloy of Copper. We fhall have occafion to be more particular on this fubject when we come to treat of lead.

Copper is foluble in all the acids, to which it communicates a green colour, and fometimes a blue. Even the neutral falts, and water itfelf, act upon this metal. With regard to water indeed, as the procuring it abfolutely pure and free from any faline mixture is next to an impoffibility, it remains a queftion whether the effect it produces on Copper be not owing to certain faline particles contained in it. It is this great facility of being diffolved that renders Copper fo fubject to ruft; which is nothing elfe but fome parts of its furface corroded by faline particles contained in the furrounding air and water.

The ruft of Copper is always green or blue, or of a colour between thefe two. Internally ufed it is very noxious, being a real poifon, as are all the folutions of this metal made by any acid whatever:

The

The blue colour, which Copper conftantly affumes when corroded by any faline fubftance, is a fure fign by which it may be difcovered wherever it exifts, even in a very fmall quantity.

Copper diffolved in the vitriolic acid forms a kind of metalline falt, which fhoots into rhomboidal cryftals of a moft beautiful blue colour. Thefe cryftals are called *Blue Vitriol*, or *Vitriol of Copper*. They are fometimes found ready formed in the bowels of the earth; and may be artificially made by diffolving Copper in the vitriolic acid; but the folution will not fucceed unlefs the acid be well dephlegmated. The tafte of this vitriol is faltifh and aftringent. It retains a confiderable quantity of water in cryftallizing, on which account it is eafily rendered fluid by fire.

It muft be obferved that, when it is expofed to a certain degree of heat in order to free it of its humidity, a great part of its acid flies off at the fame time: and hence it is that, after calcination, there remains only a kind of earth, or metalline calx, of a red colour, which contains but very little acid. This earth cannot be brought to flow but with the greateft difficulty.

A folution of copper in the nitrous acid forms a falt which does not cryftallize, but, when dried, powerfully attracts the moifture of the air. The fame thing happens when it is diffolved in fpirit of falt, or in *aqua regis*.

If the Copper thus diffolved by any of thefe acids be precipitated by an earth or an alkali, it retains nearly the colour it had in the folution: but thefe precipitates are fcarce any thing more than the earth of Copper, or Copper deprived of moft of its phlogifton; fo that if they were expofed to a violent fire, without any additament, a great part of them would be converted into an earth that could never be reduced to a metalline form. Therefore,

when we intend to reduce thefe precipitates to Copper, it is neceffary to add a certain quantity of a
fubftance capable of reftoring to them the phlogifton they have loft.

The fubftance which hath been found fitteft for
fuch reductions is charcoal-duft; becaufe charcoal
is nothing but a phlogifton clofely combined with
an earth, which renders it exceedingly fixed, and
capable of refifting a violent force of fire. But
as charcoal will not melt, and confequently is capable of preventing rather than forwarding the flux
of a metalline calx or glafs, which neverthelefs is
effentially neceffary to complete the reduction, it
hath been contrived to mix it, or any other fubftance containing the phlogifton, with fuch fixed
alkalis as eafily flow, and are fit to promote the
flux of other bodies. Thefe mixtures are called
Reducing Fluxes; becaufe the general name of *Fluxes*
is given to all falts, or mixtures of falts, which facilitate fufion.

If Sulphur be applied to Copper made perfectly
red-hot, the metal immediately runs; and thefe
two fubftances uniting form a new compound much
more fufible than pure Copper.

This compound is deftroyed by the fole force of
fire, for two reafons: the firft is, that, fulphur
being volatile, the fire is capable of fubliming a
great part of it, efpecially when it is in a great proportion to the Copper with which it is joined; the
fecond is, that the portion of fulphur which remains, being more intimately united with the Copper, though it be rendered lefs combuftible by that
union, is neverthelefs burnt and confumed in time.
Copper being combined with fulphur, and together
with it expofed to the force of fire, is found to be
partly changed into a blue vitriol; becaufe the vitriolic acid, being difengaged by burning the fulphur,
is by that means qualified to diffolve the Copper.

The

The affinity of Copper with fulphur is greater than that of Silver.

This metal, as well as the other imperfect metals and the femi-metals, being mingled with nitre and expofed to the fire, is decompofed and calcined much fooner than by itfelf; becaufe the phlogifton which it contains occafions the deflagration of the nitre, and confequently the two fubftances mutually decompofe each other. There are certain, metalline fubftances whofe phlogifton is fo abundant, and fo weakly connected with their earth, that when they are thus treated with nitre, there arifes immediately a detonation, accompanied with flame, and as violent as if fulphur or charcoal-duft had been employed; fo that in a moment the metalline fubftance lofes its phlogifton, and is calcined. The nitre, after thefe detonations, always affumes an alkaline character.

§. IV. *Of* Iron.

Iron is lighter and lefs ductile than Copper; but it is much harder, and of more difficult fufion.

It is the only body that has the property of being attracted by the magnet, which therefore ferves to difcover it wherever it is. But it muft be obferved that it hath this property only when in its metalline ftate, and lofes it when converted to an earth or calx. Hence very few Iron-ores are attracted by the load-ftone; becaufe, for the moft part, they are only forts of earths, which require a phlogifton to be added before they can be brought to the form of true Iron.

When Iron hath undergone no other preparation but the fufion which is neceffary to fmelt it from its ore, it is ufually quite brittle, and flies to pieces under the hammer: which arifes in fome meafure from its containing a certain portion of unmetallic earth
interpofed

interpofed between its parts. This we call *Pig Iron.*

By melting this a fecond time it is rendered purer, and more free from heterogeneous matters: but ftill, as its proper parts are probably not brought fufficiently near, or clofely enough united, till the Iron hath undergone fome further preparation befides that of fufion, it feldom hath any degree of malleability.

The way to give it this property is to make it juft red hot, and then hammer it for fome time in all directions; to the end that its parts may be properly united, incorporated, and welded together, and that the heterogeneous matters which keep them afunder may be feparated. Iron made by this means as malleable as poffible we call *Bar Iron,* or *Forged Iron.*

Bar Iron is ftill harder to fufe than Pig Iron: to make it flow requires the utmoft force of fire.

Iron has the property of imbibing a greater quantity of phlogifton than is neceffary to give it the metalline form. It may be made to take in this fuperabundant phlogifton two ways: the firft is by fufing it again with matters that contain the phlogifton; the fecond is, by encompaffing it with a quantity of fuch matters, charcoal-duft for inftance, and then expofing it fo encompaffed, for a certain time, to a degree of fire barely fufficient to keep it red-hot. This fecond method, whereby one fubftance is incorporated with another by means of fire, but without fufing either of them, is in general called *Cementation.*

Iron thus impregnated with an additional quantity of phlogifton is called *Steel.* The hardnefs of Steel may be confiderably augmented by *tempering* it; that is, by making it red-hot, and fuddenly quenching it in fome cold liquor. The hotter the metal, and the colder the liquor in which it is

quenched,

quenched, the harder will the Steel be. By this means tools are made, such as files and sheers, capable of cutting and dividing the hardest bodies, as glass, pebbles, and Iron itself. The colour of Steel is darker than that of Iron, and the facets which appear on breaking it are smaller. It is also less ductile and more brittle, especially when tempered.

As Iron may be impregnated with an additional quantity of phlogiston, and thereby converted into Steel, so may Steel be again deprived of that superabundant phlogiston, and brought back to the condition of Iron. This is effected by cementing it with poor earths, such as calcined bones and chalk. By the same operation Steel may be *untempered:* nay, it will lose the hardness it had acquired by tempering; if it be but made red-hot, and left to cool gradually. As Iron and Steel differ only in the respects we have here taken notice of, their properties being in all other respects the same, what follows is equally applicable to both.

Iron being exposed to the action of fire for some time, especially when divided into small particles, such as filings, is calcined and loses its phlogiston. By this means it turns to a kind of reddish yellow earth, which on account of its colour is called *Crocus Martis*, or *Saffron of Mars.*

This calx of Iron has the singular property of flowing in the fire with somewhat less difficulty than Iron itself; whereas every other metalline calx flows with less ease than the metal that produced it. It has moreover the remarkable property of uniting with the phlogiston, and of being reduced to Iron without fusion; requiring for that purpose only to be made red-hot.

Iron may be incorporated with Silver, and even with Gold, by means of certain operations. Under the article of Lead we fhall fee how it may be fe-parated from thefe metals.

The acids produce on it much the fame effects as on Copper: every one of them acts upon it. Certain neutral falts, alkalis, and even water itfelf, are capable of diffolving it; and hence it is alfo very fubject to ruft. The vitriolic acid diffolves it with the greateft eafe: but the circumftances which attend the folution thereof are different from thofe with'which the fame acid diffolves Copper: for, 1. whereas the vitriolic acid muft be concentrated to diffolve Copper, it muft on the contrary be diluted with water to diffolve Iron, which it will not touch when well dephlegmated. 2. The vapours which rife in this diffolution are inflammable; fo that if it be made in a fmall-necked bottle, and the flame of a candle be applied to the mouth thereof, the vapours in the bottle take fire with fuch rapidity as to produce a confiderable explofion.

This folution is of a beautiful green colour; and from this union of the vitriolic acid with Iron there refults a neutral metalline falt, which has the property of fhooting into cryftals of a rhomboidal figure, and a green colour. Thefe cryftals are called *Green Vitriol, Vitriol of Mars,* and *Copperas.*

Green Vitriol hath a faltifh and aftringent tafte. As it retains a great deal of water, in cryftalliz-ing it quickly flows by the action of fire: but this fluidity is owing to its water only, and is not a real fufion; for as foon as its moifture is evaporated, it refumes a folid form. Its green tranfparent colour is now changed into an opaque white: and, if the calcination be continued, its acid alfo exhales and is diffipated in vapours; and as it lofes that, it turns gradually to a yellow colour, which comes fo much the nearer to a red the longer the calcination is continued, or the higher the force of the fire is raifed; which being driven to the utmoft,

3

what

what remains is of a very deep red. This remainder is nothing but the body of the Iron, which having loſt its phlogiſton is now no more than an earth,
nearly of the ſame nature with that which is left
after calcining the metal itſelf.

Green Vitriol diſſolved in water ſpontaneouſly
lets fall a yellowiſh earthy ſediment. If this ſolution
be defecated by filtration, it ſtill continues to depoſite ſome of the ſame ſubſtance, till the vitriol be
wholly decompoſed. This ſediment is nothing but
the earth of Iron, which is then called *Ochre*.

The nitrous acid diſſolves Iron with great eaſe.
This ſolution is of a yellow colour, inclining more
or leſs to a ruſſet, or dark-brown, as it is more or
leſs ſaturated with Iron. Iron diſſolved by this acid
alſo, falls ſpontaneouſly in a kind of calx, which is
incapable of being diſſolved a ſecond time; for the
nitrous acid will not act upon Iron that has loſt its
phlogiſton. This ſolution does not cryſtallize, and
if evaporated to dryneſs attracts the moiſture of the
air.

Spirit of ſalt likewiſe diſſolves Iron, and this ſolution is green. The vapours which riſe during the
diſſolution are inflammable, like thoſe which aſcend
when this metal is attacked by the vitriolic acid.
Aqua regis makes a ſolution of Iron, which is of a
yellow colour.

Iron hath a greater affinity than either Silver or
Copper with the nitrous and vitriolic acids: ſo that
if Iron be preſented to a ſolution of either in one of
theſe two acids, the diſſolved metal will be precipitated; becauſe the acid quits it for the Iron, with
which it has a greater affinity.

On this occaſion it muſt be obſerved that if a ſolution of Copper in the vitriolic acid be precipitated
by means of Iron, the precipitate has the form and
ſplendour of a metal, and does not require the addition of a phlogiſton to reduce it to true Copper;

F 2 which

which is not the cafe, as has been fhewn, when the precipitation is effected by earths or alkaline falts.

The colour of this metalline precipitate hath deceived feveral perfons, who being unacquainted with fuch phenomena, and with the nature of blue vitriol, imagined that Iron was tranfmuted into Copper, when they faw a bit of Iron laid in a folution of that vitriol become, in form and external appearance, exactly like Copper: whereas the furface only of the Iron was crufted over with the particles of Copper contained in the vitriol, which had gradually fallen upon and adhered to the Iron, as they were precipitated out of the folution.

Among the folvents of Iron we mentioned fixed alkalis; and that they have fuch a power is proved by the following phenomenon. If a large proportion of alkaline falts be fuddenly mixed with a folution of Iron in an acid, no precipitation enfues, and the liquor remains clear and pellucid; or if at firft it look a little turbid, that appearance lafts but a moment, and the liquor prefently recovers its tranfparency. The reafon is, that the quantity of alkali is more than fufficient to faturate all the acid of the folution, and the fuperabundant portion thereof, meeting with the Iron already finely divided by the acid, diffolves it with eafe as faft as it falls, and fo prevents its muddying the liquor. To evince that this is fo in fact, let the alkali be applied in a quantity that is not fufficient, or but barely fufficient, to faturate the acid, and the Iron will then precipitate like any other metal.

Water alfo acts upon Iron; and therefore Iron expofed to moifture grows rufty. If Iron-filings be expofed to the dew, they turn wholly to a ruft, which is called *Crocus Martis Aperiens*.

Iron expofed to the fire together with nitre makes it detonate pretty brifkly, fets it in a flame, and decompofes it with rapidity.

This

This metal hath a greater affinity than any other metalline fubftance with fulphur; on which account it is fuccefsfully ufed to precipitate and feparate all metalline fubftances combined with fulphur.

Sulphur uniting with Iron communicates to it fuch a degree of fufibility, that if a mafs of this metal heated red-hot be rubbed with a bit of fulphur, it inceffantly runs into as perfect a fufion as a metal expofed to the focus of a large burning-glafs.

§. V. Of TIN.

TIN is the lighteft of all metals. Though it yields eafily to the impreffion of hard bodies, it has but little ductility. Being bent backwards and forwards it makes a fmall crackling noife. It flows with a very moderate degree of fire, and long before it comes to be red-hot. When it is in fufion, its furface foon grows dufky, and there forms upon it a thin dark-coloured dufty pellicle, which is no other than a part of the Tin that has loft its phlogifton, or a calx of Tin. The metal thus calcined eafily recovers its metalline form on the addition of a phlogifton. If the calx of Tin be urged by a ftrong fire it grows white, but the greateft violence of heat will not fufe it; which makes fome Chymifts confider it as a calcinable or abforbent earth, rather than a vitrifiable one. Yet it turns to glafs, in fome fort, when mixed with any other fubftance that vitrifies eafily. However, it always produces an imperfect glafs only, which is not at all tranfparent, but of an opaque white. The calx of Tin thus vitrified is called *Enamel.* Enamels are made of feveral colours by the addition of this or that metalline calx.

Tin unites eafily with all the metals; but it deftroys the ductility and malleability of every one of them, Lead excepted. Nay, it poffeffes this property of making metals brittle in fuch an eminent de-

gree,

gree, that the very vapour of it, when in fufion, is capable of producing this effect. Moreover, which is very fingular, the moft ductile metals, even Gold and Silver, are thofe on which it works this change with the moft eafe, and in the greateft degree. It has alfo the property of making Silver mixed with it flow over a very fmall fire.

It adheres to, and in fome meafure incorporates with, the furface of Copper and of Iron; whence arofe the practice of coating over thofe metals with Tin. Tin-plates are no other than thin plates of Iron tinned over.

If to twenty parts of Tin one part of Copper be added, this alloy renders it much more folid, and the mixed mafs continues tolerably ductile.

If on the contrary to one part of Tin ten parts of Copper be added, together with a little Zink, a femi-metal to be confidered hereafter, from this combination there refults a metalline compound which is hard, brittle, and very fonorous; fo that it is ufed for cafting bells: this compofition is called *Bronze* and *Bell-metal.*

Tin hath an affinity with the vitriolic, nitrous, and marine acids. All of them attack and corrode it; yet none of them is able to diffolve it without great difficulty: fo that if a clear folution thereof be defired, particular methods muft be employed for that purpofe; for the acids do but in a manner calcine it, and convert it to a kind of white calx or precipitate. The folvent which has the greateft power over it is *aqua regis,* which has even a greater affinity therewith than with Gold itfelf; whence it follows that Gold diffolved in *aqua regis* may be precipitated by means of Tin; but then the *aqua regis* muft be weakened. Gold thus precipitated by Tin is of a moft beautiful colour, and is ufed for a red in enameling and painting on porcelain, as alfo to give a red colour to artificial gems. If the *aqua regis*

be

be not lowered, the precipitate will not have the purple colour.

Tin hath the property of giving a great luftre to all red colours in general; on which account it is ufed by the dyers for ftriking a beautiful fcarlet, and tin veffels are employed in making fine fyrup of violets. Water does not act upon this metal, as it does upon Iron and Copper; for which reafon it is not fubject to ruft: neverthelefs when it is ex-pofed to the air its furface foon lofes its polifh and fplendour.

Tin mixed with nitre and expofed to the fire de-flagrates with it, makes it detonate, and is immedi-ately converted to a *refractory calx :* for fo all fub-ftances are called which are incapable of fufion.

Tin readily unites with fulphur, and with it be-comes a brittle and friable mafs.

§. VI. *Of* LEAD.

NEXT to Gold and Mercury Lead is the heavieft of all metalline fubftances, but in hardnefs is ex-ceeded by every one of them. Of all metals alfo it melts the eafieft except Tin. While it is in fufion there gathers inceffantly on its furface, as on that of Tin, a blackifh dufty pellicle, which is nothing but a calx of Lead.

This calx further calcined by a moderate fire, the flame being reverberated on it, foon grows white. If the calcination be continued it becomes yellow, and at laft of a beautiful red. In this ftate it is called *Minium*, and is ufed as a pigment. *Minium* is not eafily made, and the operation fucceeds well in large manufactures only.

To convert Lead into *Litharge,* which is the metal in a manner half vitrified, you need only keep it melted by a pretty ftrong fire; for then as its furface gradually calcines, it tends more and more to fufion and vitrification.

All

All thefe preparations of Lead are greatly difpofed to perfect fufion and vitrification, and for that
purpofe require but a moderate degree of fire; the
calx or earth of Lead being of all metalline earths
that which vitrifies the moft eafily.

Lead hath not only the property of turning
into glafs with the greateft facility, but it hath alfo
that of promoting greatly the vitrification of all the
other imperfect metals; and, when it is actually
vitrified, procures the ready fufion of all earths and
ftones in general, even thofe which are refractory,
that is, which could not be fufed without its
help.

Glafs of Lead, befides its great fufibility, hath
alfo the fingular property of being fo fubtile and
active as to corrode and penetrate the crucibles in
which it is melted, unlefs they be of an earth that
is exceeding hard, compact, and withal very refractory: for Glafs of Lead being one of the moft
powerful fluxes that we know, if the earth of the
crucible in which it is melted be in the fmalleft
degree fufible, it will be immediately vitrified;
efpecially if there be any metallic matter in its compofition.

The great activity of Glafs of Lead may be
weakened by joining it with other vitrifiable matters: but unlefs thefe be added in a very great proportion, it will ftill remain powerful enough to penetrate common earths, and carry off the matters
combined with it.

On thefe properties of Lead, and of the Glafs of
Lead, depends the whole bufinefs of refining Gold and
Silver. It hath been fhewn that as thefe two metals
are indeftructible by fire, and the only ones which
have that advantage, they may be feparated from
the imperfect metals, when mixed therewith, by expofing the compound to a degree of fire fufficiently
ftrong to vitrify the latter; which when once con-

3   verted

verted into glafs can no longer remain united with any metal that has its metalline form. But it is very difficult to procure this vitrification of the imperfect metals, when united with Gold and Silver ; nay, it is in a manner impoffible to vitrify them entirely, for two reafons : firft, becaufe moft of them are naturally very difficult to vitrify ; fecondly, becaufe the union they have contracted with the perfect metals defends them, in a manner, from the action of the fire, and that fo much the more effectually as the proportion of the perfect metals is greater; which being indeftructible, and in fome fort coating over thofe with which they are alloyed, ferve them as a prefervative and impenetrable fhield againft the utmoft violence of fire.

It is therefore clear that a great deal of labour may be faved, and that Gold and Silver may be refined to a much greater degree of purity than can otherwife be obtained, if to a mixture of thefe metals with Copper, for inftance, or any other imperfect metal, be added a certain quantity of Lead. For the Lead, by its known property, will infallibly produce the defired vitrification ; and as it likewife increafes the proportion of the imperfect metals, and fo leffens that of the perfect metals, in the mafs, it evidently deprives the former of a part of their guard, and fo effects a more complete vitrification. In conclufion, as the Glafs of Lead hath the property of running through the crucible, and carrying with it the matters which it has vitrified, it follows that when the vitrification of the imperfect metals is effected by its means, all thofe vitrified matters together penetrate the veffel containing the fufed metalline mafs, difappear, and leave only the Gold and Siver perfectly pure, and freed, as far as is poffible, from all admixture of heterogeneous parts.

The better to promote the feparation of fuch parts, it is ufual to employ in this procefs a particular fort

of

of ſmall crucibles, made of the aſhes of calcined bones, which are exceedingly porous and eaſily pervaded. They are called *cupels*, on account of their figure, which is that of a wide-mouthed cup : and from hence the operation takes its name; for when we refine Gold and Silver in this manner we are ſaid to *cupel* thoſe metals. It is eaſy to perceive that the more Lead is added the more accurately will the Gold and Silver be refined; and that ſo much the more Lead ought to be added as the perfect metals are alloyed with a greater proportion of the imperfect. This is the moſt ſevere trial to which a perfect metal can be put; and conſequently any metal that ſtands it may be fairly conſidered as ſuch.

In order to denote the fineneſs of Gold, it is ſuppoſed to be divided into twenty-four parts called *carats*; and Gold which is quite pure and free from all alloy is ſaid to be twenty-four *carats fine*; that which contains $\frac{1}{24}$ part of alloy is called Gold of twenty-three carats; that which contains $\frac{2}{24}$ of alloy is but twenty-two carats; and ſo on. Silver again is ſuppoſed to be divided into twelve parts only, which are called *penny-weights* : ſo that when abſolutely pure it is ſaid to be twelve *penny-weights fine*; when it contains $\frac{1}{12}$ of alloy, it is then called eleven penny-weights fine; when it contains $\frac{2}{12}$ of alloy, it is called ten penny-weights fine, and ſo on.

In treating of Copper we promiſed to ſhew, under the article of Lead, how to ſeparate it from Iron. The proceſs is founded on that property of Lead which renders it incapable of mixing and uniting with Iron, though it readily diſſolves all other metalline ſubſtances. Therefore if you have a maſs compounded of Copper and Iron, it muſt be fuſed with a certain quantity of Lead, and then the Copper, having a greater affinity with Lead than with Iron, will deſert the latter and join the former, which being incapable of any union with Iron, as

was

was faid, will wholly exclude it from the new compound. The next point is to feparate the Lead
from the Copper; which is done by expofing the
mafs compounded of thefe two metals to a degree
of fire ftrong enough to deprive the Lead of its
metalline form, but too weak to have the fame effect on the Copper: and this may be done; fince
of all the imperfect metals Lead is, next to Tin,
the eafieft to be calcined, and Copper on the contrary refifts the greateft force of fire longeft, without lofing its metalline form. Now what we gain
by this exchange, viz. by feparating Copper from
Iron and uniting it with Lead, confifts in this, that
as Lead is calcined with lefs fire than Iron, the Copper is lefs expofed to be deftroyed: For it muft be
obferved that, however moderate the fire be, it is
hardly poffible to prevent a certain quantity thereof from being calcined in the operation.

Lead melted with a third part of Tin forms a
compound, which being expofed to a fire capable
of making it thoroughly red-hot, fwells, puffs up,
feems in fome fort to take fire, and is prefently calcined. Thefe two metals mixed together are much
fooner calcined than either of them feparately.

Both Lead and Tin are in fome meafure affected
by water, and by a moift air; but they are both
much lefs fubject than Iron or Copper to be corroded by thefe folvents, and of courfe are much lefs
liable to ruft.

The vitriolic acid acts upon and diffolves Lead,
much in the fame manner as it doth Silver.

The nitrous acid diffolves this metal with much
eafe, and in great quantities; and from this folution
a fmall portion of mercury may be obtained. On
this fubject fee our *Elements of the Practice of Chymiftry*.

When this folution of Lead is diluted with a
good deal of water, the Lead precipitates in the
form

form of a white powder; which happens becaufe the acid is rendered too weak to keep the Lead diffolved.

If this folution of Lead be evaporated to a certain degree, it fhoots into cryftals formed like regular pyramids with fquare bafes. Thefe cryftals are of a yellowifh colour, and of a faccharine tafte: they do not eafily diffolve in water. This nitrous metalline falt has the fingular property of detonating in a crucible, without any additament, or the contaĉt of any other inflammable fubftance. This property it derives from the great quantity of phlogifton contained in, and but loofely connected with, the Lead which is one of its principles.

If fpirit of falt, or even fea-falt in fubftance, be added to a folution of Lead in the nitrous acid, a white precipitate immediately falls; which is no other than the Lead united with the marine acid. This precipitate is extremely like the precipitate of Silver made in the fame manner, and that being called *Luna cornea* hath occafioned this to be named *Plumbum corneum*. Like the *luna cornea* it is very fufible, and being melted hardens like it into a kind of horny fubftance: it is volatile, and may be reduced by means of inflammable matters combined with alkalis. But it differs from the *luna cornea* in this chiefly, that it diffolves eafily in water; whereas the *luna cornea*, on the contrary, diffolves therein with great difficulty, and in a very fmall quantity.

As this precipitation of Lead from its folution in fpirit of nitre is procured by the marine acid, Lead is thereby proved to have a greater affinity with the latter acid than with the former. Yet, if you attempt to diffolve Lead directly by the acid of fea-falt, the folution is not fo eafily effected as by the fpirit of nitre, and it is always imperfect; for it wants one of the conditions effential to every folution in a liquor, namely tranfparency.

If

If Lead be boiled for a long time in a lixivium of. fixed alkali, part of it will be diffolved.

Sulphur renders this metal refractory and fcarce fufible ; and the mafs they form when united together is friable. Hence it appears that fulphur acts upon Lead much in the fame manner as upon Tin ; that is, it renders both thefe metals lefs fufible, which are naturally the moft fufible of any, while it exceedingly facilitates the fufion of Silver, Copper, and Iron, metals which of themfelves flow with the greateft difficulty.

C H A P. VIII.

Of QUICK-SILVER.

WE treat of Quick-filver in a chapter apart, becaufe this metallic fubftance cannot be claffed with the metals properly fo called, and yet has fome properties which will not allow us to confound it with the femi-metals. The reafon why Quick-filver, by the Chymifts commonly called Mercury, is not reputed a metal is, that it wants one of the effential properties thereof, to wit, malleability. When it is pure and unadulterated with any mixture, it is always fluid, and of courfe unmalleable. But as, on the other hand, it eminently poffeffes the opacity, the fplendour, and above all the gravity of a metal, being next to Gold the heavieft of all bodies, it may be confidered as a true metal, differing from the reft no otherwife than by being conftantly in fufion ; which we may fuppofe arifes from its aptnefs to flow with fuch a fmall degree of heat, that be there ever fo little warmth on earth, there is ftill more than enough to keep Mercury in fufion ;

fufion; which would become folid and malleable if it
were poffible to apply to it a degree of cold confiderable enough for that purpofe.   Thefe properties
will not allow us to confound it with the femi-metals.   Add that we are not yet affured by any un-
doubted experiment that it can be wholly deprived
of its phlogifton, as the imperfect metals may.
Indeed we cannot apply the force of fire to it as
could be wifhed : for it is fo volatile that it flies off
and exhales in vapours, with a much lefs degree of
fire than is neceffary to make it red-hot.   The vapours of Mercury thus raifed by the action of fire,
being collected and united in a certain quantity, appear to be no other than true Mercury, retaining
every one of its properties ; and no experiment hath
ever been able to fhew the leaft change thus produced in its nature.

If Mercury be expofed to the greateft heat that it
can bear without fublimation, and continued in it
for feveral months, or even a whole year together,
it turns to a red powder, which the Chymifts call
Mercurius præcipitatus per fe. But to fucceed in this
operation it is abfolutely neceffary that the heat be
fuch as is above-fpecified ; for this metallic fubftance
may remain expofed to a weaker heat for a confiderable number of years, without undergoing any
fenfible alteration.

Some Chymifts fancied that by this operation they had fixed Mercury and changed its nature ; but without any reafon : for if the Mercury
thus feemingly tranfmuted be expofed to a fomewhat ftronger degree of fire, it fublimes and exhales
in vapours as ufual ; and thofe vapours collected
are nothing elfe but running Mercury, which has
recovered all its properties without the help of any
additament.   ·

Mercury has the property of diffolving all the
metals, Iron only excepted.   But it is a condition
                                        abfolutely

abfolutely neceffary to the fuccefs of fuch diffolution, that the metalline fubftances be poffeffed of their phlogifton; for if they be calcined, Mercury cannot touch them : and hence it follows that Mercury doth not unite with fubftances that are purely earthy. Such a combination of a metal with Mercury is called an *Amalgam*. Trituration alone is fufficient to effect it; however, a proper degree of heat alfo is of ufe.

Mercury amalgamated with a metal gives it a confiftence more or lefs foft, and even fluid, according to the greater or fmaller proportion of Mercury employed. All amalgams are foftened by heat, and hardened by cold.

Mercury is very volatile; vaftly more fo than the moft unfixed metals : moreover the union it contracts with any metal is not fufficiently intimate to entitle the new compound refulting from that union to all the properties of the two fubftances united; at leaft with regard to their degree of fixity and volatility. From all which it follows that the beft and fureft method of feparating it from metals diffolved by it, is to expofe the amalgam to a degree of heat fufficient to make all the Quick-filver rife and evaporate; after which the metal remains in the form of a powder, and being fufed recovers its malleability. If it be thought proper to fave the Quick-filver, the operation muft be performed in clofe veffels, which will confine and collect the mercurial vapours. This operation is moft frequently employed to feparate Gold and Silver from the feveral forts of earths and fands with which they are mixed in the ore; becaufe thefe two metals, Gold efpecially, are of fufficient value to compenfate the lofs of Mercury, which is inevitable in this procefs: befides, as they very readily amalgamate with it, this way of feparating them from every thing unmetallic is very facile and commodious.

Mercury

Mercury is diffolved by acids ; but with circumftances peculiar to each particular fort of acid.

The vitriolic acid, concentrated and made boiling hot, feizes on it, and prefently reduces it to a kind of white powder, which turns yellow by the affufion of water, but does not diffolve in it : it is called *Turbith Mineral.* However, the vitriolic acid on this occafion unites with a great part of the Mercury, in fuch a manner that the compound is foluble in water. For if to the water which was ufed to wafh the Turbith a fixed alkali be added, there falls inftantly a ruffet-coloured precipitate, which is no other than Mercury feparated from the vitriolic acid by the intervention of the alkali.

This diffolution of Mercury by the vitriolic acid is accompanied with a very remarkable phenomenon; which is, that the acid contracts a ftrong fmell of volatile fpirit of fulphur : a notable proof that part of the phlogifton of the Mercury hath united therewith. And yet, if the Mercury be feparated by means of a fixed alkali, it does not appear to have fuffered any alteration. Turbith Mineral is not fo volatile as pure Mercury.

The nitrous acid diffolves Mercury with eafe. The folution is limpid and tranfparent, and as it grows cold fhoots into cryftals, which are a nitrous mercurial falt.

If this folution be evaporated to drynefs, the Mercury remains impregnated with a little of the acid, under the form of a red powder, which hath obtained the names of *Red Precipitate,* and *Arcanum Corallinum.* This Precipitate, as well as Turbith, is lefs volatile than pure Mercury.

If this folution of Mercury be mixed with a folution of Copper, made likewife in the nitrous acid, and the mixture evaporated to drynefs, there will remain a green powder called *Green Precipitate.*

Thefe

These precipitates are cauftic and corrofive ; and are ufed as fuch in furgery.

Though Mercury be diffolved more eafily and completely by the nitrous acid than by the vitriolic, yet it has a greater affinity with the latter than with the former : for if a vitriolic acid be poured into a folution of Mercury in fpirit of nitre, the Mercury will quit the latter acid in which it was diffolved, and join the other which was added. The fame thing happens when the marine acid is employed inftead of the vitriolic. ·

Mercury combined with fpirit of falt forms a fingular body ; a metalline falt which fhoots into long cryftals, pointed like daggers. This falt is volatile, and fublimes eafily without decompofition. It is moreover the moft violent of all the corrofives hitherto difcovered by Chymiftry. It is called *Corrofive Sublimate*, becaufe it muft abfolutely be fublimed to make the combination perfect. There are feveral ways of doing this : but the operation will never fail, if the Mercury be rarified into vapours, and meet with the marine acid in a fimilar ftate.

Corrofive Sublimate is diffolved by water, but in very fmall quantities only. It is decompounded by fixed alkalis, which precipitate the Mercury in a reddifh yellow powder, called on account of its colour *Yellow Precipitate.*

If Corrofive Sublimate be mixed with tin, and the compound diftilled, a liquor comes over which continually emits abundance of denfe fumes, and from the name of its inventor is called the *Smoking Liquor of Libavius.* This liquor is no other than the tin combined with the marine acid of the Corrofive Sublimate, which therefore it hath actually decompounded : whence it follows that this acid hath a greater affinity with tin than with Mercury.

The marine acid in Corrofive Sublimate is not quite faturated with Mercury ; but is capable of

Vol. I.   G   taking

taking up a much greater quantity thereof. For if Corrosive Sublimate be mixed with fresh Mercury, and sublimed a second time, another compound will be produced containing much more Mercury, and less acrimonious; for which reason it is named *Sweet Sublimate of Mercury, Mercurius dulcis, Aquila alba.* This compound may be taken internally, and is purgative or emetic according to the dose administered. It may be rendered still more gentle by repeated sublimations, and then it takes the title of. *Panacea Mercurialis.* No way hath hitherto been found to dissolve Mercury in *aqua regis* without great difficulty, and even then it is but imperfectly dissolved.

Mercury unites easily and intimately with sulphur. If these two substances be only rubbed together in a gentle heat, or even without any heat, they will contract an union, tho' but an incomplete one. This combination takes the form of a black powder, which has procured it the name of *Æthiops Mineral.*

If a more intimate and perfect union be desired, this compound must be exposed to a stronger heat; and then a red ponderous substance will be sublimed, appearing like a mass of shining needles; this is the combination desired, and is called *Cinabar.* In this form chiefly is Mercury found in the bowels of the earth. Cinabar finely levigated acquires a much brighter red colour, and is known to painters by the name of *Vermilion.*

Cinabar rises wholly by sublimation, without suffering any decomposition; because the two substances of which it consists, viz. Mercury and Sulphur, are both volatile.

Though Mercury unites and combines very well with sulphur, as hath been said, yet it hath less affinity with that mineral than any other metal;—Gold only excepted: whence it follows that any of the other metals will decompound Cinabar, by uniting

· 3   with

with its fulphur, and fo fetting the Mercury at liberty to appear in its ufual form. Mercury thus feparated from fulphur is efteemed the pureft, and bears the name of *Mercury revivified from Cinabar.*

Iron is generally ufed in this operation, preferably to the other metals, becaufe among them all it has the greateft affinity with fulphur, and is the only one that has none with Mercury.

Cinabar may alfo be decompounded by means of fixed alkalis; the affinity of thefe falts with fulphur being generally greater than that of any metalline fubftance whatever.

CHAP. IX.

Of the SEMI-METALS.

§. I. *Of* REGULUS *of* ANTIMONY.

REGULUS of Antimony is a metallic fubftance of a pretty bright white colour. It has the fplendour, opacity, and gravity of a metal: but it is quite unmalleable, and crumbles to duft, inftead of yielding or ftretching, under the hammer; on which account it is claffed with the Semi-metals.

It begins to flow as foon as it is moderately red; but, like the other Semi-metals, it cannot ftand a violent degree of fire; being thereby diffipated into fmoke and white vapours, which adhere to fuch cold bodies as they meet with, and fo are collected into a kind of *farina* called *Flowers of Antimony.*

If Regulus of Antimony, inftead of being expofed to a ftrong fire, be only heated fo moderately that it fhall not even melt, it will calcine, lofe its phlogifton, and take the form of a greyifh

powder

powder deftitute of all fplendour : this powder is-
called *Calx of Antimony*.

This calx is not volatile like the Regulus, but
will endure a very violent fire; and being. expofed thereto will flow, and turn to a glafs of the.
yellowifh colour of a hyacinth.

It is to be obferved that the more the Regulus is
deprived of its phlogifton by continued calcination,
the more refractory is the calx obtained from it.
The glafs thereof has alfo fo much the lefs colour,
and comes the nearer to common glafs.

The calx and the Glafs of Antimony will recover
their metalline form, like every other Calx and
Glafs of a metal, if reduced by reftoring to them
their loft phlogifton. Yet if the calcination be carried too far, their reduction will become much more
difficult, and a much fmaller quantity of Regulus
will be refufcitated.

Regulus of Antimony is capable of diffolving the
metals ; but its affinities with them are various, and
differ according to the following order. It affects
Iron the moft powerfully, next Copper, then Tin,
Lead, and Silver. It promotes the fufion of metals,
but makes them all brittle and unmalleable.

It will not amalgamate with Mercury ; and
though by certain proceffes, particularly the addition of water and continued trituration, a fort of
union between thefe two fubftances may be produced, yet it is but apparent and momentary ; for
being left to themfelves and undifturbed they
quickly difunite and feparate *.

* M. Malouin, however, hath found a way to unite thefe two
metallic fubftances : but then he does it by the interpofition of
fulphur ; that is, he combines crude Antimony with Mercury.
This combination is brought about in the fame way that Æthiops Mineral is made ; viz. either by fufion, or by trituraticn
only without fire. It refembles the common Æthiops, and
M. Malouin calls it *Æthiops of Antimony*. He obferved that
Mercury unites with Antimony much more intimately, by
melting, than by rubbing them together.

The vitriolic acid, affisted by heat, and even by diftillation, diffolves Regulus of Antimony. The nitrous acid likewife attacks it ; but the folution can by no art be made clear and limpid : fo that the Regulus is only calcined, in a manner, by this acid.

The marine acid diffolves it well enough ; but then it muft be exceedingly concentrated, and applied in a peculiar manner, and efpecially by diftillation. One of the beft methods of procuring a perfect union between the acid of fea-falt and Regulus of Antimony, is to pulverize the latter, mix it with corrofive fublimate, and diftill the whole. There rifes in the operation a white matter, thick and fcarce fluid, which is no other than the Regulus of Antimony united and combined with the acid of fea-falt. This compound is extremely corrofive, and is called *Butter of Antimony*.

- It is plain that the corrofive fublimate is here decompounded ; that the Mercury is revivified, and that the acid which was combined therewith hath quitted it to join the Regulus of Antimony, with which its affinity is greater. This Butter of Antimony by repeated diftillations acquires a confiderable degree of fluidity and limpidnefs.

If the acid of nitre be mixed with Butter of Antimony, and the whole diftilled, there rifes an acid liquor, or a fort of *aqua regis*, which ftill retains fome of the diffolved Regulus, and is called *Bezoardic Spirit of Nitre*. After the diftillation there remains a white matter, from which frefh fpirit of nitre is again abftracted, and which being then wafhed with water is called *Bezoar Mineral*. This Bezoar Mineral is neither fo volatile, nor fo cauftic, as Butter of Antimony ; becaufe the nitrous acid hath not the property of volatilizing metallic fubftances, as the marine acid does, and becaufe it remains much more intimately combined with the reguline part.

G 3   If

If Butter of Antimony be mixed with water, the liquor immediately becomes turbid and milky, and a precipitate falls, which is nothing but the metal-lic matter partly feparated from its acid, which is too much weakened by the addition of water to keep it diffolved. Yet this precipitate ftill retains a good deal of acid ; for which reafon it continues to be a violent emetic, and in fome degree corro-five. It hath therefore been very improperly called *Mercurius Vitæ*.

The proper folvent of Regulus of Antimony is *aqua regis*; by means whereof a clear and limpid folution of this Semi-metal may be obtained.

Regulus of Antimony mixed with nitre, and pro-jected into a red-hot crucible, fets the nitre in a flame, and makes it detonate. As it produces this effect by means of its phlogifton, it muft needs at the fame time be calcined, and lofe its metallic pro-perties, which accordingly happens, and when the nitre is in a triple proportion to the Regulus, the latter is fo perfectly calcined as to leave only a white powder, which is fufed with great difficulty, and then turns to a faintly coloured glafs, not very dif-ferent from common glafs, and which is not redu-cible to a Regulus by the addition of inflammable matter; at leaft it yields but a very fmall quantity thereof. If lefs nitre be ufed, the calx is not fo white ; the glafs it produces is more like a metal-line glafs, and is more eafily reduced. The calx of the Regulus thus prepared by nitre is called, on account of the medicinal virtue afcribed to it, *Dia-phoretic Antimony*, or *Diaphoretic Mineral*.

Nitre always becomes an alkali by deflagration, and in the prefent cafe retains part of the calx, which it even renders foluble in water. This calx may be feparated from the alkali, if an acid be em-ployed to precipitate it; and then it is called *Mate-ria Perlata.* This pearly matter is a calx of Anti-mony

mony, fo completely deprived of its phlogifton as to be altogether incapable of reduction to a Regulus.

Regulus of Antimony readily joins and unites with fulphur, forming therewith a compound which has a very faint metallic fplendour. This compound appears like a mafs of long needles adhering together laterally; and under this form it is ufually found in the ore, or at leaft when only feparated by fufion from the ftones and earthy matters with which the ore is mixed. It is called *Crude Antimony*.

Antimony flows with a moderate heat, and becomes even more fluid than other metallic fubftances. The action of fire diffipates or confumes the fulphur it contains, and its phlogifton alfo, fo as to convert it into a calx and a glafs, as it does the Regulus.

Aqua regis, which we obferved to be the proper folvent of the Regulus, being poured on Antimony, attacks and diffolves the reguline part, but touches not the fulphur; in confequence whereof it decompofes the Antimony, and feparates its fulphur from its Regulus.

There are feveral other ways of effecting this decompofition, and obtaining the reguline part of Antimony by itfelf: they confift either in deftroying the fulphureous part of the Antimony by combuftion, or in melting the Antimony with fome fubftance, which has a greater affinity than its reguline part with fulphur. Moft metals are very fit for this latter purpofe: for though the Regulus has a confiderable affinity with fulphur, yet all the metals, except Gold and Mercury, have a greater.

If therefore Iron, Copper, Lead, Silver, or Tin, be melted with Antimony, the metal employed will unite with the fulphur, and feparate it from the Regulus.

It

It muſt be obſerved that, as theſe metals have ſome affinity with the Regulus of Antimony, the Regulus will be joined in the operation by ſome of the metal employed as a *Precipitant,* (ſo thoſe ſubſtances are called which ſerve as the means of ſeparating two bodies from each other;) and therefore the Regulus procured in this manner will not be abſolutely pure : on this account care is taken to diſtinguiſh each by adding the name of the metal employed in its precipitation ; and thence come theſe titles, *Martial Regulus of Antimony,* or only *Martial Regulus, Regulus Veneris ;* and ſo of the reſt.

Antimony is employed with advantage to ſeparate Gold from all the other metals with which it may be alloyed. It has been ſhewn that all the metals have a greater affinity, than the reguline part of Antimony, with ſulphur, Gold only excepted ; which is incapable of contracting any union therewith: and therefore, if a maſs compounded of Gold, and ſeveral other metals be melted with Antimony, every thing in that maſs which is not Gold will unite with the ſulphur of the Antimony. This union occaſions two ſeparations, to wit, that of the ſulphur of the Antimony from its reguline part, and that of the Gold from the metals with which it was adulterated ; and from the whole two new compounds ariſe ; namely, a combination of the metals with the ſulphur, which being lighteſt riſes to the ſurface in fuſion ; and a metalline maſs formed of the Gold and the reguline part of the Antimony united together, which being much the heavieſt ſinks to the bottom. There is no difficulty in parting the Gold from the Regulus of Antimony with which it is alloyed : for the metalline maſs need only be expoſed to a degree of fire capable of diſſipating into vapours all the Semi-metal it contains ; which being very volatile, the operation is much eaſier, and more expeditiouſly finiſhed, than if the metals with which

the

the Gold was debafed were to be vitrified on the
cupel; without taking into the account that if Silver were one of them, recourfe muft needs be had
to the procefs of quartation after that of the cupel.

If equal parts of nitre and Antimony be mixed
together, and the mixture expofed to the action of
fire, a violent detonation enfues; the nitre deflagrating confumes the fulphur of the Antimony, and
even a part of its phlogifton. After the detonation
there remains a greyifh matter which contains fixed
nitre, vitriolated tartar, and the reguline part of
the Antimony in fome meafure deprived of its phlogifton, and half vitrified by the action of the fire,
which is confiderably increafed by the deflagration.
This matter is called *Liver of Antimony.*

If inftead of equal parts of nitre and Antimony,
two parts of the former be ufed to one of the latter, then the reguline part lofes much more of its
phlogifton, and remains in the form of a yellowifh
powder.

Again, if three parts of nitre be taken to one of
Antimony, the Regulus is thereby entirely robbed
of its phlogifton, and converted to a white calx
which bears the name of *Diaphoretic Antimony*, or
Diaphoretic Mineral. The pearly matter may be
precipitated by pouring an acid on the faline fubftances which here remain after the detonation, in
the fame manner as we fhewed above was to be
done with regard to the Regulus.

In the two laft operations, where the nitre is in a
double or triple proportion to the Antimony, the
reguline part is found after the detonation to be converted into a calx, and not into a half vitrified
matter, which we have feen is the effect when equal
parts only of nitre and Antimony are ufed. The
reafon of this difference is, that in thefe two cafes the
reguline part, being wholly, or almoft wholly, deprived of its phlogifton, becomes, as was obferved,

more

more difficult to fuse, and consequently cannot begin to vitrify in the same degree of heat as that
which hath not loft so much of its phlogiston. If
instead of performing the operation with equal parts
of nitre and Antimony alone, a portion of some
substance which abounds with phlogiston be added,
in that case the sulphur only of the Antimony will
be consumed, and the Regulus will remain united
with its phlogiston and separated from its sulphur.

The Regulus prepared in this manner is absolutely pure, because no metalline substance being
employed, none can mix with and adulterate it.
It is called *Regulus of Antimony per se*, or only *Regulus of Antimony*.

It is true indeed that in this operation much of
the reguline part unavoidably loses its phlogiston
and is calcined, and consequently a much smaller
quantity of Regulus is obtained than when metalline precipitants are employed: but this loss is easily
repaired, if it be thought proper, by restoring to
the calcined part its loft phlogiston.

Antimony melted with two parts of fixed alkali
yields no Regulus, but is entirely dissolved by the
salt, and forms with it a mass of a reddish yellow
colour.

The reason why no precipitate is produced on
this occasion is, that the alkali uniting with the
sulphur of the Antimony forms therewith the combination called Liver of Sulphur, which by its nature is qualified to keep the reguline part dissolved.
This mass formed by the union of the Antimony
with the alkali is soluble in water. If any acid
whatever be dropt into this solution, there falls a
precipitate of a reddish yellow colour; because the
acid unites with the alkali, and forces it to quit the
matters with which it was combined. This precipitate is called *Golden Sulphur of Antimony*.

As

As in the operation for preparing *Regulus of Antimony per se*, some of the nitre is, by the inflammable matters added thereto, turned to an alkali, this alkali seizes on part of the Antimony, and therewith forms a compound like that just described. Hence it comes that if the scoria formed in this process be dissolved in water, and an acid dropped into the solution, a true golden sulphur of Antimony is thereby separated.

This union of Antimony with an alkali may also be brought about by the humid way; that is, by making use of an alkali resolved into a liquor, and boiling the mineral in it. The alkaline liquor, in proportion as it acts upon the Antimony, gradually becomes reddish and turbid. If left to settle and cool when well saturated therewith, it gradually deposites the Antimony it had taken up, which precipitates in the form of a red powder: and this precipitate is the celebrated remedy known by the name of *Kermes Mineral*. It is plain that the kermes is nearly the same thing with the golden sulphur: yet it differs from it in some respects; and especially in this, that being taken inwardly it operates much more gently than the golden sulphur, which is a violent emetic. Nitre fixed by charcoal, and resolved into a liquor, is the only alkali employed in preparing the kermes.

It was shewn above that Regulus of Antimony mixed and distilled with corrosive sublimate decompounds it, disengages the Mercury, and joining itself to the marine acid forms therewith a new combination, called Butter of Antimony. If the same operation be performed with crude Antimony instead of its Regulus, the same effects are produced: but then the Antimony itself is also decomposed; that is, the reguline part is separated from the sulphur, which being set free unites with the Mercury, now also at liberty, and these two
together

together form a true cinabar called *Cinabar of An-timony.*

§. II. *Of* BISMUTH.

BISMUTH, known alfo by the name of Tin-glafs, is a femi-metal, having almoſt the fame appearance as Regulus of Antimony; yet it has a more duſky caſt, inclining fomewhat to red, and even prefents fome changeable ſtreaks, efpecially after lying long in the air.

When expofed to the fire it melts long before it is red, and confequently with lefs heat than Regulus of Antimony, which does not flow, as was ſhewn above, till it begin to be red-hot. It becomes volatile, like all the other femi-metals, when acted on by a violent fire: being kept in fufion by a proper degree of heat it lofes its phlogiſton with its metallic form, and turns to a powder or a calx; and that again is converted into glafs by the continued action of fire. The calx and glafs of Biſmuth may be reduced, like any other metallic calx, by reſtoring their phlogiſton.

Biſmuth mixes with all the metals in fufion, and even facilitates the fufion of fuch as do not otherwife flow readily. It whitens them by its union, and deſtroys their malleability.

It amalgamates with Mercury, if they be rubbed together with the addition of water: yet after fome time thefe two metalline fubſtances defert each other, and the Biſmuth appears again in the form of a powder. Hence it is plain that the union it contracts with Mercury is not perfect; and yet it has the fingular property of attenuating Lead, and altering it in fuch a manner that it afterwards amalgamates with Mercury much more perfectly, fo as even to pafs with it through ſhamoy leather without any feparation. The Biſmuth employed in making this amalgama afterwards feparates from it fponta-neouſly,

neoufly, as ufual; but the Lead ftill, continues
united with the Mercury, and always retains the
property thus acquired.

The vitriolic acid does not diffolve Bifmuth: its
proper folvent is the nitrous acid, which diffolves
it with violence, and abundance of fumes.

Bifmuth diffolved in the nitrous acid is precipitated not only by alkalis, but even by the bare addition of water. This precipitate is extremely white,
and known by the name of *Magiftery of Bifmuth.*

The acid of fea-falt and *aqua regis* likewife act
upon Bifmuth, but with lefs violence.

This femi-metal does not fenfibly deflagrate with
nitre; yet it is quickly deprived of its phlogifton,
and turned into a vitrifiable calx, when expofed
with it to the action of fire.

It readily unites with fulphur in fufion, and forms
therewith a compound which appears to confift of
needles adhering laterally to each other.

It may be feparated from the fulphur with which
it is combined, by only expofing it to the fire,
without any additament; for the fulphur is either
confumed or fublimed, and leaves the Bifmuth behind.

§. III. *Of* ZINC.

ZINC to appearance differs but little from Bifmuth, and has even been confounded with it by
feveral authors. Neverthelefs, befides that it has
fomething of a blueifh caft, and is harder than Bifmuth, it differs from it effentially in its properties,
as will prefently be fhewn. Thefe two metallic
fubftances fcarce refemble each other in any thing,
but the qualities common to all femi-metals.

Zinc melts the moment it grows red in the fire,
and then alfo begins to turn to a calx, which, like
any other metallic calx, may be reduced by means
of the phlogifton: but if the fire be confiderably
increafed,

increafed, it fublimes, flames, and burns like an oily matter; which is a proof of the great quantity of phlogifton in its compofition. At the fame time abundance of flowers rife from it in the form of white flakes, flying about in the air like very light bodies; and into this form may the whole fubftance of the Zinc be converted. Several names have been given to thefe flowers, fuch as Pompholyx, Philofophic Wool. They are fuppofed to be no other than the Zinc itfelf deprived of its phlogifton; yet nobody has hitherto been able to refufcitate them in the form of Zinc, by reftoring their phlogifton according to the methods ufed in the reduction of metals. Though they rife in the air with very great eafe while the Zinc is calcining, yet when once formed they are very fixed; for they withftand the utmoft violence of fire, and are capable of being vitrified, efpecially if joined with a fixed alkali. They are foluble in acids.

Zinc unites with all metalline fubftances, except Bifmuth. It has this fingular property, that being mixed with Copper, even in a confiderable quantity, fuch as a fourth part, it does not greatly leffen the ductility thereof, and at the fame time communicates to it a very beautiful colour not unlike that of Gold: on which account the compofition is frequently made, and produces what is called *Brafs*. This metal melts much more eafily than Copper alone, becaufe of the Zinc with which it is alloyed. If it be expofed to a great degree of heat, the Zinc which it contains takes fire, and fublimes in white flowers, juft as when it is pure.

It is to be obferved that Brafs is ductile only while it is cold, and not then unlefs the Zinc ufed in making it was very pure; otherwife the compofition will prove but a *Tombac* or *Prince's Metal*, having very little malleability.

Zinc

Zinc is very volatile, and carries off with it any metallic fubftance with which it is fufed, making a kind of fublimate thereof. In the furnaces where they fmelt ores containing Zinc, the matter thus fublimed is called *Cadmia Fornacum*, to diftinguifh it from the native *Cadmia* called alfo *Calamine*, or *Lapis Calaminaris*; which, properly fpeaking, is an ore of Zinc, containing a great deal of that femi-metal, together with fome Iron, and a ftony fubftance. The name of *Cadmia Fornacum* is not appropriated folely to the metallic fublimates procured by means of Zinc, but is given in general to all the metallic fublimates found in fmelting houfes.

If a violent and fudden heat be applied to Zinc, it fublimes in its metalline form; there not being time for it to burn and be refolved into flowers.

This femi-metal is foluble in all the acids, but efpecially in fpirit of nitre, which attacks and diffolves it with very great violence.

Zinc has a greater affinity than iron or copper with the vitriolic acid; and therefore it decompounds the green and blue vitriols, precipitating thofe two metals by uniting with the vitriolic acid, with which it forms a metallic falt, or vitriol, called *White Vitriol*, or *Vitriol of Zinc*.

Nitre mixed with Zinc, and projected into a redhot crucible, detonates with violence, and during the detonation there rifes a great quantity of white flowers, like thofe which appear when it is calcined by itfelf.

Sulphur has no power over Zinc. Even liver of fulphur, which diffolves all other metallic fubftances, contracts no union with this femi-metal.

Meff. Hellot and Malouin have beftowed a great deal of pains on this femi-metal. An account of their experiments is to be found in the Memoirs of the Academy of Sciences.

§. IV. *Of*

§. IV. *Of* REGULUS OF ARSENIC.

REGULUS of Arſenic is the moſt volatile of all the ſemi-metals. A very moderate heat makes it wholly evaporate, and fly off in fumes; on which account it cannot be brought to fuſion, nor can any conſiderable maſſes thereof be obtained. It has a metallic colour, ſomewhat reſembling Lead; but it ſoon loſes its ſplendour when expoſed to the air.

It unites readily enough with metallic ſubſtances, having the ſame affinities with them as Regulus of Antimony hath. It makes them brittle, and unmalleable. It hath alſo the property of rendering them volatile, and greatly facilitates their ſcorification.

It very eaſily parts with its phlogiſton and its metallic form. When expoſed to the fire it riſes in a kind of ſhining cryſtalline calx, which on that account looks more like a ſaline matter than a metallic calx. To this calx or theſe flowers are given the names of *White Arſenic*, *Cryſtalline Arſenic*, and moſt commonly plain *Arſenic*.

The properties of this ſubſtance are very ſingular, and extremely different from thoſe of any other metallic calx. Hitherto it hath been but little examined; and this led me to make ſome attempts towards diſcovering its nature, which may be ſeen in the Memoirs of the Academy of Sciences.

Arſenic differs from every other metalline calx, firſt, in being volatile; whereas the calxes of all other metallic ſubſtances, not excepting thoſe of the moſt volatile ſemi-metals, ſuch as Regulus of Antimony and Zinc, are exceedingly fixed; and ſecondly, in having a ſaline character, which is not found in any other metalline calx.

The ſaline character of Arſenic appears, firſt, from its being ſoluble in water; ſecondly, from its corroſive quality, which makes it one of the moſt violent

lent

lent poifons : a quality from which the other metallic fubftances are free, when they are not combined
with fome faline matter. Regulus of Antimony
muft however be excepted. But then the beft Chymifts agree that this femi-metal is either nearly of
the fame nature with Arfenic, or contains a portion
thereof in its compofition: befides, its noxious
qualities never difcover themfelves fo plainly as
when it is combined with fome acid. Laftly, Arfenic acts juft like the vitriolic acid upon nitre; that
is, it decompounds that neutral falt, by expelling
its acid from its alkaline bafis, of which it takes
poffeffion, and therewith forms a new faline compound.

This combination is a fpecies of falt that is perfectly neutral. When the operation is performed
in a clofe veffel, the falt fhoots into cryftals in the
form of right-angled quadrangular prifms, terminated at each extremity by pyramids that are
alfo quadrangular and right-angled; fome of which
however, inftead of ending in a point, are obtufe
as if truncated. The confequence is different
when the operation is performed in an open veffel;
for then nothing is obtained but an alkaline falt
impregnated with Arfenic, which cannot be cryftallized.

The caufe of this different effect is that, when
the Arfenic is once engaged in the alkaline bafis of
the nitre, it can never be feparated from it by the
utmoft force of fire, fo long as it is kept in a clofe
veffel; whereas, if you expofe it to the fire without
that precaution, it readily feparates from it. This
property of Arfenic was never before obferved by
any Chymift, and therefore this our new fpecies of
Neutral arfenical falt was abfolutely unknown till
lately.

This new falt poffeffes many fingular properties,
the chief of which are thefe. Firft, it cannot be

decompounded by the intervention of any acid, even the ftrongeft acid of vitriol; and this, joined to its property of expelling the nitrous acid from its bafis, fhews that it has a very great affinity with fixed alkalis.

Secondly, this very falt, on which pure acids have no effect, is decompounded with the greateft eafe by acids united with metallic fubftances. The reafon of this phenomenon is curious, and furnifhes us with an inftanee of what we advanced concerning double affinities.

If to a folution of any metallic fubftance whatever, made by any acid whatever, (except that of Mercury by the marine acid, and that of Gold by *aqua regis*), a certain quantity of our New Salt diffolved in water be added, the metallic fubftance is inftantaneoufly feparated from the acid in which it was diffolved, and falls to the bottom of the liquor.

All metallic precipitates obtained in this manner are found to be a combination of the metal with Arfenic; whence it neceffarily follows that the new Neutral Salt is by this means decompounded, its arfenical part uniting with the metallic fubftance, and its alkaline bafis with the acid in which that fubftance was diffolved.

The affinities of thefe feveral bodies muft be confidered as operating on this occafion in the following manner: The acids which tend to decompound the Neutral Salt of Arfenic, by virtue of their affinity with its alkaline bafis, are not able to accomplifh it, becaufe this affinity is powerfully counteracted by that which the Arfenic has with the fame alkaline bafis, and which is equal or even fuperior to theirs. But if thefe acids happen to be united with a fubftance which naturally has a very great affinity with the arfenical part of the Neutral Salt, then, the two parts of which this Salt confifts being drawn different ways by two feveral affinities tending to fe-

7   parate

parate them from each other, the Salt will undergo a decompofition, which could not have been effected without the help of this fecond affinity. Now as metallic fubftances have a great affinity with Arfenic, it is not furprifing that the Neutral Salt of Arfenic, which cannot be decompounded by a pure acid, fhould neverthelefs yield to an acid combined with a metal. The decompofition of this Salt, therefore, and the precipitation which of courfe it produces in metallic folutions, are brought about by the means of a double affinity; namely, that of the acid with the alkaline bafis of the Neutral Salt, and that of the metal with the arfenical part of that falt.

Arfenic has not the fame effect on fea-falt as on nitre, and cannot expel its acid: a very fingular phenomenon, for which it is hard to affign a reafon; for the nitrous acid is known to have a greater affinity than the marine acid with alkalis, and even with the bafis of fea-falt itfelf.

Yet Arfenic may be combined with the bafis of fea-falt, and a Neutral falt thereby obtained, like that which refults from the decompofition of nitre by Arfenic: but for that purpofe a quadrangular nitre muft be firft prepared, and Arfenic applied thereto as to common nitre.

The Salt produced by uniting Arfenic with the bafis of fea-falt very much refembles the Neutral falt of Arfenic above treated of, as well in the figure of its cryftals as in its feveral properties.

Arfenic prefents another fingular phenomenon; both with the alkali of nitre and with that of fea-falt; which is, that if it be combined with thefe falts in a fluid ftate, it forms with them a faline compound, quite different from the Neutral falts of arfenic which refult from the decompofition of nitrous falts.

This faline compound, which I call *Liver of Arfenic*, takes up a much greater quantity of Arfenic

than

than is neceſſary for the perfect ſaturation of the alkali. It has the appearance of a glue, which is ſo
much the thicker the more Arſenic it contains. Its
ſmell is diſagreeable; it attracts the moiſture of the
air, and does not cryſtallize; it is eaſily decompounded by any acid whatever, which precipitates
the Arſenic and unites with the alkali. Laſtly, the
effects it produces on metallic ſolutions are different
from thoſe of our neutral arſenical ſalts. But the
bounds which I have ſet myſelf in this treatiſe will
not allow me to be more particular. Such as have
the curioſity to enquire further into the ſubject may
conſult my Diſſertations on Arſenic, publiſhed
among the Memoirs of the Academy of Sciences.

Arſenic is eaſily reduced to a Regulus. It need
only be mixed with any matter containing the phlogiſton, and by the help of a moderate heat a true
Regulus will ſublime. This Regulus, as was ſaid,
is very volatile, and calcines with the greateſt eaſe;
which is the reaſon why it cannot be obtained but
in ſmall quantities, and alſo why, in order to obtain maſſes of it, ſome have thought of adding
thereto ſome metal with which it has a great affinity, ſuch as Copper or Iron; becauſe by joining
with the metal it is partly fixed and reſtrained from
flying off. But it is plain the Regulus obtained by
this means is not pure, as it muſt partake conſiderably of the metal employed.

Arſenic readily unites with ſulphur, and riſes
with it in a yellow compound called *Orpiment*.

Sulphur cannot be ſeparated from Arſenic but by
the intervention of two bodies only; to wit, a fixed
alkali and Mercury.

The property which Mercury poſſeſſes of ſeparating ſulphur from Arſenic is founded on this, that
theſe two metallic ſubſtances are incapable of contracting any union; whereas though moſt of the
other metals and ſemi-metals have a greater affinity

with

with fulphur than Mercury hath, as was fhewn in treating of the decompofition of Cinabar, neverthelefs they are all unable to decompound Orpiment; becaufe fome of them have as great an affinity with Arfenic as with fulphur; others have no affinity with either; and laftly, fulphur hath as great an affinity with Arfenic as with any of them.

It muft be obferved that, if fixed alkalis be employed to purify Arfenic in this manner, no more muft be ufed than is neceffary to abforb the fulphur or the phlogifton, of which alfo it is their nature to deprive Arfenic; for otherwife, as it has been fhewn, that Arfenic readily unites with alkalis, they would abforb a confiderable quantity thereof.

CHAP. X.

Of OIL in general.

OIL is an unctuous body, which burns and confumes with flame and fmoke, and is not foluble in water. It confifts of the phlogifton united with water by means of an acid. There is moreover in its compofition a certain proportion of earth, more or lefs according to each feveral fort of Oil.

The inflammability of Oil evidently proves that it contains the phlogifton. That an acid is one of its conftituent principles many experiments demonftrate, of which thefe are the chief: If certain Oils be long triturated with an alkaline falt, and the alkali afterwards diffolved in water, cryftals of a true neutral falt will be produced: fome metals, and particularly Copper, are corroded and rufted by Oils, juft as they are by acids: again, acid cryftals

H 3   are

are found in fome Oils that have been long kept. This acid in Oil ferves undoubtedly to unite its phlogifton with its water; becaufe thefe two fub-ftances having no affinity with each other cannot be united without the intervention of fuch a medium as an acid, ·which has an affinity with both.. As to the exiftence of water in Oils, it appears plainly when they are decompofed by repeated diftillations, efpecially after mixing them with abforbent earths. Laftly, when an Oil is deftroyed by burning, a cer-tain quantity of earth is conftantly left behind.

· We are very fure that the abovementioned prin-ciples enter into the compofition of Oils; for they may be obtained from every one of them : but it is not abfolutely certain that they confift of thefe only, and that they do not contain fome other principle which may efcape our notice in decompofing them ; for hitherto it doth not appear, by any experiment we can depend on, that Oil was ever produced by combining together the principles here fpecified : yet fuch redintegrations are the only means we have of fatisfying ourfelves that we know all the principles which conftitute a body.

· Oils expofed to the fire in clofe veffels pafs over almoft wholly from the containing veffel into any other applied to receive them. There remains, however, a fmall quantity of black matter, which is extremely fixed, and continues unalterable as long as it hath no communication with the external air, be the force of the fire ever fo violent. This matter is no other than part of the phlogifton of the Oil united with its moft fixed and groffeft earth; and this is what we called *Charcoal*, or plainly a *Coal*.

§. I. *Of* CHARCOAL.

WHEN Oil happens to be united to much earth, as it is in vegetable and animal bodies, it leaves a confiderable quantity of *Coal* or Charred matter.

This

This Coal, expofed to the fire in the open air, burns and waftes, but without blazing like other combuftible matters: there appears only a fmall blueifh flame, but not the leaft fmoke. Moft commonly it only glows and fparkles, and fo gradually falls into afhes, which are nothing but the earth of the body, combined with an alkaline falt in burning. This alkaline falt may be feparated from the earth, by lixiviating the afhes with water, which diffolves all the falt, and leaves the earth quite pure.

Charcoal is unalterable and indeftructible by any other body but fire; whence it follows, that when it is not actually kindled and ignited, the moft powerful agents, fuch as the acids, though ever fo ftrong and concentrated, have not the leaft effect on it.

The cafe is otherwife when it is lighted, that is; when its phlogifton begins to feparate from its earth; for then the pure acid of vitriol being joined therewith contracts an inftantaneous union with its phlogifton, and evaporates in a volatile fulphureous fpirit. If the vitriolic acid, inftead of being applied quite pure, be firft clogged with fome bafis, efpecially an alkaline one, it quits that bafis, enters into a more intimate union with the phlogifton of the burning Coal, and fo forms an actual fulphur, with which the alkali now unites and forms a hepar.

The pure acid of fea-falt hath not been obferved to act in the leaft upon Charcoal, efpecially when it is not on fire. But when this acid is incorporated with an alkaline or metallic bafis, and combined according to a peculiar procefs with burning Charcoal, it in like manner quits its bafis, unites with the phlogifton, and therewith forms a Phofphorus, of which we have already taken notice.

Nor has the pure nitrous acid any effect on a charred Coal, even when ignited: and fo far is it from being able to kindle a cold one, that when

H 4 poured

poured on a live one, it extinguiſhes it like water. But when this acid is united with a baſis, it quits it rapidly as ſoon as it touches a burning coal, and ruſhes violently into an union with the phlogiſton thereof. From this union there probably ariſes, as we ſaid before, a kind of ſulphur or phoſphorus, which is ſo inflammable as to be deſtroyed by the fire the very moment it is generated.

The acids of nitre and vitriol act upon Oils; but very differently, according to the quantity of phlegm they contain. If they be weakened with much water, they have no effect at all upon Oils; if they contain little water, or be dephlegmated to a certain degree, they diſſolve them with heat, and with them form compounds of a thick conſiſtence. Acids thus combined in a conſiderable proportion with Oils render them ſoluble in water.

§. II. *Of* SOAP.

ALKALIS alſo have the ſame property. When an Oil is combined with an acid or an alkali in ſuch a manner that the compound reſulting from their union is ſoluble in water; ſuch a compound may in general be called a *Soap*. Soap itſelf hath the property of rendering fat bodies in ſome meaſure ſoluble in water; on which account it is very uſeful for ſcouring or cleanſing any thing greaſy.

Oily and ſaline ſubſtances, combined together, obſerve the ſame general rules as all other combinations; that is, they mutually communicate the properties belonging to each: thus Oils, which naturally are not ſoluble in water, acquire by their union with ſaline matters the property of diſſolving therein; and ſalts loſe by their conjunction with Oils part of their natural tendency to incorporate with water; ſo that while they ſerve to conſtitute Soap they do not, as before, attract the moiſture of the air, &c. and in like manner, as they are not

in-

inflammable, they confiderably leffen the inflammability of the Oils combined with them.

Acid Soaps are decompounded by alkalis, as alkaline Soaps are by acids, according to the general
rules of affinities.

The acids of nitre and vitriol, when highly concentrated, diffolve Oils with fuch violence as to heat
them, make them black, burn them, and even fet
them on fire. How fea-falt affects Oils is not yet
fufficiently afcertained.

All Oils have the property of diffolving fulphur;
which is not at all furprifing, feeing each of its component principles hath an affinity with Oil.

It is alfo a property common to all Oils to become
more fluid, fubtile, light, and limpid the oftener
they are diftilled. On the contrary, by being incorporated with faline fubftances they acquire a
greater confiftence, and fometimes form compounds
that are almoft folid.

CHAP. XI.

Of the feveral Sorts of OILS.

OILS are diftinguifhed by the fubftances from
which they are drawn: and as Oils are extracted from minerals, from vegetables, and from
animals, there are of courfe Mineral, Vegetable,
and Animal Oils.

§. I. Of MINERAL OILS.

IN the bowels of the earth we find but one fort
of Oil, called *Petroleum :* its fmell is ftrong and not
difagreeable, and its colour fometimes more fometimes

times lefs yellow. There are certain mineral fubftances which yield by diftillation a great deal of
Oil very like Petroleum. This fort of fubftance
is called a *Bitumen,* and is, indeed, nothing but an
Oil rendered confiftent and folid by being combined
with an acid; as appears from hence, that by uniting Petroleum with the acid of vitriol we can produce an artificial Bitumen very like the native.

§. II. *Of* V E G E T A B L E O I L S.

VEGETABLE fubftances yield a very great quantity and variety of Oils : for there is not a plant,
or part of a plant, that does not contain one or more
forts thereof, generally peculiar to itfelf, and different from all others.

By expreffion only, that is, by bruifing and
fqueezing vegetable fubftances, particularly certain
fruits and feeds, a fort of Oil is obtained which has
fcarce any fmell or tafte. Oils of this fort are very
mild and unctuous; and, becaufe in this refpect
they refemble animal fat more than the reft do,
they are called *Fat Oils.*

Thefe Oils, being expofed to the air for fome
time, fooner or later grow thick, acquire an acrid
tafte, and a ftrong difagreeable fmell. Some of
them congeal with the fmalleft degree of cold.
This fort of Oil is well adapted to diffolve thofe
preparations of Lead called Litharge and Minium,
with which they form a thick tenacious fubftance,
that is ufed for the bafis of almoft all plaifters.
They alfo diffolve Lead in its metalline form ; but
not fo eafily as the forts of calx abovementioned ;
probably becaufe its body is not fo much opened,
nor its parts fo divided.

By expreffion alone we alfo procure from certain
vegetable fubftances another fort of Oil, which is
thin, limpid, volatile, of a pungent tafte,. and retains the fmell of the vegetable that yielded it ; on
which

which account it is called an *Essential Oil.* Of this there are several sorts, differing from one another like the Fat Oils, according to the subjects from which they are obtained.

We must observe that it is very difficult, or rather in most cases impossible, to force from the greatest part of vegetables, by expression only, all the essential Oil they contain. For this purpose therefore recourse must be had to fire: a gentle heat, not exceeding that of boiling water, will extract all the essential Oils of vegetables; and this is the most usual and most convenient way of procuring them.

The fat Oils cannot be obtained by the same method: these being much less volatile than the essential Oils require a much greater degree of heat to raise them; which nevertheless they cannot bear without being much spoiled and entirely changed in their nature, as shall presently be shewn. All Oils, therefore, which rise with the heat of boiling water, and such alone, should be called Essential Oils.

Essential Oils, in a longer or shorter time, according to the nature of each, lose the fragrant smell they had when newly distilled, and acquire another, which is strong, rancid, and much less agreeable: they also lose their tenuity, becoming thick and viscid; and in this state they greatly resemble those substances abounding in Oil which flow from certain trees, and which are called *Balsams* or *resins*, according as they are less or more consistent.

Balsams and Resins are not soluble in water. But there are other Oily compounds which likewise run from trees; and, though not unlike Resins, are however soluble in water. These are called *Gums*; and their property of dissolving in water arises from their containing more water and more salt than Resins have; or at least their saline parts are less clogged and more disengaged.

Balsams

Balfams and Refins diftilled with the heat of boiling water yield great quantities of a limpid, fubtle, odoriferous, and, in one word, effential Oil. In the ftill there remains a fubftance thicker and more confiftent than the Balfam or Refin was before diftillation. The fame thing happens to effential Oils which by length of time have acquired a confiftence and are grown refinous. If they be rediftilled, they recover their former tenuity, leaving behind them a remainder thicker and more refinous than they themfelves were. This fecond diftillation is called the *Rectification* of an Oil.

It muft be obferved that an effential Oil, combined with an acid ftrong enough to diffolve it, immediately becomes as thick and refinous, in confequence of this union, as if it had been long expofed to the air: which proves the confiftence an Oil acquires by long keeping to be owing to this, that its lighteft and lefs acid parts being evaporated, the proportion of its acid to the remainder is fo increafed, that it produces therein the fame change, as an additional acid mixed with the Oil would have wrought before the evaporation.

⸱ This alfo fhews us that Balfams and Refins are only effential Oils combined with a great proportion of acid, and thereby thickned.

If vegetable fubftances, from which no more effential Oil can be drawn by the heat of boiling water, be expofed to a ftronger heat, they yield an additional quantity of Oil; but it is thicker and heavier than the effential Oil. Thefe Oils are black, and have a very difagreeable burnt fmell, which hath made them be called *Fetid*, or *Empyreumatic* Oils. They are moreover very acrid.

It muft be obferved that, if a vegetable fubftance be expofed to a degree of heat greater than that of boiling water, before the fat or the effential Oil is extracted from it, an empyreumatic Oil only will then

then be obtained ; becaufe both the fat and effential Oils, when expofed to the force of fire, are thereby burnt, rendered acrid, acquire a fmell of the fire, and, in a word, become truly empyreumatic. There is ground to think that an empyreumatic Oil is nothing elfe but an effential or fat Oil burnt and fpoiled by the fire, and that no other Oil befides thefe two exifts naturally in vegetables.

Empyreumatic Oils, diftilled and rectified feveral times by a gentle heat, acquire by every diftillation a greater degree of tenuity, lightnefs, and limpidity. By this means alfo they lofe fomething of their difagreeable odour; fo that they gradually come nearer and nearer to the nature of effential Oils : and if the rectifications be often enough repeated, ten or twelve times for inftance, they become perfectly like thofe Oils ; except that their fmell will never be fo agreeable, nor like that of the fubftances from which they were obtained.

Fat Oils may alfo be brought by the fame means to refemble effential Oils : but neither effential nor empyreumatic Oils are capable of acquiring the properties of fat Oils.

§. III. *Of* Animal Oils.

Distillation procures us confiderable quantities of Oil from all the parts of animal bodies, and efpecially from their fat.. This Oil at firft is not very fluid, and is extremely fetid : but by many rectifications it gradually acquires a great degree of clearnefs and tenuity, and at the fame time lofes much of its difagreeable odour. Animal Oils, thus rendered thin and fluid by a great number of rectifications, have the reputation of being an excellent medicine, and a fpecific in the epilepfy.

CHAP.

CHAP. XII.

Of FERMENTATION *in general.*

BY Fermentation is meant an inteſtine motion, which, ariſing ſpontaneouſly among the inſenſible parts of a body, produces a new diſpoſition and a different combination of thoſe parts.

To excite a Fermentation in a mixt body it is neceſſary firſt, that there be in the compoſition of that mixt a certain proportion of watery, ſaline, oily, and earthy parts: but this proportion is not yet ſufficiently aſcertained. Secondly, it is requiſite that the body to be fermented be placed in a certain degree of temperate heat: for much cold obſtructs fermentation; and too much heat decompoſes bodies. Laſtly, the concurrence of the air is alſo neceſſary to fermentation.

All vegetable and animal ſubſtances are ſuſceptible of Fermentation, becauſe all of them contain in a due proportion the principles above ſpecified. However, many of them want the proper quantity of water, and cannot ferment while they remain in ſuch a ſtate of dryneſs. But it is eaſy to ſupply that defect, and ſo render them very apt to ferment.

With reſpect to minerals properly ſo called, (that is, excluding ſuch vegetable and animal ſubſtances as may have lain long buried in the earth,) they are not ſubject to any Fermentation; at leaſt that our ſenſes can perceive.

There are three ſorts of Fermentation, diſtinguiſhed from one another by their ſeveral productions. The firſt produces wines and ſpirituous liquors; for

which

which reason it is called the *Vinous* or *Spirituous Fermentation* : the result of the second is an acid liquor ; and therefore it is called the *Acetous Fermentation* : and the third generates an alkaline salt ; which, however, differs from the alkaline salts hitherto treated of, in this respect chiefly that, instead of being fixed, it is extremely volatile : this last fort takes the name of the *Putrid* or *Putrefactive Fermentation.* We shall now consider these three forts of Fermentation and their effects a little more particularly.

These three forts of Fermentation may take place successively in the same subject ; which proves them to be only three different degrees of fermentation, all proceeding from one and the same cause, rather than three distinct fermentations. These degrees of fermentation always follow the order in which we have here placed them.

C H A P. XIII.

Of the SPIRITUOUS FERMENTATION.

THE juices of almost all fruits, all saccharine vegetable matters, all farinaceous feeds and grains of every kind, being diluted with a sufficient quantity of water, are proper subjects of Spirituous Fermentation. If such liquors be exposed, in vessels slightly stopped, to a moderate degree of heat, they begin in some time to grow turbid ; there arises insensibly a small commotion among their parts, attended with a hissing noise ; this by little and little increases, till the grosser parts appear, like little feeds or grains, moving to and fro, agitated among themselves, and thrown up to the surface. At the same time some air bubbles rise, and the
liquor

liquor acquires a pungent, penetrating fmell, occafioned by the very fubtile vapours which exhale
from it.

Thefe vapours have never yet been collected, in
order to examine their nature; and they are known
only by their noxious effects. They are fo actively
pernicious, that if a man comes rafhly into a clofe
place, where large quantities of liquors are fermenting, he fuddenly drops down and expires, as
if he were knocked on the head.

When thefe feveral phenomena begin to go off,
it is proper to ftop the fermentation, if a very fpirituous liquor be required : for if it be fuffered to
continue longer, the liquor will become acid, and
from thence proceed to its laft ftage, that is, to putrefaction. This is done by ftopping the containing
veffels very clofe, and removing them into a cooler
place. Then the impurities precipitate, and fet-
tling at the bottom leave the liquor clear and
tranfparent: and now the palate difcovers that the
fweet faccharine tafte it had before fermentation is
changed to an agreeable pungency which is not
acid.

Liquors thus fermented are in general called
Wines : for though in common life that word properly fignifies the fermented juice of grapes only,
and particular names are given to the fermented
juices of other vegetable fubftances; as that obtained from Apples is called *Cyder*; that made from
malt is called *Beer :* yet in Chymiftry it is of ufe to
have one general term denoting every liquor that
has undergone this firft degree of fermentation.

By diftillation we draw from Wine an inflammable liquor, of a yellowifh white colour, light,
and of a penetrating, pleafant fmell. This liquor
is the truly fpirituous part of the wine, and the
product of fermentation. That which comes off in
the firft diftillation is commonly loaded with much
phlegm

phlegm and some oily parts, from which it may be afterwards freed. In this ftate it goes by the name of *Brandy*; but when freed from thefe heterogeneous matters by repeated diftillations, it becomes ftill clearer, lighter, more fragrant, and much more inflammable, and then is called *Spirit of Wine*, and *Rectified Spirit of Wine*, or an *Ardent Spirit*, if confiderably purified. The properties which diftinguifh an Ardent Spirit from all other fubftances are its being inflammable; its burning and confuming entirely, without the leaft appearance of fmoke or fuliginofity; its containing no particles reducible to a coal; and its being perfectly mifcible with water. Ardent Spirits are lighter and more volatile than any of the principles of the mixts from which they were produced, and confequently more fo than the phlegm, the acid, and the oil of which they themfelves confift. This arifes from a particular difpofition of thefe principles, which are in a fingular manner attenuated by fermentation, and thereby rendered more fufceptible of expanfion and rarefaction.

Ardent fpirits are fuppofed to be the natural folvents of oils and oily matters. But it is very remarkable that they diffolve effential oils only, without touching the fat of animals, or the fat oils obtained from vegetables by expreffion; yet when thefe oils have once undergone the action of fire, they become foluble in fpirit of wine, and even acquire a new degree of folubility every time they are diftilled. It is not fo with effential oils, which can never be rendered more foluble in ardent fpirits than they are at firft; and are fo far from acquiring a new degree of folubility every time they are diftilled, that on the contrary they even in fome meafure lofe that property by repeated rectifications.

I have taken fome pains to find out the caufes of thefe fingular effects, and the refult of my enquiries is publiſhed among the Memoirs of the Academy of Sciences for the year 1745. I therein conſider ardent ſpirits as conſiſting of an oil, or at leaſt a phlogiſton, mixed with a portion of water, in which it is rendered ſoluble by means of an acid. This being laid down, I ſhew that the inability of ſpirit of wine to diſſolve ſome oils muſt be imputed to its aqueous part, in which oils are not naturally ſoluble without the intervention of a ſalt: and that the power which this ſpirit exerts in diſſolving other oils with eaſe, ſuch as eſſential oils, muſt in all probability be owing to this, that in theſe oils it meets with the neceſſary ſaline medium, that is, with an acid, which numberleſs experiments ſhew they actually contain.

On the other hand, I there prove that the acid in eſſential oils is ſuperabundant, and in ſome ſort foreign to their nature, or that it is but ſlightly connected with them, and in part deſerts them every time they are diſtilled; which renders them leſs ſoluble after every new rectification: whereas, on the contrary, the fat expreſſed oils in their natural ſtate give not the leaſt ſign of acidity; but the action of fire upon them diſcovers an acid which was not perceivable before. Hence I conjecture that theſe oils contain no more acid than is juſt neceſſary to conſtitute them oils; that this acid is intimately blended with their other component parts; that it is ſo ſheathed and entangled by theſe parts as to be incapable of exerting any of its properties; and that on this account theſe oils in their natural ſtate are not ſoluble in ſpirit of wine: but that the diſpoſition of their parts being gradually changed by the fire, and their acid, being by that means ſet more and more at liberty, at length recovers its properties, and particularly that of rendering the oily parts ſoluble in an aqueous menſtruum: and hence it follows that the fat oils become ſo much the more ſoluble

luble

luble in fpirit of wine the oftner they are expofed to the action of fire.

Spirit of wine doth not diffolve fixed alkalis; or at leaft it takes up but a very fmall quantity thereof; and hence ardent fpirits may be freed from much of their phlegm by means of thefe falts thoroughly dried: for as they ftrongly imbibe moifture, and have even a greater affinity than ardent fpirits with water, if a fixed alkali, well exficcated, be mixed with fpirit of wine that is not perfectly dephlegmated, the alkali immediately attracts its fuperfluous moiftures, and is thereby refolved into a liquor, which on account of its gravity defcends to the bottom of the veffel. The fpirit of wine, which fwims at top, is by this means as much dephlegmated, and as dry, as if it had been rectified by feveral diftillations. As it takes up fome alkaline particles in this operation, it is thereby qualified to diffolve oily matters with the greater facility. When rectified in this manner it is called *Alcoholized Spirit of Wine.*

Yet fpirit of wine, even when rectified to an alcohol, is not capable of diffolving all oily matters. Thofe named Gums will by no means enter into any fort of union therewith; but it readily diffolves moft of thofe which are known by the appellation of Refins. When it has diffolved a certain proportion of refinous particles it acquires a greater confiftence, and forms what is called a *Spirit-Varnifh*, or a *Drying Varnifh*, becaufe it foon dries. This Varnifh is fubject to be damaged by water. Many forts thereof are prepared, differing from each other according to the different refins employed, or the proportions in which they are ufed. Moft of thefe Varnifhes are tranfparent and colourlefs.

Such bitumens or refins, as fpirit of wine will not touch, are diffolved in oils, by means of fire, and then form another kind of Varnifh, which water does

not

not hurt. Thefe Varnifhes are ufually coloured, and require much longer time to dry than the Spirit-Varnifhes : they are called *Oil-Varnifhes.*

Spirit of wine hath a much greater affinity with water than with oily matters : and therefore if a folution of any oil or refin in fpirit of wine be mixed with water, the liquor immediately grows turbid, and acquires a whitifh milky colour, owing entirely to the oily parts being feparated from the fpirituous menftruum by the acceffion of water, and too finely divided to appear in their natural form. But if the liquor ftand fome time quiet, feveral of thefe particles unite together, and gradually acquire a bulk fufficient to render them very perceptible to the eye.

Acids have an affinity with fpirit of wine, and may be combined with it. By this union they lofe moft of their acidity, and on that account are faid to be *Dulcified.* But as thefe combinations of acids, efpecially of the vitriolic acid, with fpirit of wine, furnifh fome new productions of very fingular properties, and as an examination thereof may throw much light on the nature of ardent fpirits, it will not be amifs to take notice of them in this place, and confider each of them particularly.

One part of highly concentrated oil of vitriol being mixed with four parts of well dephlegmated fpirit of wine, there arifes immediately a confiderable ebullition, and effervefcence, attended with great heat, and abundance of vapours, which fmell pleafantly, but are hurtful to the lungs. At the fame time is heard a hiffing like that produced by a piece of red-hot iron plunged into water. Indeed it is proper to mix the liquors very gradually ; for otherwife the veffels in which the operation is performed will be in great danger of breaking.

If the two liquors thus mixed be diftilled with a very gentle heat, there rifes firft a fpirit of wine of

a moſt penetrating and grateful odour: when about
half thereof is come over, what follows has a quicker
and more ſulphureous ſmell, and is alſo more loaded
with phlegm. When the liquor begins to boil a
little, there comes off a phlegm which ſmells very
ſtrong of ſulphur, and grows gradually more acid.
On this phlegm floats a ſmall quantity of a very
light and very limped oil. In the ſtill there remains
a thick, blackiſh ſubſtance, ſomewhat like a reſin
or bitumen. From this ſubſtance may be ſeparated
a good deal of a vitriolic but ſulphureous acid.
When that is extracted, there remains a black maſs
like a charred coal, which being put into a crucible
and expoſed to a violent heat leaves a ſmall portion
of earth, very fixed, and even vitrifiable.

By rectifying the ardent ſpirit, which came over
in diſtilling the abovementioned mixture, a very
ſingular liquor is obtained, which differs eſſentially
both from oils and from ardent ſpirits, though in
certain reſpects it reſembles them both. This liquor is known in Chymiſtry by the name of *Æther*,
and its chief properties are as follow.

Æther is lighter, more volatile, and more inflammable, than the moſt highly rectified ſpirit of
wine. It quickly flies off when expoſed to the
air, and ſuddenly catches fire when any flame approaches it. It burns like ſpirit of wine without
the leaſt ſmoke, and conſumes entirely without
leaving the ſmalleſt appearance of a coal or of aſhes.
It diſſolves oils and oily matters with great eaſe and
rapidity. Theſe properties it has in common with
an ardent ſpirit. But it reſembles an oil in that it is
not miſcible with water; and this makes it eſſentially different from ſpirit of wine, the nature of
which is to be miſcible with all aqueous liquors.

Another very ſingular property of Æther is its
great affinity with gold, exceeding even that of *aqua
regis*. It does not indeed diſſolve gold when in a

I 3

maſs,

mafs, and in its metalline form: but if a fmall quantity of Æther be added to a folution of gold in *aqua regis*, and the whole fhaken together, the gold feparates from the *aqua regis*, joins the Æther, and remains diffolved therein.

The reafon of all the phenomena above-mentioned, refulting from the mixture of fpirit of wine with oil of vitriol, is founded on the great affinity between this acid and water. For if the vitriolic acid be weak, and as it were over-dofed with watery parts, neither oil nor Æther can be obtained by means thereof: but when highly concentrated, it attracts the aqueous parts very powerfully; and therefore being mixed with fpirit of wine lays hold of moft of the water contained in it, and even robs it of fome portion of that which is effential to its nature, and neceffary to conftitute it fpirit of wine: whence it comes to pafs that a certain quantity of the oily particles in its compofition being feparated from the watery particles, and fo brought nearer to each other, they unite and affume their natural form; and thus the oil that fwims at top of the fulphureous phlegm is produced.

The vitriolic acid moreover thickens and even burns fome of this oil; and hence comes the bituminous refiduum left at the bottom of the ftill, which looks like the refult of a vitriolic acid combined with common oil. Laftly, the vitriolic acid becomes fulphureous, as it always doth when united with oily matters, and alfo very aqueous, on account of the quantity of phlegm which it attracts from the fpirit of wine.

Æther may be confidered as a fpirit of wine exceedingly dephlegmated, even to fuch a degree that its nature is thereby changed; fo that the few aqueous particles left in it are not fufficient to diffolve the oily particles and keep them afunder; which therefore being now much nearer to one another

than

than in common spirit of wine, the liquor hath loft its property of being mifcible with water.

Spirit of nitre well dephlegmated, and combined with spirit of wine, prefents likewife fome very fingular appearances,

Firft, in the very inftant of its mixture with fpirit of wine, it produces a greater and more violent effervefcence than the vitriolic acid occafions.

Secondly, this mixture, without the help of diftillation, and only by ftopping the bottle in which the liquors are contained, affords a fort of Æther, produced probably by the vapours which afcend from, and fwim at top of the mixture. This is a very fingular liquor. Dr. Navier was the firft that took notice of it, and gave a defcription thereof, which may be feen in the Memoirs of the Academy of Sciences.

Thirdly, fome authors pretend that, by diftilling the mixture under confideration, an oil is obtained greatly refembling that which, as we obferved above, rifes from fpirit of wine combined with the vitriolic acid : others again deny this. For my part, I believe the thing depends on the different concentration of the fpirit of nitre, as well as on the quality of the fpirit of wine, which is fometimes more fometimes lefs oily.

Fourthly, the two liquors we are fpeaking of, being intimately mixed by diftillation, form a liquor flightly acid, ufed in medicine, and known by the name of *Sweet* or *Dulcified Spirit of Nitre* : a very proper name, feeing the nitrous acid, by uniting with the fpirit of wine, actually lofes almoft all its acidity and corrofive quality.

Fifthly and laftly, when the diftillation is finifhed, there remains in the bottom of the veffel a thick, blackifh fubftance, nearly refembling that which is found after diftilling oil of vitriol and fpirit of wine,

Spirit of Salt hath likewife been combined with
fpirit of wine; but it does not unite therewith fo-
eafily or fo intimately as the two acids abovementioned. To mix them thoroughly, the fpirit of
falt muft be highly concentrated, and fmoaking,
and moreover the affiftance of the ftill muft be
called in. Some authors pretend that from this
mixture alfo a fmall quantity of oil may be obtained;
which probably happens when the liquors have the
qualities above fpecified. The marine acid likewife,
by uniting with fpirit of wine, lofes moft of its
acidity; on which account it is in like manner called
Sweet or *Dulcified Spirit of Salt*. A thick refiduum
is alfo found here after diftillation.

C H A P. XV.

Of the ACETOUS FERMENTATION.

BESIDES an ardent fpirit, wine affords a
great deal of water, oil, earth, and a fort of
acid which fhall be confidered prefently. When
the fpirituous part is feparated from thefe other
matters, they undergo no further change. But if
all the conftituent parts of wine remain combined
together, then, after fome time, fhorter or longer
as the degree of heat in which the wine ftands is
greater or lefs, the fermentation begins afrefh, or
rather arrives at its fecond ftage. The liquor once
more grows turbid, a new inteftine motion arifes,
and after fome days it is found changed into an
acid; which, however, is very different from thofe
hitherto treated of. The liquor then takes the name
of *Vinegar*. The acetous fermentation differs from
the fpirituous, not only in its effect, but alfo in feveral of its concomitant circumftances. Moderate
motion

motion is of fervice to this, whereas it obftructs
the fpirituous; and it is attended with much more
warmth than the fpirituous. The vapours it produces are not noxious, like thofe of fermenting wine.
Laftly, Vinegar depofits no tartar, even when the
wine employed in this operation is quite new, and
hath not had time to difcharge its tartar : inftead
of tartar Vinegar depofites a vifcid matter which is
very apt to putrify.

It muft be obferved that wine is not the only
fubftance that is fufceptible of the acetous fermentation : for feveral vegetable and even animal
matters, which are not fubject to the fpirituous fermentation, turn four before they putrify. But as
vinous liquors poffefs in a very eminent degree the
property of being fufceptible of the acetous fermentation, and likewife of producing the ftrongeft acids
that can refult from fuch fermentation, their acid
fhall be more particularly confidered in this place.

§. I. *Of* Vinegar.

If wine, which has gone through this fecond
ftage of fermentation, be diftilled, inftead of an ardent fpirit, only an acid liquor is obtained, which
is called *Diftilled Vinegar*.

This acid has the fame properties as the mineral
acids of which we have already treated ; that is, it
unites with alkaline forts, abforbent earths, and metallic fubftances, and therewith forms neutral faline
combinations.

Its affinity with thefe fubftances obferves the fame
order as that obferved by the mineral acids with regard to the fame fubftances ; but in general it is
weaker ; that is, any mineral acid is capable of expelling the acid of Vinegar out of all matters with
which it is united.

Vinegar hath likewife a greater affinity than fulphur with alkalis ; whence it follows that it is capable

pable of decompounding that combination of fulphur with an alkali called Liver of Sulphur, and of
precipitating the fulphur it contains.

The acid of Vinegar is always clogged with a
certain proportion of oily parts, which greatly weaken
it, and deprive it of much of its activity ; and for
this reafon it is not near fo ftrong as the mineral
acids, which are not entangled with any oil.   By
diftillation, indeed, it may be freed from this oil,
and at the fame time from the great quantity of
water which in a manner fuffocates it, and by that
means it may be brought much nearer to the nature
of the mineral acids : but this attempt hath not
yet been profecuted with the affiduity it deferves.
Befides diftillation, there is another way of freeing
Vinegar from a good deal of its phlegm ; and that
is, by expofing it to a hard froft, which readily con-
geals the watery part into ice, while the acid retains its fluidity.

Vinegar, faturated with a fixed alkali, forms a
neutral oily falt, of a dark colour, which is femi-
volatile, melts with a very gentle heat, flames when
thrown upon burning coals, and diffolves in fpirit
of wine, of which, however, it requires fix parts to
complete the folution.   This folution being evaporated to drynefs leaves a matter in the form of leaves
lying on each other ; on which account it hath obtained the name of *Terra Foliata.* The fame foli-
ated matter will be obtained, though the falt be not
previoufly diffolved in fpirit of wine ; but not fo
readily. This falt is alfo called *Regenerated Tartar.*
Under the head of Tartar we fhall fee the reafon of
thefe different appellations. Regenerated Tartar is
alfo in fome degree capable of cryftallizing : for
this purpofe a folution thereof in water muft be
flowly evaporated to the confiftence of a fyrup, and
then fuffered to ftand quiet in a cool place ; by
which means it will fhoot into clufters of cryftals,
lying

lying one upon another, not unlike the feathers on a quill.

With Vinegar and several abforbent earths, fuch as calcined pearls, coral, fhells of fifh, &c. are alfo formed neutral faline compounds, each of which takes the name of the particular earth employed in its compofition.

Vinegar perfectly diffolves Lead, and converts it to a neutral metallic falt, which fhoots into cryftals, and has a fweet faccharine tafte. This compound is called *Sugar of Lead*, or *Sal Saturni*.

If Lead be expofed to the bare vapour of Vinegar, it will be thereby corroded, calcined, and converted into a white matter much ufed in painting, and known by the name of *Cerufe*; or, when it is finer than ordinary, *White Lead*.

Vinegar corrodes Copper likewife, and converts it into a beautiful green ruft, which alfo is ufed in painting, and diftinguifhed by the name of *Verdegris*. However, Vinegar is not commonly employed to make Verdegris : for this purpofe they ufe wine, or the rape of wine, from which fire extricates an acid analogous to that of Vinegar.

In treating of the feveral fubftances which conftitute wine we mentioned an acid matter, but did not then enter into a particular examination thereof; becaufe as that matter greatly refembles the acid of Vinegar, we thought it more proper to defer the confideration of its properties till we had treated of the acetous fermentation, and its effects.

§. II. *Of* TARTAR.

THIS fubftance is a faline compound, confifting of earthy, oily, and efpecially acid parts. It is found in the form of crufts, adhering to the inner fides of veffels in which wines have ftood for fome time, particularly acid wines, fuch as thofe of Germany.

Tartar

Tartar derives its origin from the fuperabundant quantity of acid contained in the juice of the grape. This fuperfluous acid, being more than is requifite to conftitute the ardent fpirit, unites with fome of the oil and earth contained in the fermented liquor, and forms a kind of falt; which for fome time continues fufpended in that liquor, but, when the wine ftands undifturbed in a cool place, is depofited, as hath been faid, on the fides of the cafk.

Tartar in this ftate contains many earthy parts, which are fuperfluous, and foreign to its nature. From thefe it may be freed by boiling it repeatedly with a fort of earth found in the neighbourhood of Montpelier, as may be feen in the Memoirs of the Academy of Sciences.

When it is purified, there appears on the furface of the liquor a fort of white cryftalline pellicle, which is fkimmed off as it forms. This matter is called *Cream of Tartar*. The fame liquor which produces this Cream, and in which the purified Tartar is diffolved, being fet to cool yields a great number of white femi-tranfparent cryftals, which are called *Cryftals of Tartar*. The Cream and the Cryftals of Tartar are therefore no other than purified Tartar, and differ from each other in their form only.

-Though the Cryftals of Tartar have every appearance of a neutral falt, yet they are far from being fuch; for they have all the properties of a true acid, which fcarce differs from that of vinegar, except that it contains lefs water, and more earth and oil; to which it owes its folid form, as well as its property of not being foluble in water without much difficulty: for a very great quantity of water is requifite to keep the Cryftals of Tartar in folution; and it muft moreover be boiling hot; otherwife as foon as it cools moft of the

Tartar

Tartar diffolved in it feparates from the liquor,
and falls to the bottom in the form of a white
powder.

Tartar is decompofed by calcination in the open
fire. All its oily parts are confumed or diffipated
in fmoke, together with moft of its acid. The
other part of its acid, uniting intimately with its
earth, forms a very ftrong and very pure fixed alkali, called *Salt of Tartar.*

It will be fhewn in its proper place that almoft
every vegetable matter, as well as Tartar, leaves a
fixed alkali in its afhes : yet Tartar has thefe peculiar properties ; firft, it affumes an alkaline character even when burnt or calcined in clofe veffels, whereas other fubftances acquire it only by
being burnt in the open air ; fecondly, the alkali
of Tartar is ftronger and more faline than almoft
any that is obtained from other matters.

This alkali, when thoroughly calcined, powerfully attracts the moifture of the air, and melts
into an unctuous alkaline liquor, improperly called
Oil of Tartar per deliquium. This is the alkali
generally ufed in making the *Terra Foliata*, mentioned under the head of Vinegar ; for which reafon this combination is called *Terra Foliata Tartari*;
a name fuitable enough. But the fame cannot be
faid of the other name, *Regenerated Tartar*, which
is alfo given it. 'Tis true that on this occafion
an oily acid is reftored to the earth of the Tartar,
analogous to that of which the fire had deprived
it : but the compound thence refulting is a neutral falt which very readily diffolves in water ;
whereas Tartar is manifeftly acid, and not foluble,
or at leaft hardly foluble, in water.

Cryftals of Tartar combined with alkali of Tartar produce a great effervefcence while they are
mixing, as all acids ufually do ; and if the combination be brought exactly up to the point of faturation,

a per-

a perfectly neutral falt is formed, which fhoots into
cryftals, and eafily diffolves in water; and this
hath procured it the name of *Soluble Tartar*. It is
alfo called the *Vegetable Salt*, as being obtained
from vegetables only; and again, *Tartarifed Tartar*, becaufe it confifts of the acid and the alkali
of Tartar combined together.

Cryftals of Tartar combined with alkalis procured from the afhes of maritime plants, fuch as Soda,
which alkalis refemble the bafis of fea-falt, form
likewife a neutral falt, which cryftallizes well, and
diffolves eafily in water. This falt is another fort
of foluble Tartar. It is called *Saignette's Salt*, from
the inventor's name.

Both the Vegetable Salt and Saignette's Salt are
gently purgative foaps, and much ufed in Medicine.

Tartar likewife diffolves the abforbent earths, as
lime, chalk, &c. and with them forms neutral
falts which are foluble in water *. It even attacks
metallic bodies, and when combined with them
becomes foluble. A foluble Tartar for medical
ufe is prepared with Cryftals of Tartar and Iron:
the metallic falt thereby produced hath the name
of, *Chalybeated Soluble Tartar*. This falt attracts
the moifture of the air, and is one of thofe which
do not cryftallize.

Cryftallized Tartar acts alfo upon feveral other
metallic fubftances: for inftance, it diffolves the
Regulus, Liver, and Glafs of Antimony, and
thence acquires an emetic quality: it is then called
Stibiated or *Emetic Tartar.* It likewife diffolves
Lead, and therewith forms a falt which, in the
figure of its cryftals, refembles Tartarifed Tartar.

It is very extraordinary that Tartar, which of
itfelf is not foluble in water, fhould be foluble

* See Mr. Duhamel's Effays on this fubject in the Memoirs
of the Academy of Sciences.

therein

therein when become a neutral falt by uniting either with alkalis or with abforbent earths, or even with metals. With refpect to alkalis, indeed, it may be urged that, having themfelves a great affinity with water, they communicate to Tartar fome of that facility with which they naturally unite therewith : but the fame cannot be alledged concerning abforbent earths, and metallic fubftances, which water diffolves not at all, or at leaft with great difficulty, and in fmall quantity. This effect, therefore, muft be attributed wholly to fome change in the difpofition of its parts which is to us unknown.

All the Soluble Tartars are eafily decompounded by expofing them to a certain degree of heat. In diftillation they yield the fame principles which are obtained from Tartar ; and what remains fixed in the fire, after they are thoroughly burnt, is a compound of the alkali which Tartar naturally produces, and of the alkaline or metallic fubftance with which it was converted into a neutral falt.

As Cryftal of Tartar is the weakeft of all acids, on account of the oily and earthy matters with which it is combined, Soluble Tartars are decompounded by all the acids ; by any of which cryftal of Tartar may be feparated from the fubftance that ferves it for a bafis and renders it a neutral falt.

The other acids which are procured from vegetables, and even thofe which are obtainable from fome animal fubftances, may all be referred to and compared with either Vinegar or Tartar, according to the quantities of oil or earth with which they are combined.

After all, thefe acids have not yet been thoroughly examined. There is great reafon to think that they are no other than the mineral acids, which in paffing through the bodies of vegetables, and even of animals, undergo a confiderable change, efpecially by contracting an union with oily matters.

For,

For, as we faid before in treating of Vinegar, by freeing them from their oil they are brought very near to the nature of mineral acids : and fo likewife the mineral acids acquire many of the properties of vegetable acids by being combined with oils.

C H A P. XV.

Of the PUTRID FERMENTATION, *or* PUTREFACTION.

EVERY body which hath gone through the two ftages of fermentation above defcribed, that is, the fpirituous and the acetous fermentation, being left to itfelf in a due degree of warmth, which varies according to the fubject, advances to the laft ftage of fermentation ; that is, to putrefaction.

It is proper to obferve, before we go any further, that the converfe of this propofition is not true ; that is, it is not neceffary that a body fhould fucceffively pafs through the fpirituous and the acetous fermentation, before it can arrive at the putrid ; but that, as certain fubftances fall into the acetous without having gone through the fpirituous fermentation, fo others begin to putrify without having undergone either the fpirituous or the acetous fermentation ; of which laft kind are, for inftance, moft animal fubftances. When therefore we reprefented thefe three forts of fermentation as three different degrees or ftages of one and the fame fermentation, we fuppofed it to be excited in a body fufceptible of fermentation in its full extent.

However, there is ftill room to think that every fubftance which is capable of fermenting always

passes

paffes neceffarily through thefe three different ftages;
but that the fubftances moft difpofed thereto pafs
with fuch rapidity through the firft, and even the
fecond, that they arrive at the third before our
fenfes can perceive the leaft figns of either of the
two former. This opinion is not deftitute of probability; yet it is not fupported by proofs fufficiently ftrong and numerous to compel our affent.

When a body is in a putrefying ftate it is eafy to
difcover, (as in the two forts of fermentation already treated of) by the vapours which rife from it,
by the opacity which invades it, if a pellucid liquor,
and frequently even by a greater degree of heat
than is found in the two other forts of fermentation,
that an inteftine motion is begun among its conftituent parts, which lafts till the whole be entirely
putrefied.

The effect of this inteftine motion is in this, as in
the two other forts of fermentation, to break the
union, and change the difpofition, of the particles
conftituting the body in which it is excited, and
to produce a new combination. This is brought
about by a mechanifm to which we are ftrangers,
and concerning which nothing beyond conjectures
can be advanced : but thefe we neglect, refolving
to keep wholly to facts, as the only things in Natural Philofophy that are pofitively certain.

If, then, we examine a fubftance that has undergone putrefaction, we fhall foon perceive that
it contains a principle which did not exift in it before. If this fubftance be diftilled, there rifes firft,
by means of a very gentle heat, a faline matter
which is exceedingly volatile, and affects the organ
of fmelling brifkly and difagreeably. Nor is the
aid of diftillation neceffary to difcover the prefence
of this product of putrefaction : it readily ma-
nifefts itfelf in moft fubftances where it exifts, as
any one may foon be convinced by obferving the

different finell of frefh and of putrefied urine; for
the latter not only affects the nofe, but even makes
the eyes finart, and irritates them fo as to draw tears
from them in abundance.

· This faline principle which is the product of putrefaction, when feparated from the other principles
of the body which affords it, and collected by itfelf, appears either in the form of a liquor, or in-
that of a concrete falt, according to the different
methods ufed to obtain it. In the former ftate it is
called a *Volatile Urinous Spirit*; and in the latter a
Volatile Urinous Salt. The qualification of urinous
is given it, becaufe, as was faid, a great deal-
thereof is generated in putrefied urine, to which it
communicates its finell. It goes alfo by the general
name of a *Volatile Alkali*, whether in a concrete or
in a liquid form. The enumeration of its properties will fhew why it is called an Alkali.

Volatile Alkalis, from whatever fubftance obtained, are all alike, and have all the fame properties; differing only according to their degrees of
purity. The Volatile Alkali, as well as the Fixed,
confifts of a certain quantity of acid combined with
and entangled by a portion of the earth of the mixt
body from which it was obtained; and on that account it has many properties like thofe of a Fixed
Alkali. But there is moreover in its compofition a
confiderable quantity of fat or oily matter, of
which there is none in a Fixed Alkali; and on this
account again there is a great difference between
them. Thus the Volatility of the Alkali produced
by putrefaction, which is the principal difference
between it and the other kind of Alkali whofe nature it is to be Fixed, muft be attributed to the
portion of oil which it contains: for there is a
certain method of volatilizing Fixed Alkalis by
means of a fatty fubftance.

I   *Volatile*

Volatile Alkalis have a great affinity with acids, unite therewith ,rapidly and with ebullition, and form with them neutral falts, which fhoot into cryftals, but differ from one another according to the kind of acid employed in the combination.

The neutral falts which have a Volatile Alkali for their bafis are in general called *Ammoniacal Salts.* That whofe acid is the acid of fea-falt is called *Sal Ammoniac.* As this was the firft known, it gave name to all the reft. Great quantities of this falt are made in Egypt, and thence brought to us. They fublime it from the foot of cow's-dung, which is the fuel of that country, and contains fea-falt, together with a Volatile Alkali, or at leaft the materials proper for forming it, and confequently all the ingredients that enter into the compofition of Sal Ammoniac. See the Memoirs of the Academy of Sciences.

The neutral falts formed by combining the acids of nitre and of vitriol with a Volatile Alkali are called, after their acids, *Nitrous Sal Ammoniac,* and *Vitriolic Sal Ammoniac:* the latter, from the name of its inventor, is alfo called Glauber's *Secret Sal Ammoniac.*

A Volatile Alkali, then, has the fame property as a Fixed Alkali with regard to acids : yet they differ in this, that the affinity of the former with acids is weaker than that of the latter : and hence it follows that any Sal Ammoniac may be decompounded by a Fixed Alkali, which will lay hold of the acid, and difcharge the Volatile Alkali.

A Volatile Alkali will decompound any neutral falt which has not a Fixed Alkali for its bafis ; that is, all fuch as confift of an acid combined with an abforbent earth or a metallic fubftance. By joining with the acids in which they are diffolved, it difengages the earths or metallic fubftances, takes their

place,

place, and in conjunction with their acids, forms Ammoniac Salts.

Hence it might be concluded that, of all substances, next to the Phlogiston and the Fixed Alkalis, Volatile Alkalis have the greatest affinity with acids in general. Yet there is some difficulty in this matter: for absorbent earths, and several metallic substances, are also capable of decompounding Ammoniacal Salts, discharging their Volatile Alkali, and forming new compounds by uniting with their acids. This might induce us to think that these substances have nearly the same affinity with acids.

But it is proper to observe, that a Volatile Alkali decompounds such neutral salts as have for their basis either an absorbent earth or a metallic substance, without the aid of fire; whereas absorbent earths or metallic substances will not decompound an Ammoniacal Salt, unless they be assisted by a certain degree of heat.

Now as all these matters are extremely fixed, at least in comparison with a Volatile Alkali, they have the advantage of being able to resist the force of fire, and so of acting in conjunction therewith; and fire greatly promotes the natural action of substances upon one another: whereas the Volatile Alkali in the Ammoniacal Salt, being unable to abide the force of fire, is compelled to desert its acid; and that so much the more quickly, as its affinity therewith is considerably weakened by the presence of an earthy or metallic substance, both of which have a great affinity with acids.

These considerations oblige us to conclude, that Volatile Alkalis have a somewhat greater affinity, than absorbent earths and metallic substances, with acids.

Ammoniacal Salts projected upon nitre in fusion make it detonate: and the Nitrous Sal Ammoniac

I   detonates

detonates by itself, without the addition of any in-flammable matter. This fingular effect evidently demonftrates the exiftence of an oily matter in Volatile Alkalis; for it is certain that nitre will never deflagrate without the concurrence, and even the immediate contact, of fome combuftible matter.

This oily fubftance is often found combined with Volatile Alkalis in fuch a large proportion as to difguife it, in fome meafure, and render it exceeding foul. The falt may be freed from its fuperfluous oil by repeated fublimations; and particularly by fubliming it from abforbent earths, which readily drink up oils. This is called the *Rectification* of a Volatile Alkali. The falt, which before was of a yellowifh or dirty colour, by being thus rectified becomes very white, and acquires an odour more pungent and lefs fetid than it had at firft, that is, when obtained by one fingle diftillation from a putrid fubftance.

It is proper to obferve that the rectification of a Volatile Alkali muft not be carried too far, or repeated too often; for by that means it may be entirely decompofed at length; and particularly if an abforbent earth, and efpecially chalk, be employed for that purpofe, the falt may be converted into an oil, an earth, and water.

Volatile Alkalis act upon feveral metallic fubftances, and particularly on copper; of which they make a moft beautiful blue folution. On this property depends a pretty fingular effect, which happens fometimes when we attempt, by means of a Volatile Alkali, to feparate copper from any acid with which it is combined. Inftead of feeing the liquor grow turbid, and the metal fall, both which generally happen when any Alkali whatever is mixed with a metallic folution, we are furprifed to obferve the folution of copper, upon adding a Volatile Alkali, retain its limpidity, and let fall no precipitate; or at

leaft

leaſt if the liquor does grow turbid, it remains ſo but for a moment, and inſtantly recovers its tranſparency.

This is occaſioned by adding ſuch a quantity of Volatile Alkali as is more than ſufficient fully to ſaturate the acid of the ſolution, and conſiderable enough to diſſolve all the copper as faſt as it is ſeparated from the acid. On this occaſion the liquor acquires a deeper blue than it had before; which ariſes from the property which Volatile Alkalis have of giving this metal, when combined with them, a fuller blue than any other ſolvent can: hence we have a touchſtone to diſcover copper wherever it is; for let the quantity of this metal combined with other metals be ever ſo ſmall, a Volatile Alkali never fails to diſcover it, by making it appear of a blue colour.

Though a Volatile Alkali be conſtantly the reſult of putrefaction, yet it muſt not therefore be imagined that none can be produced by any other means; on the contrary, moſt of thoſe ſubſtances which contain the ingredients neceſſary to form it, yield no inconſiderable quantity thereof in diſtillation. Tartar, for example, which by being burnt in an open fire is converted, as was ſhewn, into a Fixed Alkali, yields a Volatile Alkali when it is decompoſed in cloſe veſſels; that is, when it is diſtilled: becauſe in this latter caſe the oily part is not diſſipated or burnt, as it is by calcination in a naked fire, but has time to unite with ſome of the earth and acid of the mixt, in ſuch a manner as to form a true Volatile Alkali.

To prove that on this occaſion, as well as on all others, where unputrefied bodies yield a Volatile Alkali, this ſalt is the product of the fire, we need only obſerve, that in theſe diſtillations it never riſes till after ſome part of the phlegm, of the acid, and even of the thick oil of the mixt, is come over;

which

which never is the cafe when it is formed before-
hand in the body which is the fubject of the operation, as it is in thofe which have undergone putrefaction : for this falt, being much lighter and
more volatile than thofe other fubftances, rifes of
courfe before them in diftillation.

CHAP. XVI.
A General View of CHYMICAL
DECOMPOSITION.

THOUGH we have confidered all the fubftances which enter into the compofition of
Vegetables, Animals, and Minerals, whether as primary or as fecondary principles, it will not be improper to fhew in what order we obtain thefe principles from the feveral mixts; and efpecially from
Vegetables and Animals, becaufe they are much
more complicated than Minerals. This is called
Analyfing a compound.

The method moft commonly taken to decompofe
bodies is by applying to them fucceffive degrees of
heat, from the gentleft to the moft violent, in appropriated veffels, fo contrived as to collect what
exhales from them. By this means the principles
are gradually feparated from each other ; the moft
volatile rife firft, and the reft follow in order, as
they come to be acted on by the proper degree of
heat : and this is called *Diftillation.*

But it being obferved that fire, applied to the decompofition of bodies, moft commonly alters their
fecondary principles very fenfibly, by combining
them in a different manner with each other, or
even partly decompofing them, and reducing them

K 4
to

to their primitive principles; other means have been ufed to feparate thofe principles without the help of fire.

With this view the mixts to be decompofed are forcibly compreffed, in order to fqueeze out of them all fuch parts of their fubftance as they will by this means part with; or elfe thofe mixts are for a long time triturated, either along with water, which carries off all their faline and faponaceous contents; or with folvents, fuch as ardent fpirits, capable of taking up every thing in them that is of an oily or refinous nature.

We fhall here give a fuccinct account of the effects of thefe different methods, as applied to the principal fubftances among Vegetables and Animals, and likewife to fome Minerals.

§. I. *The* ANALYSIS *of* VEGETABLE SUBSTANCES.

A vaft many vegetable fubftances, fuch as kernels and feeds, yield by ftrong compreffion great quantities of mild, fat, unctuous Oils, which are not foluble in ardent fpirits : thefe are what we called *Expreffed Oils.* They are alfo fometimes called *Fat Oils,* on account of their unctuoufnefs, in which they exceed all other forts of Oil. As thefe Oils are obtained without the aid of fire, it is certain that they exifted in the mixt juft as we fee them, and that they are not in the leaft altered; which could not have been the cafe had they been obtained by diftillation; for that never produces any Oils but fuch as are acrid and foluble in fpirit of wine.

Some vegetable matters, fuch as the rind of citrons, lemons, oranges, &c. alfo yield, only by being fqueezed between the fingers, a great deal of Oil. This fpirts out in fine fmall jets, which being received upon any polifhed furface, fuch as a looking-glafs, run together and form a liquor that is a real Oil.

But

But it muſt be carefully noted that this ſort of Oil, though obtained by expreſſion only, is nevertheleſs very different from the Oils mentioned before, to which the title of *Expreſſed Oils* peculiarly belongs : for this is far lighter and thinner ; moreover, it retains the perfect odour of the fruit which yields it, and is ſoluble in ſpirit of wine ; in a word, it is a true eſſential Oil, but abounds ſo in the fruits which produce it, and is lodged therein in ſuch a manner, occupying a vaſt number of little cells provided in the peel for its reception, that a very ſlight preſſure diſcharges it ; which is not the caſe with many other vegetables that contain an eſſential Oil.

Succulent and green plants yield by compreſſion a great deal of liquor or juice, which conſiſts of moſt of the phlegm, of the ſalts, and a ſmall portion of the oil and earth of the plant. Theſe juices, being ſet in a cool place for ſome time, depoſite ſaline cryſtals, which are a combination of the acid of the plant with part of its oil and earth, wherein the acid is always predominant. Theſe ſalts, as is evident from the deſcription here given, bear a great reſemblance to the tartar of wine treated of above. They are called *Eſſential Salts* ; ſo that Tartar might likewiſe be called the *Eſſential Salt of Wine*.

Dried plants, and ſuch as are of a ligneous, or acid nature, require to be long triturated with water, before they will yield their eſſential ſalts. Trituration with water is an excellent way to get out of them all their ſaline and ſaponaceous contents.

A vegetable matter that is very oily yields its eſſential ſalt with much difficulty, if at all ; becauſe the exceſſive quantity of oil entangles the ſalt ſo that it cannot extricate itſelf or ſhoot into cryſtals. Mr. Gerike, in his *Principles of Chymiſtry*, ſays, that if part of the oil of a plant be extracted by ſpirit of wine, its eſſential ſalt may be afterwards obtained with more eaſe and in greater quantity. This

muſt

muft be a very good method for fuch plants as have an exceffive proportion of effential oil; but will not fucceed if the effential falt be hindered from cryftallizing by a redundancy of fat oil, becaufe fat oils are not foluble in fpirit of wine.

Effential Salts are among thofe fubftances which cannot be extracted from mixts by diftillation: for the firft impreffion of fire decompofes them.

Though the acid which predominates in the Effential Salts of plants, be moft commonly analogous to the vegetable acid, properly fo called, that is, to the acid of vinegar and tartar, which is probably no other than the vitriolic acid difguifed; yet it fometimes differs therefrom, and fomewhat refembles the nitrous or the marine acid. This depends on the places where the plants grow which produce thefe falts: if they be maritime plants, their acid is akin to the acid of fea-falt; if on the contrary they grow upon walls, or in nitrous grounds, their acid is like that of nitre. Sometimes one and the fame plant contains falts analogous to all the three mineral acids; which fhews that the vegetable acids are no other than the mineral acids varioufly changed by circulating through plants.

Liquors containing the Effential Salts of plants being evaporated by a gentle heat to the confiftence of honey, or even further, are called *Extracts*. Hence it is plain, that an Extract is nothing but the Effential Salt of a plant, combined with fome particles of its oil and earth, that remained fufpended in the liquor, and are now incorporated by evaporation.

Extracts of plants are alfo prepared by boiling them long in water, and then evaporating fome part of it. But thefe Extracts are of inferior virtue; becaufe the fire diffipates many of the oily and faline parts.

EMULSIONS

EMULSIONS.

Subftances which abound much in Oil, being bruifed and triturated with water for fome time, afford a liquor of an opaque dead-white colour, like milk. This liquor confifts of fuch juices as the water is capable of diffolving, together with a portion of the oil, which being naturally indiffoluble in water, is only divided and difperfed in the liquor, the limpidity whereof is by that means deftroyed. This fort of oily liquor, in which the oil is only divided, not diffolved, is called an *Emulfion.* The oily particles in Emulfions fpontaneoufly feparate from the water, when left at reft, and uniting into greater maffes rife, on account of their lightnefs, to the furface of the liquor, which by that means recovers a degree of tranfparency.

If vegetables abounding in effential oils and refins be digefted in fpirit of wine, the menftruum takes up thefe oily matters, as being capable of diffolving them; and they may afterwards be eafily feparated from it by the affufion of water. The water, with which fpirit of wine has a greater affinity than with oily matters, feparates them by this means from their folvent, agreeably to the common laws of affinities.

Without the help of fire fcarce any thing, befides the fubftances already mentioned, can be obtained from a plant: but by the means of diftillation we are enabled to analyfe them more completely. In profecuting this method of extracting from a plant the feveral principles of which it confifts, the following order is to be obferved.

A plant being expofed to a very gentle heat, in a diftilling veffel fet in the *balneum mariæ*, yields a water which retains the perfect fmell thereof. Some Chymifts, and particularly the illuftrious Boerhaave, have called this liquor the *Spiritus Rector.* The

nature

nature of this odoriferous part of plants is not yet thoroughly known; becaufe it is fo very volatile that it is difficult to fubject it to the experiments neceffary for difcovering all its properties.

If inftead of diftilling the plant in the *balneum maria*, it be diftilled over a naked fire, with the precaution of putting a certain quantity of water into the diftilling veffel along with it, to prevent its fuffering a greater heat than that of boiling water, all the effential oil contained in that plant will rife together with that water, and with the fame degree of heat.

On this occafion it muft be obferved that no effential oil can be obtained from a plant after the *Spiritus Rector* hath been drawn off; which gives ground to think that the volatility of thefe oils is owing to that fpirit,

The heat of boiling water is alfo fufficient to feparate from vegetable matters the fat oils which they contain. That, however, is to be done by the way of decoction only, and not by diftillation: becaufe though thefe oils will fwim on water, yet they will not rife in vapours without a greater degree of heat.

When the effential oil is come over, if the plant be expofed to a naked fire, without the addition of water, and the heat be increafed a little, a phlegm will rife that gradually grows acid; after which, if the heat be increafed as occafion requires, there will come over a thicker and heavier oil; from fome a volatile alkali; and laft of all, a very thick, black, empyreumatic oil.

When nothing more rifes with the ftrongeft degree of heat, there remains of the plant a mere coal only, called the *Caput Mortuum*, or *Terra Damnata*. This coal when burnt falls into afhes, which being lixiviated with water, gives a fixed alkali.

It

It is obfervable that in the diftillation of plants which yield an acid and a volatile alkali, thefe two falts are often found quite diftinct and feparate in the fame receiver; which feems very extraordinary, confidering that they are naturally difpofed to unite, and have a great affinity with one another. The reafon of this phenomenon is that they are both combined with much oil, which embarraffes them fo that they cannot unite to form a neutral falt, as they would not fail to do were it not for that impediment.

All vegetables, except fuch as yield a great deal of volatile alkali, being burnt in an open fire, and fo as to flame, leave in their afhes a large quantity of an acrid, cauftic, fixed alkali. But if care be taken to fmother them, fo as to prevent their flaming while they burn, by covering them with fomething that may continually beat down again what exhales, the falt obtained from their afhes will be much lefs acrid and cauftic; the caufe whereof is, that fome part of the acid and oil of the plant being detained in the burning, and ftopped from being diffipated by the fire, combines with its alkali. Thefe falts cryftallize, and being much milder than the common fixed alkalis, may be ufed in medicine, and taken internally. They are called *Tachenius's Salts*, becaufe invented by that Chymift.

Marine plants yield a fixed alkali analogous to that of fea-falt. As for all other plants or vegetable fubftances, the fixed alkalis obtained from them, if rightly prepared and thoroughly calcined, are all perfectly alike, and of the very fame nature.

The laft obfervation I have to make on the production of fixed alkalis is, that if the plant you intend to work upon be fteeped or boiled in water before you burn it, a much fmaller quantity of falt will be obtained from it; nay, it will yield none at all, if repeated boilings have robbed it entirely of
thofe

thofe faline particles which muft neceffarily concur
with its earth to form a fixed Alkali.

§. II. *The* ANALYSIS *of* ANIMAL SUBSTANCES.

SUCCULENT animal fubftances, fuch as new-killed
flefh, yield by expreffion a juice or liquid, which
is no other than the phlegm, replete with all the
principles of the animal body, except the earth, of
which it contains but little. The hard or dry parts,
fuch as the horns, bones, &c. yield a fimilar juice,
by boiling them in water. Thefe juices become
thick, like a glue or jelly, when their watery parts
are evaporated; and in this ftate they are true extracts of animal matters. Thefe juices afford no
cryftals of effential falt, like thofe obtained from
vegetables, and fhew no fign either of an acid or
an alkali.

Great part of the oil which is in the flefh of animals may be eafily feparated without the help of
fire; for it lies in a manner by itfelf: it is commonly in a concrete form, and is called *Fat*. This
oil fomewhat refembles the fat oils of vegetables;
for like them it is mild, unctuous, indiffoluble in
fpirit of wine, and is fubtilized and attenuated by
the action of fire. But there is not in animals, as
in vegetables, any light effential oil, which rifes
with the heat of boiling water; fo that properly
fpeaking, animals contain but one fort of oil.

Few animal fubftances yield a perceptible acid.
Ants and bees are almoft the only ones from which
any can be obtained: and indeed the quantity
they yield is very fmall, as the acid itfelf is extremely weak.

The reafon thereof is, that as animals do not
draw their nourifhment immediately from the earth,
but feed wholly either on vegetables or on the flefh
of other animals, the mineral acids, which have
already undergone a great change by the union contracted

tracted between them and the oily matters of the vegetable kingdom, enter into a closer union and combination with these oily parts while they are passing through the organs and strainers of animals; whereby their properties are destroyed, or at least so impaired that they are no longer sensible.

Animal matters yield in distillation, first, a phlegm, and then, on increasing the fire, a pretty clear oil, which gradually becomes thicker, blacker, more fetid, and empyreumatic. It is accompanied with a great deal of volatile alkali; and if the fire be raised and kept up till nothing more comes over, there will remain in the distilling vessel a coal like that of vegetables; except that when it is reduced to ashes, no fixed alkali, or at least very little, can be obtained from them, as from the ashes of vegetables. This arises from hence that, as we said before, the saline principle in animals being more intimately united with the oil than it is in plants, and being consequently more attenuated and subtilized, is too volatile to enter into the combination of a fixed alkali; on the contrary it is more disposed to join in forming a volatile alkali, which on this occasion does not rise till after the oil, and therefore must certainly be the production of the fire. It must be observed, that all we have hitherto said concerning the analysis of bodies must be understood of such matters only as have not undergone any sort of Fermentation.

The chyle and the milk of animals which feed on plants still retain some likeness to vegetables; because the principles of which these liquors are composed have not gone through all the changes which they must suffer before they enter into the animal combination.

Urine and sweat are excrementitious aqueous liquors, loaded chiefly with the saline particles which are of no service towards the nourishment of the

animal,

animal, but pass through its strainers without receiving any alteration; such as the neutral salts
which have a fixed alkali for their basis, and particularly the sea-salt, which happens to be in the
food of animals, whether it exist therein naturally,
as it does in some plants, or whether the animals
eat it to please their palates.

The saliva, the pancreatic juice, and especially
the bile, are saponaceous liquors, that is, they consist of saline and oily particles combined together:
so that being themselves dissolved in an aqueous
liquor, they are capable of dissolving likewise the
oily parts, and of rendering them miscible with
water.

Lastly, the blood being the receptacle of all
these liquors partakes of the nature of each, more
or less in proportion to the quantity thereof which
it contains.

§. III. *The* Analysis *of* Mineral Substances.

Minerals differ greatly from vegetables, and
from animals; they are not near so complex as
those organized bodies, and their principles are
much more simple; whence it follows that these
principles are much more closely connected, and
that they cannot be separated without the help of
fire; which not having on their parts the same action and the same power as on organized bodies,
hath not the same ill effect on them; I mean the
effect of changing their principles, or even destroy-
ing them entirely.

I do not here speak of pure, vitrifiable, or refractory earths; of mere metals and semi-metals;
of pure acids; or even of their simplest combinations, such as sulphur, vitriol, allum, sea-salt : of
all these we have said enough.

We are now to treat of bodies that are more complex, and therefore more susceptible of decomposition.

These

Thefe bodies are compound maffes, or combinations of thofe above-mentioned; that is, metallic fubftances as they are found in the bowels of the earth, united with feveral forts of fand, ftones, earths, femi-metals, 'fulphur, &c. When the metallic matter is combined with other matters, in fuch a proportion to the reft that it may be feparated from them with advantage and profit, thefe compounds are called *Ores:* when the cafe is otherwife, they are called *Pyrites*, and *Marcafites*; efpecially if fulphur or arfenic be predominant therein, which often happens.

In order to analyfe an ore, and get out of it the metal it contains, the firft ftep is to free it from a great deal of earth and ftones, which commonly adhere to it very flightly and fuperficially. This is effected by pounding the ore, and then wafhing it in water; to the bottom of which the metalline parts prefently fink, as being the heavieft, while the fmall particles of earth and ftone remain fufpended fome time longer.

Thus the metallic part is left combined with fuch matters only as are moft intimately complicated with it. Thefe fubftances are moft commonly fulphur and arfenic. Now as they are much more volatile than other mineral matters, they may be diffipated in vapours, or the fulphur may be confumed, by expofing the ore which contains them to a proper degree of heat. If the fulphur and arfenic be defired by themfelves, the fumes thereof may be catched and collected in proper veffels and places. This operation is called *Roafting* an Ore.

The metal thus depurated is now fit to be expofed to a greater force of fire, capable of melting it.

On this occafion the femi-metals and the imperfect metals require the addition of fome matter abounding in phlogifton, particularly charcoaldduft; becaufe thefe metallic fubftances lofe their phlogifton by the action of the fire, or of the fluxes joined with them, and therefore without this precaution would never acquire either the fplendour or the ductility of a metal. By this means the metallic fubftance is more accurately feparated from the earthy and ftony parts, of which fome portion always remains combined therewith till it is brought to fufion. For, as we obferved before, a metallic glafs or calx only will contract an union with fuch matters; a metal poffeffed of its phlogifton and metalline form being utterly incapable thereof.

We took notice of the caufe of this feparation above, where we fhewed that a metal poffeffed of its phlogifton and metalline form will not remain intimately united with any calcined or vitrified matter, not even with its own calx or glafs.

The metal therefore on this occafion gathers into a mafs, and lies at the bottom of the veffel, as being moft ponderous; while the hetcrogeneous matters float upon it in the form of a glafs, or a femivitrification. Thefe floating matters take the name of *Scoria*, and the metalline fubftance at bottom is called the *Regulus*.

It frequently happens that the metalline regulus thus precipitated is itfelf a compound of feveral metals mixed together, which are afterwards to be feparated. We cannot at prefent enter into a detail of the operations neceffary for that purpofe: they will appear in our Treatife of *Practical Chymiftry:* but the principles on which they are founded may be deduced from what we have faid above, concerning the properties of the feveral metals and of acids.

It is proper to obferve, before we quit this fubject, that the rules here laid down for analyfing ores are not abfolutely general: for example, it is often advifable to roaft the ore before you wafh it; for by that means fome ores are opened, attenuated, and made very friable, which would coft much trou-

ble

ble and expence, on account of their exceffive hardnefs, if you fhould attempt to pound them without
a previous torrefaction.

It is alfo frequently neceffary to feparate the ore
from part only of its ftone ; fometimes to leave the
whole ; and fometimes to add more to it, before
you fmelt it.   This depends on the quality of the
ftone, which always helps to promote fufion when
it is in its own nature fufible and vitrifiable.   It is
then called the *Fluor* of the ore : but of this we muft
fay as we did of the preceding article ; it is fufficient
for our prefent purpofe to lay down the fundamental
principles on which the reafon of every procefs is
built ; the defcription of the operations themfelves
being referved for our fecond Part.

We fhall now give a fuccinct account of the principal ores and mineral bodies, contenting ourfelves
with juft pointing out the particulars of which they
feverally confift.

Of *the* P Y R I T E S.

The yellow Pyrites.

THE yellow Pyrites is a mineral confifting of
fulphur, iron, an unmetallic earth, and frequently a
little copper: the fulphur, which is the only one
of thefe principles that is volatile, may be feparated
from the reft by fublimation : it ufually makes a
fourth, and fometimes a third, of the whole weight
of thefe Pyrites.   The other principles are feparated from one another by fufion and reduction
with the phlogifton, which, by metallizing the ferruginous and cupreous earths, parts them from the
unmetallic earth : for this earth vitrifies, and cannot afterwards continue united with metallic matters poffeffed of their metalline form, as hath been
faid before.

There is yet another way of decompofing the yellow Pyrites, which is to let it lye till it efflorefces, or begins to fhoot into flowers; which is nothing but a fort of flow accenfion of the fulphur it contains. The fulphur being by this means decompofed, its acid unites with the ferruginous and cupreous parts of the Pyrites, and therewith forms green and blue vitriols; which may be extracted by fteeping in water the Pyrites which has efflorefced or been burnt, and then evaporating the lixivium to a pellicle; for by this means the vitriol will fhoot into cryftals.

Sometimes the Pyrites contains alfo an earth of the fame nature with that of alum: a Pyrites of this fort, after flowering, yields alum as well as vitriol.

The White Pyrites.

The white Pyrites contains much arfenic, a ferruginous earth, and an unmetallic earth. The arfenic being a volatile principle may be feparated by fublimation or diftillation from the reft, which are fixed: and thefe again may be disjoined from each other by fufion and reduction, as was faid in relation to the yellow Pyrites.

The Copper Pyrites.

The Copper Pyrites contains fulphur, copper, and an unmetallic earth. A great deal thereof likewife holds arfenic, and its colour approaches more or lefs to Orange, yellow, or white, according to the quantity of arfenic in it. It may be decompofed by the fame means as the yellow and white Pyrites.

Of

OF ORES.

Of Gold Ores.

GOLD being conftantly found in its metalline form, and never combined with fulphur and arfenic, its matrices are not, properly fpeaking, ores; becaufe the metal contained in them is not mineralized. The gold is only lodged between particles of ftone, earth, or fand, from which it is eafily feparated by lotion, and by amalgamation with quickfilver. The gold thus found is feldom pure, but is frequently alloyed with more or lefs filver, from which it is to be feparated by quartation.

It is alfo very common to find gold in moft ores of other metals or femi-metals, and even in the Pyrites; but the quantity contained therein is generally fo fmall, that it would not pay the coft of extracting it. However, if any fhould incline to attempt it, merely out of curiofity, it would be neceffary to begin with treating thefe ores in the manner proper for feparating their metalline part; then to cupel the metalline regulus fo obtained; and laftly to refine it by quartation.

Of Silver Ores.

IT is no rare thing to find filver, as well as gold, in its metalline form, only lodged in fundry earths and ftony matters, from which it may be feparated in the fame manner as gold. But the greateft quantities of this metal are ufually dug out of the bowels of the earth in a truly mineral ftate; that is, combined with different fubftances, and particularly with fulphur and arfenic.

Several filver ores are diftinguifhed by peculiar characterifticks, and are accordingly denoted by particular names. That which is called the *Vi-*

treous

treous Silver Ore, is scarce any thing else but a combination of silver and sulphur. Another is known by the name of the *Horny Silver Ore*, because when in thin plates it is semi-transparent : in this ore the silver is mineralized by sulphur and a little arsenic. The *Red Silver Ore* is of the colour which its name imports, sometimes more, sometimes less vivid ; and is chiefly composed of silver, arsenic, and sulphur : it also contains a little iron.

These three ores are very rich in silver : the first contains nearly three fourths of its weight, and the others about two thirds of theirs.

There is a fourth, called the *White Silver Ore*, which though it be heavier, is not so rich in silver, because it contains much copper. Many other minerals contain silver, yet are not, properly speaking, silver ores ; because a much greater quantity of other metals than of silver is found in them.

When a silver ore is to be decomposed, in order to have the silver pure, or when silver is to be extracted out of any ore that contains it, the first thing to be done is to roast the ore, in order to clear it of the volatile minerals : and as silver cannot be had pure without the operation of the cupel, which requires more or less lead to be joined with it, it is usual to mix with the torrified silver ore a quantity of lead, proportioned to that of the heterogeneous matters combined with the silver, and to melt the whole together. Part of the added lead vitrifies during the fusion, and at the same time converts some of the heterogeneous matters also into glass, with which it forms a scoria that rises to the surface of the matter. The other part of the lead, with which the silver is mixed, falls to the bottom in the form of a regulus, which must be cupelled in order to have the silver pure.

Of

Of *Copper Ores.*

COPPER is much feldomer found in a metalline form, than gold or filver: it is commonly in a mineral ftate: it is mineralized by fulphur and arfenic: almoft all its ores contain alfo more or lefs iron; fometimes a little filver or even gold, together with unmetallic earths and ftones, as all ores do.

Moft copper ores are of a beautiful green or blue, or elfe in fhades blended of thefe two colours. The minerals called *mountain green,* and *mountain blue,* are true copper ores; not in the form of hard ftones, like other ores, but crumbly and friable like earth.

Neverthelefs there are feveral copper ores of different colours, as afh-coloured, whitifh, and fhaded with yellow or orange; which colours arife from the different proportions of arfenic, fulphur, and iron, which thefe ores contain.

In order to decompofe a copper ore, and to extract the copper it contains, it is firft of all to be freed from as many of its earthy, ftony, fulphureous and arfenical parts, as is poffible, by roafting and wafhing; then what remains is to be mixed with a flux compounded of a fixed alkali and fome inflammable matter; a little fea-falt is to be put over all, and the whole melted by a ftrong fire. The falts facilitate the fufion and fcorification of the unmetallic matters, and therewith form a flag, which being the lighteft rifes to the furface. The metalline matters are collected below in the form of a fhining regulus of copper; which, however, is not ufually fine copper, but requires to be purified in the manner to be fhewn in our fecond part.

In order to feparate the copper from the unmetallic matters, it is abfolutely neceffary to melt its ore along with inflammable fubftances abounding in phlogifton.

·phlogifton. For, as this metal is not poffeffed of its metalline form while it is in a mineral ftate, as it is deftitute of the true quantity of phlogifton, and, though it were not, would lofe it by the action · of the fire, it would come to pafs that if its ore were melted without the addition of any inflammable matter, the cupreous earth or calx would be fcorified and confounded with the unmetallic matters ; and as all metallic matters, except gold and filver, are fubject to this inconvenience as well as copper, the addition of an inflammable fubftance, in fluxing all ores that contain them, is a general rule that ought conftantly to be obferved.

Of Iron Ores.

IRON is feldom found pure and malleable in the earth; yet it is much feldomer found in the mineral ftate, properly fo called, than any of the other metals : for moft iron ores are fcarce any thing more than a ferruginous earth mixed· in different proportions with unmetallic earths and ftones. Some of them, however, contain alfo volatile minerals, fuch as fulphur and arfenic ; and therefore it is neceffary to roaft the iron ores, like all others, before you attempt to extract the metal out'of them. That being done, they are to be fmelted with a flux confifting of fufible and inflammable matters, as the general rule directs.

Iron is the commoneft of all metals; nay, it is fo univerfally diffufed through the earth, that it is difficult to find any ftone, earth, or fand, that does not contain fome of it; and therefore none of thefe are ufually confidered and treated as iron ores, except · fuch as contain a great deal of that metal, and melt eafily.' The hematites, emery, yellow pyrites, calamine, all contain a pretty confiderable quantity of iron ; but nobody attempts to extract it from them, becaufe they are very hard to melt.

5   Ferru-

Ferruginous earth being naturally of an orange colour, a ftone or earth may be juged to contain iron, if either naturally, or after roafting, it appears to have one fhade of yellow or red.

The fingular property which iron has of being attracted by the magnet, and of being the only body, exclufive of all others, that is fo, likewife affords us an eafy method of difcovering the prefence of this metal among other matters, where it often exifts in fuch a fmall quantity that it could not otherwife be found out. For this purpofe the body in which iron is fufpected to lurk, muft be pulverifed and torrefied with fome inflammable matter; and then the powder thus roafted being touched with a magnet, or an animated bar, if it contains any particles of iron they will infallibly adhere to the magnet or bar.

Of Tin Ores.

TIN is never found in the earth pure and malleable, but always in a mineral ftate, and always mineralized by arfenic. Tin ores are not fulphureous; whence it comes that though tin be the lighteft of all metals, its ores are neverthelefs heavier than thofe of other metals, as arfenic greatly exceeds fulphur in gravity. Some tin ores contain alfo a little iron. The ores of tin are to be wafhed, roafted, and fmelted with a reducing flux, according to the general rules.

Of Lead Ores.

LEAD, like tin, is never found but in a mineral ftate. It is moft commonly mineralized by fulphur; yet there are fome lead ores which alfo contain arfenic.

Lead ores, as well as others, muft be roafted and fmelted with a reducing flux: however, as it is difficult

ficult to free them from all their fulphur by torrefaction only, the reducing flux employed in their
fufion may be made up with a quantity of iron
filings, which being incapable of any union with
lead, and having a much greater affinity than that
metal with fulphur, will on this occafion be of great
fervice by interpofing between them.

Of *Quick-filver Ores.*

RUNNING Mercury is fometimes found in certain
earths, or grey, friable ftones; but moft commonly
in a mineral ftate. It is always mineralized by fulphur, and by fulphur alone: fo that cinabar is the
only ore of quick-filver that we know of: and a
very rich one it is, feeing it contains fix or feven
times as much mercury as fulphur.

Roafting can be of no ufe towards decompofing
the ore of mercury, and feparating its fulphur; becaufe mercury being itfelf very volatile would be
carried off by the fire together with the fulphur. In
order therefore to part the two fubftances of which
cinabar confifts, recourfe muft neceffarily be had
to fome third body, which will unite with one of
them, and by that means feparate it from the other.
Now all the metals, except Gold, having a greater
affinity than mercury with fulphur, fuch a body
is eafily found: any metal but Gold may be employed with fuccefs in this decompofition; but as
iron hath a greater affinity with fulphur than any
of the reft, and is moreover the only one that cannot unite with mercury, it muft on account of
thefe two qualities be preferred to all the reft.

Fixed alkalis are alfo well qualified to abforb
the fulphur of cinabar. Cinabar muft be decompofed in clofe veffels, and by the way of diftillation; otherwife the mercury, as foon as it feparates

from

from the fulphur, will be diffipated in vapours and entirely loft.

In this operation it is needlefs to add either flux or phlogifton ; becaufe the cinabar is decompofed without melting, and the mercury, though in a mineral ftate, contains, like gold and filver, all the phlogifton requifite to fecure its metalline properties.

Of the Ores of Regulus of Antimony.

Regulus of Antimony is always found in a mineral ftate : it is mineralized by fulphur ; but fometimes, though rarely, it is alfo combined with a little arfenic.

When the ore of regulus of antimony is to be decompofed, the firft thing to be done is to expofe it to a degree of heat too weak to melt its earthy and ftony parts, but ftrong enough to fufe its reguline, together with its fulphureous parts, which by this means are feparated from the earth, and united into one mafs, known by the name of Antimony.

It is plain that this firft operation, which is founded on the great fufibility of antimony, produces, with regard to the ore of regulus of antimony, the fame effect that wafhing hath on other ores : fo that after this firft fufion nothing more is requifite to the obtaining of a pure regulus of antimony, but to feparate it from its fulphur by roafting, and to melt it with fome matter abounding in phlogifton, in the fame manner as other metallic matters are treated. The term *Calcination* is generally ufed to exprefs this torrefaction of antimony, by means whereof the metallic earth of the regulus of antimony is feparated from its fulphur.

As regulus of antimony hath, like mercury, much lefs affinity with fulphur than the other metals have, it follows that antimony may be decompofed by the fame means as cinabar ; but the regu-

lus

lus fo obtained is adulterated with a portion of the additament made ufe of, which combines therewith.

There is ftill another procefs employed for obtaining the regulus of antimony : it confifts, as was mentioned in its place, in detonating the mineral with a mixture of nitre and tartar, applied in fuch a proportion-that, after the detonation has confumed the fulphur, there may remain fo much inflammable matter as will be fufficient to furnifh the metalline earth of the antimony with the phlogifton neceffary to preferve its metallic properties. But by this method lefs regulus is produced, than by calcining, or torrefying, and reducing as ufual.

Of the Ores of Bifmuth.

The ore of Bifmuth confifts of the femi-metal mineralized by arfenic, and of an unmetallic earth. It is very eafy to decompofe this ore, and to extract the bifmuth it contains : for this purpofe it need only be expofed to a moderate heat, whereby the arfenic will be diffipated in vapours, and the bifmuth melted, which will then feparate from the unmetallic earth. This earth, at leaft, in feveral ores of bifmuth, poffeffes the property of tinging all vitrifiable matters, with which it is melted, of a beautiful blue colour.

To decompofe the ore of bifmuth no flux or inflammable matter is ufed ; becaufe this femi-metal is poffeffed, even in its mineral ftate, of all the phlogifton requifite to maintain its metalline properties ; and its great fufibility makes it unneceffary to melt the unmetallic earth contained in its ore.

Of the Ores of Zinc.

Zinc is not generally obtained from a particular ore of its own ; but fublimes during the fufion of a mineral, or rather a confufed mafs of minerals,

that

that contains this femi-metal together with iron,
copper, lead, fulphur, arfenic, and, like all other
ores, an unmetallic earth.

Neverthelefs there is a fubftance which may be
confidered as the proper ore of zinc, becaufe it
contains a pretty large quantity of that femi-metal,
a little iron, and an unmetallic earth. It is called
Calamine, or *Lapis Calaminaris* : but hitherto the
art of procuring zinc directly from this mineral
hath no where been practifed. Calamine is commonly employed only to convert copper into brafs,
or a yellow metal, by cementing it therewith. Indeed till lately no eafy or practicable method of obtaining pure zinc from calamine was publickly
known ; for that femi-metal being volatile and very
inflammable, its ore cannot be fufed like others.
Mr. Margraaf was the firft who, by mixing powdered charcoal with calamine in clofe veffels, obtained a perfect zinc from it, by the means of diftillation or fublimation, as fhall be fhewn in our
Practical Chymiftry.

Of *Arfenical Minerals.*

ARSENIC, as well as fulphur, is naturally combined with almoft all ores, or minerals containing
metallic fubftances. As it is very volatile, while
the matters with which it is united are fixed, at
leaft in comparifon therewith, it is eafily feparated
by fublimation.

The minerals that contain moft arfenic are the
white pyrites, orpiment, and cobalt. We have already confidered the white pyrites : as to Orpiment,
it confifts of fulphur and arfenic. Both thefe fubftances being very volatile, it is difficult to feparate
them by fublimation : yet, with proper management, and a due regulation of the fire, this feparation may be effected; becaufe fulphur fublimes a little more eafily than arfenic. But it is more convenient

nient, as well as more expeditious, to make ufe of fome additament that hath a greater affinity with one of thofe fubftances than with the other. Fixed alkalis and mercury, both of which have more affinity with fulphur than with arfenic, may be very properly employed on this occafion.

Cobalt is a mineral compofed of arfenic, an unmetallic earth, and frequently bifmuth : and as none of thefe are very volatile, except the arfenic, this may be eafily feparated from the reft by fublimation. The unmetallic earth which remains has, like that of the ore of bifmuth, the property of giving a blue colour to any vitrifiable matters melted with it ; whence it is conjectured that cobalt and the ore of bifmuth have a great refemblance, or are often blended with each other. Neverthelefs Mr. Brant, an ingenious Swedifh Chymift, infifts that they are very different : he pretends that the metallic fubftance contained in the true cobalt is a femi-metal of a peculiar nature, which hath been erroneoufly confounded with bifmuth : and indeed he proves by a great number of curious experiments, related in the Memoirs of the Academy of Upfal, that thefe two metallic fubftances have properties that are effentially different : to that which is obtained from cobalt, he gives the name of *Regulus of Cobalt.*

Befides the minerals already recited, there is found in the bowels of the earth another fpecies of compound body, of which we have already taken notice ; but which is fuppofed, with fome degree of proba-bility, to belong as much to the vegetable as to the mineral kingdom : I mean the *Bitumens* ; which the beft obfervations oblige us to confider as vegetable oils, that by lying long in the earth have contracted an union with the mineral acids, and by that means acquired the thicknefs, confiftence, and other properties obfervable in them.

By

By diftillation they yield an oil, and an acid not unlike a mineral acid. Mr. Bourdelin has even demonftrated, by a very artful and ingenious procefs, that amber contains a manifeft acid of fea-falt. See the Memoirs of the Royal Academy of Sciences.

C H A P. XVII.

Explanation of the Table of Affinities.

IT hath been fhewn in the courfe of this work that the caufes of almoft all the phenomena, which Chymiftry exhibits, are deducible from the mutual affinities of different fubftances, efpecially the fimpleft. We have already explained, (Chap. II.) what is meant by affinities, and have laid down the principal laws to which the relations of different bodies are fubject. The late Mr. Geoffroy, one of the beft Chymifts we have had, being convinced of the advantages which all who cultivate Chymiftry would receive from having conftantly before their eyes a ftate of the beft afcertained relations between the chief agents in Chymiftry, was the firft who undertook to reduce them into order, and unite them all in one point of view, by means of a table. We are of opinion, with that great man, that this Table will be of confiderable ufe to fuch as are beginning to ftudy Chymiftry, in helping them to form a juft idea of the relations which different fubftances have with one another; and that the practical Chymift will thereby be enabled to account for what paffes in feveral of his operations, otherwife difficult to be underftood, as well as to judge what may be expected to refult from mixtures of different compounds. Thefe reafons have induced us to

infert

infert it at the end of this Elementary Treatife, and
to give a fhort explanation of it here; efpecially
as it will ferve, at the fame time, for a recapitula-
tion of the whole work, in which the feveral axioms
of this Table are difperfed.

You have it here juft as it was drawn up by Mr.
Geoffroy, without any addition or alteration. I
own, however, that it might be improved both
ways: for fince the death of that great Chymift
many experiments have been made, fome of which
have difcovered new affinities, and others have
raifed exceptions to fome of thofe laid down by
him. But feveral reafons diffuade me from publifh-
ing a new Table of Affinities, containing all the
emendations and innovations that might be made
in the old one.

The firft is, that many of the affinities lately difcovered are not yet fufficiently verified, but, on the
contrary, fubject to be contefted: in fhort, they
are perhaps liable to more confiderable objections
and exceptions than the other.

The fecond is, that as Mr. Geoffroy's Table
contains all the fundamental affinities, it is more
fuitable to an Elementary Treatife than a much
fuller one would be; feeing this would neceffarily
fuppofe the knowledge of many things not treated
of by us, and of which it was not proper to fay
any thing in fuch a book as this.

However, as it is effential to our purpofe that we
lead none into error, we fhall take care in explain-
ing the affinities delivered by Mr. Geoffroy, to
mention the principal objections and exceptions to
which they are liable: we fhall moreover add a
very few new ones, confining ourfelves to fuch
only as are elementary and well afcertained.

The upper line of Mr. Geoffroy's Table, comprehends feveral fubftances ufed in Chymiftry. Under
each of thofe fubftances are ranged in diftinct co-
lumns

lumns feveral matters compared with them, in the order of their relation to that firft fubftance; fo as that which is the neareft to it is that which hath the greateft affinity with it, or that which none of the fubftances ftanding below it can feparate therefrom; but which, on the contrary, feparates them all when they are combined with it, and expels them in order to join itfelf therewith. The fame is to be underftood of that which occupies the fecond place of affinity; that is, it has the fame property with regard to all below it, yielding only to that which is above it: and fo of all the reft.

At the top of the firft column ftands the character which denotes an Acid in general. Immediately under this ftands the mark of a Fixed Alkali, being placed there as the fubftance which has the greateft affinity with an Acid. After the Fixed Alkali appears the Volatile Alkali, whofe affinity with Acids yields only to the Fixed Alkali. Next come the Abforbent Earths; and laft of all Metallic Subftances. Hence it follows that when a Fixed Alkali is united with an Acid it cannot be feparated therefrom by any other fubftance; that a Volatile Alkali united with an Acid cannot be feparated from it by any thing but a Fixed Alkali; that an Abforbent Earth combined with an Acid may be feparated from it either by a Fixed or by a Volatile Alkali; and laftly, that any Metallic Subftance combined with an Acid may be feparated from it by a Fixed Alkali, a Volatile Alkali, or an Abforbent Earth.

There are many important remarks to be made on this firft column. Firft, it is making the rule too general to fay that any Acid whatever has a greater affinity with a Fixed Alkali, than with any other fubftance. And indeed Mr. Geoffroy himfelf hath made an exception with refpeft to the Vitriolic. Acid: for in the fourth column, at the head of which ftands that Acid, we find the fign of the

Phlogifton placed above that of the Fixed Alkali, as having a greater affinity than the Fixed Alkali with the Vitriolic Acid. This is founded on the famous experiment, wherein Vitriolated Tartar and Glauber's Salt are decompounded by means of the Phlogifton, which feparates the Fixed Alkalis of thefe Neutral Salts, and uniting with the Vitriolic Acid contained in them forms therewith a Sulphur.

Secondly, Nitre deflagrates, and is decompofed, by the contact of any inflammable matter whatever that is actually ignited; and the operation which produces Phofphorus is no other than a decompofition of fea-falt, whofe Acid quits its Alkaline bafis to join with the Phlogifton : now thefe facts furnifh very ftrong reafons for believing that both thefe Acids, as well as the Vitriolic, have a ftronger affinity with the Phlogifton than with a Fixed Alkali. Laftly, as feveral experiments fhew the Vegetable Acids to be only the Mineral Acids difguifed and mortified, there are fufficient grounds for fufpecting that Acids in general have a greater affinity with the Phlogifton than with Fixed Alkalis : fo that inftead of making an exception with regard to the Vitriolic Acid, it would perhaps be better to lay down this greater affinity as common to all Acids whatever, and to place the Phlogifton in the firft column, immediately under the character which denotes an Acid in general. This theory, however, ftands in need of confirmation from other experiments *.

---

* Mr. Margraaf, an able German Chymift, has made feveral experiments, which induce him to think that the Acid of Phofphorus is of a particular kind, and different from that of fea-falt. May it not be the Marine Acid, but altered by the union it has contracted with the phlogifton? Or may it not be, with refpect to Phofphorus, what the volatile fulphureous fpirit is, with refpect to Sulphur? See the Memoirs of the Royal Academy of Sciences of Berlin.

Thirdly,

Thirdly, in this fame column the character of a Volatile Alkali is fet above that of an Abforbent Earth, as having a greater affinity with Acids; and yet thefe Abforbent Earths decompofe the Ammoniacal falts, drive away the Volatile Alkali from the Acids, and affume its place. This is one of the firft objections made againft Mr. Geoffroy's Table. His anfwer thereto is printed in the Memoirs of the Academy of Sciences for 1718, where his Table alfo is to be found. We have already declared our opinion about this matter in treating of a Volatile Alkali.

Fourthly, in 1744, Mr. Geoffroy, brother to the author of the Table, who hath done no lefs honour to Chymiftry than that eminent phyfician, gave in a Memoir containing an exception to the laft affinity in the firft column; namely, that which places Abforbent Earths above Metallic Subftances. He therein fhews that Alum may be converted into Copperas by boiling it in iron veffels; that on this occafion the iron precipitates the Earth of the Alum, feparates it from its Acid, and affumes its place; fo that of courfe it muft have a greater affinity, than the Abforbent Earth of Alum, with the Vitriolic Acid.

At the head of the fecond column ftands the character of the Marine Acid; which fignifies that the affinities of this Acid are the fubject of the column. Immediately below it is placed the mark of Tin. As this is a metalline fubftance, and as the firft column places metalline fubftances in the loweft degree of affinity with all Acids, it is plain we muft fuppofe Fixed Alkalis, Volatile Alkalis, and Abforbent Earths, to be placed here in order after the Marine Acid, and before Tin. Tin, then, is of all Metalline fubftances that which has the greateft affinity with the Marine Acid; and then follow Regulus of Antimony, Copper, Silver, Mercury. Gold

comes

comes laſt of all; and there are no leſs than two
vacant places above it. By this means it is in ſome
ſort excluded from the rank of ſubſtances that have
an affinity with the Marine Acid. The reaſon
thereof is that this Acid alone is not capable of
diſſolving Gold and combining therewith, neceſſa-
rily requiring for that purpoſe the aid of the Nitrous Acid, or at leaſt of the Phlogiſton. ·

   The third column exhibits the affinities of the
Nitrous Acid, the character whereof ſtands at its
head. Immediately below it is the ſign of Iron, as
the metal which has the greateſt affinity with this
Acid; and then follow other metals, each according to the degree of its relation; to wit, Copper,
Lead, Mercury, and Silver. In this column, as
in the preceding one, we muſt ſuppoſe the ſubſtances, which in the firſt column ſland above
Metallic Subſtances, to be placed in their proper
order before Iron.

   The fourth column is intended to repreſent the
Affinities of the Vitriolic Acid. Here Mr. Geoffroy
has placed the Phlogiſton as the ſubſtance which
has the greateſt affinity with this Acid, for the
reaſon given in our explanation of the firſt column.
Below it he has ranked Fixed Alkalis, Volatile
Alkalis, and Abſorbent Earths, to ſhew that this is
an exception to the firſt column. As to Metalline
ſubſtances, he has ſet down but three, being
thoſe with which the Vitriolic Acid has the moſt
perceptible affinity: theſe metals, placed in the
order of their affinities, are Iron, Copper, and
Silver.

   The fifth column ſhews the affinities of Abſorbent
Earths. As theſe Earths have no ſenſible affinity
but with Acids, this column contains only the characters of the Acids ranked according to the degree
of their ſtrength, or affinity with the Earths; to
wit, the Vitriolic, the Nitrous, and the Marine
Acids.

Acids. Underneath this laſt might be placed the Acid of Vinegar, or the Vegetable Acid.

· The ſixth column expreſſes the Affinities of Fixed Alkalis with Acids, which are the ſame with thoſe of Abſorbent Earths. Moreover, we find Sulphur placed here below all the Acids; becauſe Liver of Sulphur, which is a combination of Sulphur with a Fixed Alkali, is actually decompounded by any Acid : for any Acid precipitates the Sulphur and unites with the Alkali.

Immediately over the Sulphur, or in the ſame ſquare with it, might be ſet a mark denoting the Volatile Sulphureous Spirit; becauſe, like Sulphur, it has leſs affinity than any other Acid with Fixed Alkalis. Oils might alſo be ranked with Sulphur, becauſe they unite with Fixed Alkalis, and there-with form Soaps, which are decompounded by any Acid whatever.

· The ſeventh column points out the affinities of Volatile Alkalis, which are likewiſe the ſame as thoſe of Abſorbent Earths ; and the Vegetable Acid might be placed here alſo under the Marine Acid.

· The eighth column ſpecifies the affinities of Me-tallic ſubſtances with Acids. The affinities of the Acids, which with reſpect to Fixed Alkalis, Vola-tile Alkalis, and Abſorbent Earths, ſucceeded each other uniformly, do not appear in the ſame order here. The Marine Acid, inſtead of being placed below the Vitriolic and Nitrous Acids, ſtands, on the contrary, at their head ; becauſe, in fact, this Acid ſeparates Metalline ſubſtances from all the other Acids with which they happen to be united, and, forcing theſe Acids to quit poſſeſſion, intrudes into their place. Nevertheleſs, this is not a general rule ; for ſeveral Metalline ſubſtances muſt be ex-cepted, particularly Iron and Copper.

The ninth column declares the affinities of Sul-phur, Fixed Alkalis, Iron, Copper, Lead, Silver, Re-

· M 3   gulus

gulus of Antimony, Mercury, and Gold, ftand below it in the order of their affinities. With regard
to Gold it muft be obferved, that it will not unite
with pure Sulphur: it fuffers itfelf to be diffolved
only by the Liver of Sulphur, which is known to
be a compofition of Sulphur and Fixed Alkali.

At the head of the tenth column appears Mercury, and beneath it feveral Metalline fubftances, in
the order of their affinities with it. Thofe Metalline
fubftances are Gold, Silver, Lead, Copper, Zinc,
and Regulus of Antimony.

It is proper to remark on this column that Regulus of Antimony, which ftands the loweft, unites
but very imperfectly with Mercury; and that
after a feeming union of thefe two Metallic fubftances hath been obtained, by a tedious triture
with the addition of water, they do not continue
long united, but fpontaneoufly feparate from each
other in a fhort time. Iron and Tin are here excluded; the former with great reafon, becaufe hitherto it hath not been clearly proved, by any
known experiment, that ever Mercury was united
with Iron: but the fame objection cannot be made
to Tin, which amalgamates very well with Mercury, and might therefore be placed in this column
nearly between Lead and Copper. I ufe the word
nearly, becaufe the different degrees of affinity between Metalline fubftances and Mercury are not fo
exactly determined, as the other relations before
confidered; feeing they generally unite with it,
without excluding one another. We can therefore
fcarce judge of the degree of affinity that belongs
to each, but by the greater or lefs readinefs of each
to amalgamate therewith.

The eleventh column fhews that Lead has a
greater affinity with Silver than with Copper,

The twelfth, that Copper has a greater affinity
with Mercury than with Calamine.

The

The thirteenth, that Silver has a greater affinity with Lead than with Copper.

The fourteenth contains the affinities of Iron. Regulus of Antimony ftands immediately underneath it, as being the Metallic fubftance which has the greateft affinity with it. Silver, Copper, and Lead are placed together in the next fquare below, becaufe the degrees of affinity which thofe Metals have with Iron are not exactly determined.

The fame is to be faid of the fifteenth column: Regulus of Antimony ftands at its head; Iron is immediately below it; and below the Iron the fame three Metals occupy one fquare as before.

Laftly, the fixteenth column indicates that Water has a greater affinity with Spirit of Wine than with Salts. By this general expreffion muft not be underftood any Saline fubftance whatever; but only the Neutral Salts, which Spirit of Wine frees from the water that kept them in folution. Fixed Alkalis, on the contrary, as well as the Mineral Acids, have a greater affinity than Spirit of Wine with Water: fo that thefe Saline fubftances, being well dephlegmated, and mixed with Spirit of Wine, imbibe the water it contains and rectify it.

To thefe might be added another fhort column, having Spirit of Wine at its head: immediately below it fhould be the character of Water, and below that the mark of Oil. This column would fhew that Spirit of Wine has a greater affinity with Water than with Oils; becaufe any Oily matter whatever, that is diffolved in Spirit of Wine, may be actually feparated from it by the affufion of Water. This rule admits of no exception but in one cafe; which is when the oily fubftance partakes of the nature of foap, by having contracted an union with fome faline matter. But as this muft be imputed wholly to that adventitious faline matter being fuperadded to the oily fubftance, it is no juft foundation

'for

for an exception, and the affinity in queftion is nevertheless general.

We have now delivered every thing material that
we had to fay concerning Mr. Geoffroy's Table of
Affinities. It is, as we obferved before, of exceeding great fervice, as it collects into one view the
principal truths laid down in this Treatife. Indeed
the moft advantageous way of ufing it is, not to
delay confulting it till you have read the book
through, but to turn to it while you are reading, as
oft as any affinity between bodies is treated of;
which it will imprint more ftrongly on your mind,
by reprefenting it in a manner before your eyes.

C H A P. XVIII.

The T H E O R Y *of* C O N S T R U C T I N G
the V E S S E L S *moft commonly ufed*
in C H Y M I S T R Y.

CHYMISTS cannot perform the operations
of their art without the help of a confiderable
number of veffels, inftruments, and furnaces,
adapted to contain the bodies on which they intend
to work, and to apply to them the feveral degrees
of heat required by different proceffes. It is therefore proper, before we advance to the operations
themfelves, to confider particularly and minutely
what relates to the inftruments with which they are
to be performed.

Veffels intended for Chymical Operations fhould,
to be perfect, be able to bear, without breaking,
the fudden application of great heat and great cold;
be impenetrable to every thing, and unalterable by
any folvent; unvitrifiable, and capable of enduring
the

the moſt violent fire without melting : but hitherto no veſſels have been found with all theſe qualities united.

They are made of ſundry materials ; namely, of metal, of glaſs, and of earth. Metalline veſſels, eſpecially thoſe made of Iron or Copper, are apt to be corroded by almoſt every ſaline, oily, or even aqueous ſubſtance. For this reaſon, in order to render the uſe of them a little more extenſive, they are tinned on the inſide. But, notwithſtanding this precaution, they are on many occaſions not to be truſted ; and ſhould never be employed in any nice operations which require great accuracy : they are, moreover, incapable of reſiſting the force of fire.

Earthen veſſels are of ſeveral ſorts. Some, that are made of a refractory earth, are capable of being ſuddenly expoſed to a ſtrong fire without break- ing, and even of ſuſtaining a great degree of heat for a conſiderable time : but they generally ſuffer the vapours of the matters which they contain, as well as vitrified metals, to paſs through them, eſpe- cially the glaſs of lead, which eaſily penetrates them and runs through their pores as through a ſieve. There are others made of an earth that, when well baked, looks as if it were half vitrified : theſe be- ing much leſs porous, are capable of retaining the vapours of the matters which they contain, and even glaſs of lead in fuſion ; which is one of the ſevereſt trials a veſſel can be put to : but then they are more brittle than the other ſort.

Good glaſs veſſels ſhould conſtantly be employed in preference to all others, whenever they can poſ- ſibly be uſed : and that not only becauſe they are no way injured by the moſt active ſolvents, nor ſuffer any part of what they contain to paſs through, but alſo becauſe their tranſparency allows the Chy- miſt to obſerve what paſſes within them : which is

always

always both curious and ufeful. But it is pity that veffels of this fort fhould not be able to endure a fierce fire without melting. We fhall take care, when we come to defcribe the feveral forts of chymical inftruments, and the manner of ufing them, to note what veffels are to be preferred to others on different occafions.

Diftillation, as hath been already faid, is an operation by which we feparate from a body, by the help of a gradual heat, the feveral principles of which it confifts.

There are three methods of diftilling. The firft is performed by applying the heat over the body whofe principles are to be extracted. In this cafe, as the liquors, when heated and converted into vapours, conftantly endeavour to fly from the center of heat, they are forced to re-unite in the lower part of the veffel, that contains the matter in diftillation, and fo paffing through the pores or holes of that veffel, they fall into another cold veffel applied underneath to receive them. This way of diftilling is on this account called Diftilling *per Defcenfum.* It requires no other apparatus than two veffels figured like fegments of hollow fpheres, whereof that which is pierced with little holes, and intended to contain the matter to be diftilled, fhould be much lefs than the other, which is to contain the fire, and to fill its aperture exactly; the whole together being fupported vertically upon a third veffel, which is to ferve the purpofe of a recipient, admitting into its mouth the convex bottom of the veffel containing the matter to be diftilled, which muft accurately fill it. This method of diftilling is but little ufed.

The fecond method of diftilling is performed by applying the heat underneath the matter to be decompofed. On this occafion the liquors being heated, rarefied, and converted into vapours, rife,

and

and are condenfed in a veffel contrived for that
purpofe, which we fhall prefently defcribe. This
way of diftilling is called Diftilling *per Afcenfum*,
and is much ufed.

The veffel in which this diftillation *per Afcenfum*
is performed we call an *Alembic*.

There are feveral forts thereof differing from one
another both in the matter of which, and the manner in which, they are made.

Thofe employed to draw the odoriferous waters
and effential oils of plants are generally made of
copper, and confift of feveral pieces. The firft,
which is defigned to contain the plant, is formed
nearly like a hollow cone, the vertex whereof is
drawn out in the fhape of a hollow cylinder or tube:
this part is named the *Cucurbit*, and its tube the
Neck of the *Alembic*. To the upper end of this
tube another veffel is foldered : this is called the
Head, and commonly has likewife the form of a
cone, joined to the neck of the alembic by its bafe,
round which, on the infide, is hollowed a fmall
groove, communicating with an orifice that opens at
its moft depending part. To this orifice is foldered a
fmall pipe in a direction floping downwards, which
is called the *Nofe*, *Spout*, or *Beak* of the alembic.

As foon as the matters contained in the alembic
grow hot, vapours begin to arife from them, and
afcending through the neck of the alembic into the
head, are by the fides thereof ftopped and condenfed :
from thence they trickle down in little ftreams to
the groove, which conveys them to the fpout; and
by that they pafs out of the alembic into a glafs
veffel with a long neck, the end of the fpout being
introduced into that neck, and luted thereto.

To facilitate the refrigeration and condenfation of
the vapours circulating in the head, all alembics of
metal are moreover provided with another piece,
which is a kind of large pan of the fame metal,
fitted

fitted and foldered round the head. This piece ferves to keep cold water in, which inceffantly cools the head, and therefore it is called the *Refrigeratory*. The water in the refrigeratory itfelf grows hot after fome time, and mult therefore be changed occafionally; the heated water being firft drawn off by means of a cock fixed near the bottom of the refrigeratory. All copper alembics fhould be tinned on the infide for the reafons already given.

When faline fpirits are to be diftilled, alembics of metal mult not be ufed; becaufe the faline vapours would corrode them. In this cafe recourfe mult be had to alembics of glafs. Thefe confift of two pieces only; namely, a *Cucurbit*, whofe fuperiour orifice is admitted into and exactly luted with its *Head*, which is the fecond piece.

In general, as alembics require that the vapours of the matter to be diftilled fhould rife to a confiderable heighth, they ought to be ufed only when the molt volatile principles are to be drawn from bodies: and the lighter and more volatile the fubftances to be feparated by diftillation are, the taller mult the alembic be; becaufe the moft ponderous parts, being unable to rife above a certain height, fall back again into the cucurbit as foon as they arrive there, leaving the lighter to mount alone, whofe volatility qualifies them to afcend into the head.

When a matter is to be diftilled, that requires a very tall alembic, and yet does not admit of a metalline veffel, the end will be beft anfwered by a glafs veffel of a round or oval fhape, having a very long neck, with a fmall head fitted to its extremity. Such a veffel ferves many purpofes: it is fometimes employed as a receiver, and at other times as a digefting veffel; on which laft occafion it goes under the name of a *Matrafs*.

When

When one of thefe provided with a head is applied to the purpofe of diftilling, it forms a fort of
alembic.

There are fome alembics of glafs, blown in fuch
a manner by the workmen, that the body and head
form but one continued piece. As thefe alembics
do not ftand in need of having their feveral pieces.
luted together, they are very ufeful on fome occá-
fions, when fuch exceeding fubtile vapours rife as
are capable of tranfpiring through lutes. The
head muft have an aperture at the top, provided
with a fhort tube, through which by means of a
funnel with a long pipe, the matter to be diftilled
may be introduced into the cucurbit. This is to
be exactly clofed with a glafs ftopple, the furface
whereof muft be made to fit the infide of the tube
in every point, by rubbing thofe two pieces well
together with emery.

Another fort of alembic hath alfo been invented,
which may be ufed with advantage when *Cohobation*
is required; that is, when the liquor obtained by
diftillation is to be returned upon the matter in the
cucurbit; and efpecially when it is intended that
this cohobation fhall be repeated a great number of
times. The veffel we are fpeaking of is conftructed
exactly in the fame manner as that laft defcribed;
except that its beak, inftead of being in a ftraight
line, as in the other alembics, forms a circular arch,
and re-enters the cavity of the cucurbit, in order
to convey back again the liquor collected in the
head. This inftrument hath commonly two beaks
oppofite to each other, both turned in this manner,
and is called a *Pelican:* it faves the artift the trouble
of frequently unluting and reluting his veffels, as
well as the lofs of a great many vapours.

There are certain fubftances which in diftillation
afford matters in a concrete form, or rife wholly in
the form of a very light powder, called *Flowers.*

When

When fuch fubftances are to be diftilled, the cucurbit which contains them is covered with a head
without a nofe, which is named a *Blind-head.*

When the flowers rife in great quantities and very
high, a number of heads is employed to collect
them; or rather a number of a kind of pots, confifting of a body only without any bottom, which
fitting one into the other form a canal, that may be
lengthened or fhortened at pleafure, according as
the flowers to be fublimed are more or lefs volatile.
The laft of the heads, which terminates the canal,
is quite clofe at one end, and makes a true blind-
head. Thefe veffels are called *Aludels:* they are
ufually of earthen or ftone ware.

All the veffels above mentioned are fit only for
diftilling fuch light volatile matters as can be eafily
raifed and brought over; fuch as phlegm, effential
oils, fragrant waters, acid oily fpirits, volatile alkalis, *&c.* But when the point is to procure by
diftillation principles that are much lefs volatile,
and incapable of rifing high, fuch as the thick
fetid oils, the vitriolic, the nitrous, and the marine
acids, *&c.* we are under a neceffity of having
recourfe to other veffels, and another manner of
diftilling.

It is eafy to imagine that fuch a veffel muft be
much lower than the alembic. It is indeed no more
than a hollow globe, whofe upper part degenerates
into a neck or tube, that is bent into a horizontal
pofition; for which reafon this inftrument is called
a *Retort:* it is always of one fingle piece.

The matter to be diftilled is introduced into the
body of the retort by means of a ladle with a long
tubular fhank. Then it is fet in a furnace built pur-
pofely for this ufe, and fo that the neck of the retort
coming out of the furnace may, like the nofe of the
alembic, ftand in a floping pofition, to facilitate
the egrefs of the liquors, which by its means are
conveyed

conveyed to a receiver, into which it is introduced, and with which it is luted. This way of diftilling, in which the vapours feem rather to be driven out of the veffel horizontally and laterally, than raifed up and fublimed, is for that reafon called Diftilla-tion *per Latus.*

Retorts are, of all the inftruments of diftillation, thofe that muft fuftain the greateft heat, and refift the ftrongeft folvents; and therefore they muft not be made of metal. Some however, which are made of iron may do well enough on certain occafions: the reft are either of glafs or earth. Thofe of glafs, for the reafons above given, are preferable to the other fort, in all cafes where they are not to be expofed to fuch a force of fire as may melt them. The beft glafs, that which ftands both heat and folvents beft, is that in which there are feweft al-kaline falts. Of this fort is the green German glafs: the beautiful white cryftal glafs is far from being equally ferviceable.

Retorts, as well as alembics, may be of different forms. For example, fome matters are apt to fwell, and rife over the neck of the retort in fubftance, without fuffering any decompofition; when fuch matters are to be diftilled in a retort, it is proper that the body of the veffel, inftead of being globu-lar, be drawn out into the form of a pear, fo as nearly to refemble that of a cucurbit. In a retort of this kind, the diftance between the bottom and the neck being much greater than in thofe whofe bodies are fpherical, the matters contained have much more room for expanfion; fo that the in-convenience here mentioned is thereby prevented. Retorts of this form are called Englifh retorts. As they hold the middle place between alembics, and common retorts, they may be ufed to diftill fuch matters as have a mean degree of volatility be-tween the greateft and the leaft.

It

It is moreover proper to have, in a laboratory, fundry retorts with necks of different diameters. Wide necks will be found the fittest for conveying thick matters, and fuch as readily become fixed; for inftance, fome very thick fetid oils, butter of antimony, &c.: for as thefe matters acquire a confiftence as foon as they are out of the reach of a certain degree of heat, they would foon choak a narrow neck, and by ftopping the vapours, which rife at the fame time from the retort, might occafion the burfting of the veffels.

Some retorts are alfo made with an opening on their upper fide, like that of tubulated glafs alembics, which is to be clofed in the fame manner with a glafs ftopple. Thefe retorts are alfo called Tubulated retorts, and ought always to be ufed whenever it is neceffary to introduce frefh matter into the retort during the operation; feeing it may be done by means of this invention, without unluting and reluting the veffels; which ought always to be avoided as much as poffible.

One of the things that moft perplexes the Chymifts, is the prodigious elafticity of many different vapours, which are frequently difcharged with impetuofity during the diftillation, and are even capable of burfting the veffels with explofion, and with danger to the artift. On fuch occafions it is abfolutely neceffary to give thefe vapours vent, as we fhall direct in its proper place: but as that can never be done without lofing a great many of them; as fome of them in particular are fo elaftic that fcarce any at all would remain in the veffel; for inftance, thofe of the fpirit of nitre, and efpecially thofe of the fmoking fpirit of falt; the practice is to make ufe of very large receivers, of about eighteen or twenty inches diameter, that the vapours may have fufficient room to circulate in, and by applying to the wide furface prefented them by the exten-

five

five infide of fuch a large veffel, may be condenfed into drops. Thefe huge receivers are commonly in the form of hollow globes, and are called Ballons.

To give thefe vapours ftill more room, ballons have been contrived with two open gullets in each, diametrically oppofite to one another; whereof one admits the neck of the retort, and the other is received by one of the gullets of a fecond ballon of the fame form, which is joined in like manner to a third, and fo on. By this artifice the fpace may be enlarged at pleafure. Thefe ballons with two necks are called Adopters.

Operations on bodies that are abfolutely fixed, as metals, ftones, fand, &c. require only fuch veffels as are capable of containing thofe bodies, and refifting the force of fire. Thefe veffels are little hollow pots, of different dimenfions, which are called Crucibles. Crucibles can hardly be made of any thing but earth; they ought to have a cover of the fame material fitted to fhut them clofe. The beft earth we know is that whereof thofe pots are made in which butter is brought from Bretagne: thefe pots themfelves are exceeding good crucibles; and they are almoft the only ones that are capable of holding glafs of lead in fufion, without being penetrated by it.

For the roafting of ores, that is, freeing them, by the help of fire, from their fulphureous and arfenical parts, little cups made of the fame material with crucibles are ufed; but they are made flat, fhallow, and wider above than below, that thefe volatile matters may the more freely exhale. Thefe veffels are called Tefts, or Scorifiers: they are fcarce ever ufed but in the Docimaftic art, that is, in making fmall Affays of ores.

CHAP. XIX.

The THEORY *of* CONSTRUCTING *the* FURNACES *moft commonly ufed in* CHYMISTRY.

SKILL in conducting and applying fire properly, and determining its different degrees, is
of very great confequence to the fuccefs of Chymical operations.

As it is exceeding difficult to govern and moderate the action of fire, when the veffels in which any
operation is performed are immediately expofed to
it, Chymifts have contrived to convey heat to their
veffels, in nice operations, through different me-
diums, which they place occafionally between thofe
veffels and the fire.

Thofe intermediate fubftances in which they
plunge their veffels are called Baths. They are
either fluid or folid : the fluid baths are water or its
vapours. When the diftilling veffel is fet in water,
the bath is called *Balneum Mariæ*, or the *Water-
Bath*; and the greateft degree of heat of which it
is fufceptible is that of boiling water. When the
veffel is expofed only to the vapours which exhale
from water, this forms the *Vapour-Bath*; the heat
of which is nearly the fame with that of the *Balneum Mariæ*. Thefe baths are ufeful for diftilling
effential oils, ardent fpirits, fweet-fcented waters;
in a word, all fuch fubftances, as cannot bear a
greater heat, without prejudice either to their
odour, or to fome of their other qualities.

Baths

Baths may also be made of any other fluids, such as oils, mercury, &c. which are capable of receiving and communicating much more heat: but they are very feldom ufed. When a more confiderable degree of heat is required, a bath is prepared of any folid matter reduced to a fine powder, fuch as fand, afhes, filings of iron, &c. The heat of thefe baths may be pufhed fo far as to make the bottom of the veffel become faintly red. By plunging a thermometer into the bath, by the fide of the veffel, it is eafy to obferve the precife degree of heat applied to the fubftance on which you are working. It is neceffary that the thermometers employed on this occafion be conftructed on good principles, and fo contrived as to be eafily compared with thofe of the moft celebrated natural philofophers. Thofe of the illuftrious Réaumur are moft ufed and beft known, fo that it would not be amifs to give them the preference. When a greater heat is required than any of thofe baths can give, the veffels muft be fet immediately on live coals, or in a flaming fire: this is called working with a naked fire; and in this cafe it is much more difficult than in the other to determine the degrees of heat.

There are feveral ways of applying a naked fire. When the heat or flame is reflected upon the upper part of a veffel which is expofed to the fire, this is called a Reverberated heat. A Melting heat is that which is ftrong enough to fufe moft bodies. A Forging heat is that of a fire which is forcibly excited by the conftant blaft of a pair of bellows, or more.

There is alfo another fort of fire which ferves very commodioufly for many operations, becaufe it does not require to be fed or frequently mended: this is afforded by a lamp with one or more wicks, and may be called a Lamp-heat. It is fcarce ever employed but to heat baths, in operations which require a gentle and long continued

warmth:

warmth : if it hath any fault, it is that of growing gradually hotter.

All the different ways of applying fire require Furnaces of different conftructions : we fhall therefore defcribe fuch as are of principal and moft neceffary ufe.

Furnaces muft be divided into different parts or ftories, each of which has its particular ufe and name.

The lower part of the furnace defigned for receiving the afhes, and giving paffage to the air, is called the Afh-hole. The afh-hole is terminated above by a Grate, the ufe of which is to fupport the coals and wood, which are to be burnt thereon : this part is called the Fire-place. The fire-place is in like manner terminated above by feveral iron bars, which lie quite a-crofs it from right to left, in lines parallel to each other : the ufe of thefe bars is to fuftain the veffels in which the operations are to be performed. The fpace above thefe bars to the top of the furnace is the upper ftory, and may be called the Laboratory of the furnace. Laftly, fome furnaces are quite covered above by means of a kind of vaulted roof called the dome.

Furnaces have moreover feveral apertures : one of thefe is at the afh-hole, which gives paffage to the air, and through which the afhes that fall through the grate are raked out ; this aperture is called the afh-hole door : another is at the fire-place, through which the fire is fupplied with fuel, as occafion requires ; this is called the mouth or door of the fire-place, or the ftoke-hole : there is a third in the upper ftory, through which the neck of the veffel paffes ; and a fourth in the dome for carrying off the fuliginofities of combuftible matters, which is called the chimney.

To conclude, there are feveral other openings in the feveral parts of the furnace, the ufe whereof

is

is to admit the air into thofe places, and alfo, as
they can be eafily fhut, to incite or flacken the
activity of the fire, and fo to regulate it; which has
procured them the title of regifters. All the other
openings of the furnace fhould be made to fhut
very clofe, the better to affift in governing the fire;
by which means they likewife do the office of
regifters.

In order to our forming a juft and general idea of
the conftruction of furnaces, and of the difpofition
of the feveral apertures in them, with a view to increafe or diminifh the activity of the fire, it will
be proper to lay down, as our ground-work, certain principles of natural philofophy, the truth of
which is demonftrated by experience.

And firft, every body knows that combuftible
matters will not burn or confume unlefs they have a
free communication with the air; infomuch that if
they be deprived thereof, even when burning moft
rapidly, they will be extinguifhed at once: that
confequently combuftion is greatly promoted by the
frequent acceffion of frefh air, and that a ftream of
air, directed fo as to pafs with impetuofity through
burning fuel, excites the fire to the greateft poffible
activity.

Secondly, it is certain that the air which touches,
or comes near ignited bodies is heated, rarefied, and
rendered lighter than the air about it, that is, further
diftant from the center of heat; and confequently
that this air, fo heated and become lighter, is neceffarily determined thereby to afcend and mount aloft,
in order to make room for that which is lefs heated
and not fo light, which by its weight and elafticity tends to occupy the place quitted by the other.
Another confequence hereof is, that if fire be kindled in a place inclofed every where but above and
below, a current of air will be formed in that place,
running in a direction from the bottom to the top;

fo

fo that if any light bodies be applied to the opening below, they will be carried up towards the fire; but, on the contrary, if they be held at the opening above, they will be impelled by a force which will drive them up and carry them away from the fire.

Thirdly and laftly, it is a truth demonftrated in hydraulics that the velocity of a given quantity of any fluid, determined to flow in any direction whatever, is fo much the greater the narrower the channel is to which that fluid is confined; and confequently that the velocity of a fluid will be increafed by making it run from a wider through a narrower paffage.

Thefe principles being eftablifhed, it is eafy to apply them to the conftruction of furnaces. Firft, if a fire be kindled in the fire-place of a furnace, which is open on all fides, it burns nearly as if it were in the open air. It has with the furrounding air a free communication; fo that frefh air is continually admitted to facilitate the entire combuftion of the inflammable matters employed as fuel. But there being nothing to determine that air to pafs with rapidity through the fire in this cafe, it does not at all augment the activity thereof, but fuffers it to wafte away quietly.

Secondly, if the afh-hole or dome of a furnace, in which a fire is burning, be fhut quite clofe, then there is no longer any free communication between the air and the fire: if the afh-hole be fhut, the air is debarred from having free accefs to the fire; if the dome be ftopt, the egrefs of the air rarefied by the fire is prevented; and confequently the fire muft in either cafe burn very faintly and flowly, gradually die away, and at laft go quite out.

Thirdly, if all the openings of the furnace be wholly clofed, it is evident that the fire will be very quickly extinguifhed.

Fourthly,

Fourthly, if only the lateral openings of the fire-place be shut, leaving the ash-hole and upper part of the furnace open; it is plain that the air entering by the ash-hole will necessarily be determined to go out at top, and that consequently a current of air will be formed, which will pass through the fire, and make it burn briskly and vigorously.

Fifthly, if both the ash-hole and the upper story of the furnace be of some length, and form canals either cylindric or prismatic, then the air being kept in the same direction through a longer space, the course of its stream will be both stronger and better determined, and consequently the fire will be more animated by it.

Sixthly and lastly, if the ash-hole and the upper part of the furnace, instead of being cylindric or prismatic canals, have the form of truncated cones or pyramids, standing on their bases, and so ordered that the upper opening of the ash-hole, adjoining to the fire place, may be wider than the base of the superiour cone or pyramid, then the stream of air, being forced to pass incessantly from a larger channel through a smaller, must be considerably accelerated, and procure to the fire the greatest activity which it can receive from the make of a furnace.

The materials fittest for building furnaces are, 1. Bricks, joined together with potters clay mixed with sand and moistened with water. 2. Potters clay mingled with potsherds, moistened with water, and baked in a violent fire. 3. Iron; of which all furnaces may be made, with this precaution, that the inside be provided with a great many prominent points, as fastenings for a coat of earth, with which the internal parts of the furnace must necessarily be covered to defend it from the action of the fire.

The

The reverberating furnace is one of thofe that are moft employed in Chymiftry: it is proper for diftillations by the retort, and fhould be conftructed in the following manner.

Firft, the ufe of the afh-hole being, as was faid, to give paffage to the air and to receive the afhes, no bad confequence can attend its being made pretty high: it may have from twelve to twenty or twenty-four inches in heighth. Its aperture fhould be wide enough to admit billets of wood, when a great fire is to be made.

Secondly, the afh-hole muft be terminated at its upper part by an iron grate, the bars of which fhould be very fubftantial, that they may refift the action of the fire: this grate is the bottom of the fire-place, and deftined to fupport the coals. In the lateral part of the fire-place, and nearly about the fame heighth with the grate, there fhould be a hole of fuch a fize that it may eafily admit charcoal, as well as little tongs and fhovels for managing the fire. This aperture or mouth of the fire-place fhould be perpendicularly over the mouth of the afh-hole.

Thirdly, from fix to eight or ten inches high above the grate over the afh-hole, little apertures muft be made in the walls of the furnace, of eight or ten lines in diameter, an inch from one another, and thofe in one fide muft be diametrically oppofite to thofe in the other. The ufe of thefe holes is to receive bars of iron for the retort to reft on; which fhould be, as I faid, at different heights, in order to accommodate retorts of different fizes. At the upper extremity of this part of the furnace, which reaches from the iron bars to the top, the heighth whereof fhould be fomewhat lefs than the width of the furnace, muft be cut a femi-circular aperture for the neck of the retort to come through. This hole muft by no means be over the doors of the fire-place and afh-hole; for then, as it gives paffage to the neck of the retort, it muſt of courſe be oppoſite to the receiver, and in that caſe the receiver itſelf would ſtand over againſt thoſe two apertures; which would be attended with this double inconvenience, that the receiver would not only grow very hot, but greatly embarraſs the operator, whoſe free acceſs to the fire-place and aſh-hole would be thereby obſtructed. It is proper therefore that the ſemi-circular cut we are ſpeaking of be ſo placed that when the greateſt ballons are luted to the retort they may leave an open paſſage to the fire-place and aſh-hole.

Fourthly, in order to cover in the laboratory of the reverberating furnace, there muſt be a roof made for it in the form of a cupola, or concave hemiſphere, having the ſame diameter as the furnace. This dome ſhould have a ſemicircular cut in its rim anſwering to that above-directed to be made in the upper extremity of the furnace, ſo that, when adjuſted to each other, the two together may form a circular hole for the neck of the retort to paſs through. At the top of this dome there muſt alſo be a circular hole of three or four inches diameter, carrying a ſhort tapering funnel of the ſame diameter, and three inches high, which will ſerve for a chimney to carry off all fuliginoſities, and accelerate the current of the air. This paſſage may be ſhut at pleaſure with a flat cover. Moreover, as it is neceſſary that the dome ſhould be taken off and put on with eaſe, it ſhould have two ears or handles for that purpoſe: a portative or moveable furnace ſhould alſo have a pair of handles, fixed oppoſite to each other, between the aſh-hole and the fire-place.

Sixthly and laſtly, a conical canal muſt be provided of about three feet long, and ſufficiently wide at its lower end to admit the funnel of the aperture at the top of the dome. This conical tube is to be applied to the dome when the fire is required to be **extremely**

extremely active: it tapers gradually from its bafe upwards, and breaks off as if truncated at top, where it fhould be about two inches wide.

Befides the apertures already mentioned as necef-fary to a reverberating furnace, there muft alfo be many other fmaller holes made in its afh-hole, fire-place, laboratory, and dome, which muft all be fo contrived as to be eafily opened and fhut with ftop-ples of earth: thefe holes are the regifters of the furnace, and ferve to regulate the activity of the fire, according to the principles before laid down.

When the action of the fire is required to be exactly uniform and very brifk, it is neceffary to ftop carefully with moift earth all the little chinks in the juncture of the dome with the furnace, between the neck of the retort and the circular hole through which it paffes, and which it never fills exactly; and laftly, the holes which receive the iron bars that fuftain the retort.

It is proper to have, in a laboratory, feveral rever-berating furnaces of different magnitudes; becaufe they muft be proportioned to the fize of the retorts employed. The retort ought to fill the furnace, fo as to leave only the diftance of an inch between it and the infide of the furnace.

Yet when the retort is to be expofed to a moft violent fire, and efpecially when it is required that the heat fhall act with equal force on all parts of the furnace, and as ftrongly on its vault as on its bot-tom, a greater diftance muft be left between the re-tort and the infide of the furnace; for then the furnace may be filled with coals, even to the upper part of the dome. If moreover fome pieces of wood be put into the afh-hole, the conical canal fitted on to the funnel of the dome, and all the apertures of the furnace exactly clofed, except the afh-hole and the chimney, the greateft heat will then be excited that this furnace can produce.

The

The furnace now defcribed may alfo be employed in many other chymical operations. If, the dome be laid afide, an alembic may very well be placed therein : but then the fpace, which will be left between the body of the alembic and the top of the upper part of the furnace, muft be carefully filled up with Windfor-loam moiftened; for without that precaution the heat would foon reach the very head, which ought to be kept as cool as poffible, in order to promote the condenfation of the vapours. On this occafion therefore it will be proper, to leave no holes open in the fire-place, but the lateral ones; of which alfo thofe over-againft the receiver muft be flopped.

A pot, or broad-brimmed earthen pan, may be placed over this furnace, and being fo fitted to it as to clofe the upper part thereof accurately, and filled with fand, may ferve for a fand-heat to diftil with.

The bars defigned to fupport diftilling veffels being taken out, a crucible may ftand therein, and many operations be performed that do not require the utmoft violence of fire. In a word, this furnace is one of the moft commodious that can be, and more extenfively ufeful than any other.

The Melting furnace is defigned for applying the greateft force of heat to the moft fixed bodies, fuch as metals and earths. It is never employed in diftilling : it, is of no ufe but for calcination and fufion ; and confequently need not admit any veffels but crucibles.

The afh-hole of this furnace differs from that of the reverberating furnace only in this, that it muft be higher, in order to raife the fire-place to a level with the artift's hand; becaufe in that all the operations of this furnace are performed. The afh-hole therefore muft be about three feet high: and this heighth procures it moreover, the advantage of a good

draught

draught of air. For the fame reafon, and in confequence of the principles we laid down, it fhould be
fo built that its width leffening infenfibly from the
bottom to the top, it may be narrower where it
opens into the fire-place than any where below.

The afh-hole is terminated at its upper end, like
that of the reverberating furnace, by a grate, which
ferves for the bottom of the fire-place, and ought to
be very fubftantial that it may refift the violence of
the fire. The infide of this furnace is commonly an
elliptic curve; becaufe it is demonftrated by mathe-
maticians that furfaces having that curvature reflect
the rays of the fun, or of fire, in fuch a manner
that meeting in a point, or a line, they produce there
a violent heat. But to anfwer this purpofe thofe
furfaces muft be finely polifhed; an advantage
hardly procurable to the internal furface of this furnace, which can be made of nothing but earth :
befides, if it were poffible to give it a polifh, the
violent action of the fire that muft be employed
in this furnace would prefently deftroy it. Yet
the elliptical figure muft not be entirely difre-
garded: for, if care be taken to keep the internal
furface of the furnace as fmooth as poffible, it will
certainly reflect the heat pretty ftrongly, and collect it about the center.

The fire-place of this furnace ought to have but
four apertures.

Firft, that of the lower grate, which communicates with the afh-hole.

Secondly, a door in its fore-fide, through which
may be introduced coals, crucibles, and tongs for
managing them : this aperture fhould be made to
fhut exactly with a plate of iron, having its infide
coated with earth, and turning on two hinges fixed
to the furnace.

Thirdly, over this door a hole flanting downwards, towards the place where the crucible is to
ftand.

ſtand. The uſe of this hole is to give the operator an opportunity of examining the condition of the matters contained in his crucible, without opening the door of the fire-place: this hole ſhould be made to open and ſhut eaſily, by means of a ſtopple of earth.

Fourthly, a circular aperture of about three inches wide in the upper part or vault of the furnace, which ſhould gradually leſſen and terminate, like that of the dome of the reverberating furnace, in a ſhort conical funnel of about three inches long, and fitted to enter the conical pipe before deſcribed, which is applied when the activity of the fire is to be increaſed.

When this furnace is to be uſed, and a crucible to be placed in it, care muſt be taken to ſet on the grate a cake of baked earth, ſomewhat broader than the foot of the crucible. The uſe of this ſtand is to ſupport the crucible, and raiſe it above the grate, for which purpoſe it ſhould be two inches thick. Were it not for this precaution the bottom of the crucible, which would ſtand immediately on the grate, could never be thoroughly heated, becauſe it would be always expoſed to the ſtream of cold air which enters by the aſh-hole. Care ſhould alſo be taken to heat this earthen bottom red-hot before it be placed in the furnace, in order to free it from any humidity, which might otherwiſe happen to be driven againſt the crucible during the operation, and occaſion its breaking.

We omitted to take notice in ſpeaking of the aſh-hole that, beſides its door, it ſhould have about the middle of its heighth a ſmall hole, capable of receiving the noſel of a good perpetual bellows, which is to be introduced into it and worked, after the door is exactly ſhut, when it is thought proper to excite the activity of the fire to the utmoſt violence.

The

The Forge is only a mafs of bricks of about three feet high, along whofe upper furface is directed the nofe or pipe of a pair of large perpetual bellows, fo placed that the operator may eafily blow the fire with one hand. The coals are laid on the hearth of the forge, near the nofe of the bellows; they are confined, if neceffary, to prevent their being carried away by the wind of the bellows, within a fpace inclofed by bricks; and then by pulling the bellows the fire is continually kept up in its greateft activity. The forge is of ufe when there is occafion to apply a great degree of heat fuddenly to any fubftance, or when it is neceffary that the operator be at liberty to handle frequently the matters which he propofes to fufe or calcine.

The Cupelling furnace is that in which gold and filver are purified, by the means of lead, from all alloy of other metallic fubftances. This furnace muft give a heat ftrong enough to vitrify lead, and therewith all the alloy which the perfect metals may contain. This furnace is to be built in the following manner.

Firft, of thick iron-plates, or of fome fuch compofition of earth as we recommended for the conftruction of furnaces, muft be formed a hollow quadrengular prifm, whofe fides may be about a foot broad, and from ten to eleven inches high; and extending from thence upwards may converge towards the top, fo as to form a pyramid truncated at the heighth of feven or eight inches, and terminated by an aperture of the width of feven or eight inches every way. The lower part of the prifm is terminated, and clofed, by a plate of the fame materials of which the furnace is conftructed.

Secondly, in the fore-fide or front of this prifm there is an opening of three or four inches in heighth

5   by

by five or fix inches in breadth : this opening, which fhould be very near the bottom, is the door of the afh-hole. Immediately over this opening is placed an iron grate, the bars of which are quadrangular prifms of half an inch fquare, laid parrallel to each other, and about eight or nine inches afunder, and fo difpofed that two of their angles are laterally oppofite, the two others looking one directly upwards and the other downwards. As in this fituation the bars of the grate prefent to the fire-place very oblique furfaces, the afhes and very fmall coals do not accumulate between them, or hinder the free entrance of the air from the afh-hole. This grate terminates the afh-hole at its upper part, and ferves for the bottom of the fire-place.

Thirdly, three inches, or three and a half, above the grate, there is in the fore-fide of the furnace another opening terminated by an arch for its upper part, which confequently has the figure of a femi-circle : it ought to be four inches wide at bottom, and three inches and an half high at its middle. This opening is the door of the fire-place ; yet it is not intended for the fame ufes as the door of the fire-place in other furnaces : the purpofe for which it is actually deftined fhall be explained when we come to fhew how the furnace is to be ufed. An inch above the door of the fire-place, ftill in the fore-fide of the furnace, are two holes of about an inch diameter, and at the diftance of three inches and a half from each other, to which anfwer two other holes of the fame fize, made in the hinder part, directly oppofite to thefe. There is moreover a fifth hole of the fame width about an inch above the door of the fire-place. The defign of all thefe holes fhall be explained when we defcribe the manner in which thefe furnaces are to be ufed.

Fourthly,

Fourthly, the fore-part of the furnace is bound by three iron braces, one of which is fixed juft be-low the door of the afh-hole; the fecond occupies the whole fpace between the afh-hole door and the door of the fire-place, and has two holes in it, anfwering to thofe which we directed to be made in the furnace itfelf about this place; and the third is placed immediately over the door of the fire-place. Thefe braces muft extend from one corner of the front of the furnace to the other, and be faftened thereto with iron pins, in fuch a manner that their fides next to the doors may not lie quite clofe to the body of the furnace, but form a kind of grooves for the iron plates to flide in, that are defigned to fhut the two doors of the furnace when it is necef-fary. Each of thefe iron plates fhould have a han-dle, by which it may be conveniently moved; and to each door there fhould be two plates, which meeting each other, and joining exactly in the mid-dle of the door-place, may fhut it very clofe. Each of the two plates belonging to the door of the fire-place ought to have a hole in its upper part; one of thefe holes fhould be a flit of about two lines wide, and half an inch long; the other may be a femi-circular opening of one inch in height and two in breadth. Thefe holes fhould be placed fo that neither of them may open into the fire-place when the two plates are joined together in the middle of the door to fhut it clofe.

Fifthly, to terminate the furnace above, there muft be a pyramid formed of the fame materials with the furnace, hollow, quadrangular, three inches high on a bafe of feven inches, which bafe muft exactly fit the upper opening of the furnace: the top of this pyramidal cover muft end in a tube of three inches in diameter and two in heighth, which muft be almoft cylindrical, and yet a little inclining to the conical form. This tube ferves,

as

as in the furnaces already defcribed, to carry the conical funnel, which is fitted to the upper part when a fire of extraordinary activity is wanted.

The furnace thus conftructed is fit to ferve all the purpofes for which it is defigned: yet before it can be ufed another piece muft be provided, which, though it does not properly belong to the furnace, is neverthelefs neceffary in all the operations performed by it; and that is a piece contrived to contain the cupels, or other veffels which are to be expofed to the fire in this furnace. It is called a Muffle, and is made in the following manner.

On an oblong fquare, of four inches in breadth, and fix or feven in length, a concave femi-cylinder is erected, in the form of a vault, which makes a femi-circular canal, open at both ends. One of thefe is almoft entirely clofed, except that near the bottom two fmall femi-circular holes are left. In each of its fides likewife two fuch holes are made, and the other end is left quite open.

The Muffle is intended to bear and communicate the fierceft heat; and therefore it muft be made thin, and of an earth that will refift the violence of fire, fuch as that of which crucibles are made. The Muffle being thus conftructed, and then well baked, is fit for ufe.

When it is to be ufed it muft be put into the furnace by the upper opening, and fet upon two iron bars, introduced through the holes made for that purpofe below the door of the fire-place. The Muffle muft be placed on thefe bars in the fire-place in fuch a manner that its open end fhall ftand next to, and directly againft, the door of the fire-place, and may be joined to it with lute. Then the cupels are ranged in it, and the furnace is filled up, to the heighth of two or three inches above the Muffle, with fmall coals not bigger than a walnut, to the end that they may lie clofe round the Muffle, and procure it an equal heat on every fide. The chief ufe of the Muffle is to prevent the coals and afhes from falling into the cupels, which would be very prejudicial to the operations carrying on in them : for the lead would not vitrify as it ought, becaufe the immediate contact of the coals would continually reftore its phlogifton; or elfe the glafs of lead, which ought to penetrate and pafs through the cupels, would be rendered incapable of fo doing; becaufe the afhes mixing therewith would give it fuch a confiftence and tenacity as would deftroy that property, or at leaft confiderably leffen it. The openings therefore, which are left in the lower part of the Muffle, fhould not be fo high as to admit coals or afhes to get into the cupels; the ufe of them is to procure an eafier paffage for the heat and the air to thofe veffels. The Muffle is left quite open in its fore-part, that the operator may be at liberty to examine what paffes in the cupels, to ftir their contents, to remove them from one place to another, to convey new matters into them, &c. and alfo to promote the free accefs of the air, which muft concur with the fire towards the evaporation neceffary to the vitrification of lead; which air, if frefh were not often enough admitted, would be incapable of producing that effect; becaufe it would foon be loaded with fuch a quantity of vapours that it could not take up any more.

The government of the fire in this furnace is founded on the general principles above laid down for all furnaces. Yet as there are fome little differences, and as it is very effential to the fuccefs of the operations for which this furnace is intended, that the artift fhould be abfolutely mafter of his degree of heat, we fhall in few words fhew how that may be raifed or lowered.

When the furnace is filled with coals and kindled, if the deor of the afh-hole be fet wide open, and that of the fire-place fhut very clofe, the force of

the

the fire is increafed ; and if moreover the pyramidal cover be put on the top, and the conical funnel
added to it, the fire will become ftill more fierce.

Seeing the matters contained in this furnace are
encompaffed with fire on all fides, except in the
fore part oppofite to the door of the fire-place, and
as there are occafions which require that the force
of the fire fhould be applied to this part alfo, an
iron box, of the fhape and fize of the door, hath
been contrived to anfwer that purpofe. This box
is filled with lighted coals, and applied immediately
to the door-place, by which means the heat there
is confiderably augmented. This help may be
made ufe of at the beginning of the operation, in
order to accelerate it, and bring the heat fooner to
the defired degree ; or in cafe a very fierce heat be
required ; or at a time when the air being hot and
moift will not make the fire burn with the neceffary
vigour.

The heat may be leffened by removing the iron
box, and fhutting the door of the fire-place quite
clofe. It may be ftill further and gradually diminifhed, by taking off the conical funnel from the
top ; by fhutting the door of the fire-place with one
of its plates only, that which has the leaft, or that
which has the greateft aperture in it ; by taking off
the pyramidal cover ; by fhutting the afh-hole door
wholly or in part ; and laftly, by fetting the door of
the fire-place wide open : but, in this laft cafe, the
cold air penetrates into the cavity of the Muffle, and
refrigerates the cupels more than is almoft ever neceffary. If it be obferved, during the operation,
that the Muffle grows cold in any particular part,
it is a fign there is a vacuity left by the coals
in that place : in this cafe an iron wire muft be
thruft into the furnace, through the hole which is
over the door of the fire-place, and the coals ftirred

O 2   there-

therewith, fo as to make them fall into their places and fill up the vacant interftices.

It is proper to obferve that, befides what has been faid concerning the ways of increafing the activity of the fire in the cupelling furnace, feveral other caufes alfo may concur to procure to the matters contained in the Muffle a greater degree of heat: for example, the fmaller the Muffle is, the wider and more numerous the holes in it are; the nearer to its bottom, or further end, the cupels are placed; the more will the matters therein contained be affected with heat.

Befides the operations to be performed by the cupel, this furnace is very ufeful and even neceffary, for many chymical experiments; fuch, for inftance, as thofe relating to fundry vitrifications and enamelling. As it is pretty low, the beft way is to place it, when it is to be ufed, on a bafe of brick-work that may raife it to a level with the operator's hand.

A Lamp-furnace is exceeding ufeful for all operations that require only a moderate, but long continued, degree of heat. The furnace for working with a lamp-heat is very fimple: it confifts only of a hollow cylinder, from fifteen to eighteen inches high, and five or fix in diameter, having at its bottom an aperture large enough for a lamp to be introduced and withdrawn with eafe. The lamp muft have three or four wicks, to the end that by lighting more or fewer of them a greater or lefs degree of heat may be produced. The body of the furnace muft moreover have feveral fmall holes in it, in order to fupply the flame of the lamp with air enough to keep it alive.

On the top of this furnace ftands a bafon five or fix inches deep, which ought to fill the cavity of the cylinder exactly, and to be fupported at its circumference by a rim which may entirely cover and clofe the furnace: the ufe of this bafon is to con-

I.   tain

tain the fand through which the lamp-heat is ufu‑
ally conveyed.

Befides this there muft be a kind of cover or dome
made of the fame material with the furnace, and of
the fame diameter with the fand-bath, without any
other opening than a hole, nearly circular, cut in
its lower extremity.   This dome is a fort of rever‑
beratory, which ferves to confine the heat and di‑
rect it towards the body of the retort; for it is ufed
only when fomething is to be diftilled in a veffel of
this fafhion; and then the hole at its bottom ferves
for a paffage to the neck of the retort.   This dome
fhould have an ear or handle, for the conveniency
of putting it on and taking it off with eafe.

Of Lutes.

CHYMICAL veffels, efpecially fuch as are made of
glafs, and the earthen veffels commonly called ftone‑
ware, are very fubject to break when expofed to
fudden heat or cold; whence it comes that they
often crack when they begin to heat, and alfo when
being very hot they happen to be cooled, either by
frefh coals thrown into the furnace, or by the accefs
of cold air.   There is no way to prevent the former
of thefe accidents, but by taking the pains to warm
your veffel very flowly, and by almoft infenfible de‑
grees.   The fecond may be avoided by coating the
body of the veffel with a pafte or lute, which be‑
ing dried will defend it againft the attacks of cold.

The fitteft ftuff for coating veffels is a compofi‑
tion of fat earth, Windfor-loam, fine fand, filings of
iron, or powdered glafs, and chopped cow's hair,
mixed and made into a pafte with water.   This lute
ferves alfo to defend glafs veffels againft the vio‑
lence of the fire, and to prevent their melting eafily.

In almoft all diftillations it is of great confequence,
as hath been faid, that the neck of the diftilling vef‑
fel be exactly joined with that of the receiver into

which

which it is introduced, in order to prevent the vapours from efcaping into the air and fo being loft:
and this junction is effected by means of a lute.

A few flips of paper applied round the neck of
the veffels with common fize will be fufficient to
keep in fuch vapours as are aqueous, or not very
fpirituous.

If the vapours are more acrid, or more fpirituous, recourfe may be had to flips of bladder
long fteeped in water, which containing a fort of
natural glue clofe the junctures of the veffels very
well.

If it be required to confine vapours of a ftill more
penetrating nature, it will be proper to employ a lute
that quickly grows very hard; particularly a pafte
made with quick-lime and any fort of gelly, whether
vegetable or animal; fuch as the white of an egg,
ftiff fize, &c. This is an excellent lute and not
eafily penetrated. It is alfo ufed to ftop any cracks
or fractures that happen to glafs veffels. But it is
not capable of refifting the vapours of mineral acid
fpirits, efpecially when they are ftrong and fmoking:
for that purpofe it is neceffary to incorporate the
other ingredients thoroughly with fat earth foftened
with water; and even then it frequently happens
that this lute is penetrated by acid vapours, efpecially thofe of the fpirit of falt, which of all others
are confined with the greateft difficulty.

In fuch cafes its place may be fupplied with another, which is called Fat Lute, becaufe it is actually worked up with fat liquors. This lute is compofed of a very fine cretaceous earth, called tobacco-
pipe clay, moiftened with equal parts of the drying
oil of lint-feed, and a varnifh made of amber
and gum copal. It muft have the confiftence of a
ftiff pafte. When the joints of the veffels are clofed
up with this lute, they may, for greater fecurity,
be

be covered over with flips of linen fmeared with the lute made of quick-lime and the white of an egg.

Chymical veffels are liable to be broken in an operation by other caufes befides the fudden application of heat or cold. It frequently happens that the vapours of the matters expofed to the action of fire rufh out with fuch impetuofity, and are fo elaftic, that finding no paffage through the lute with which the joints of the veffels are clofed, they burft the veffels themfelves, fometimes with explofion and danger to the operator.

To prevent this inconvenience, it is neceffary that in every receiver there be a fmall hole, which being ftopped only with a little lute may eafily be opened and fhut again as occafion requires. It ferves for a vent-hole to let out the vapours, when the receiver begins to be too much crowded with them. Nothing but practice can teach the artift when it is requifite to open this vent. If he hits the proper time, the vapours commonly rufh out with rapidity, and a confiderable hiffing noife; and the vent fhould be ftopped again as foon as the hiffing begins to grow faint. The lute employed to ftop this fmall hole ought always to be kept fo ductile, that by taking the figure of the hole exactly it may entirely ftop it. Befides, if it fhould harden upon the glafs, it would ftick fo faft that it would be very difficult to remove it without breaking the veffel. This danger is eafily avoided by making ufe of the fat lute, which continues pliant for a long time, when it is not expofed to an exceffive heat.

This way of ftopping the vent-hole of the receiver has yet another advantage: for if the hole be of a proper width, as a line and half, or two lines, in diameter, then when the vapours are accumulated in too great a quantity, and begin to

O 4　　　　make

make a great effort againſt the ſides of the receiver, they puſh up the ſtopple, force it out, and make their way through the vent-hole : ſo that by this means the breaking of the veſſels may always be certainly prevented.   But great care muſt be taken that the vapours be not ſuffered to eſcape in this manner, except when abſolute neceſſity requires it ; for it is generally the very ſtrongeſt and moſt ſubtile part of a liquor which is thus diſſipated and loſt.

Heat being the chief cauſe that puts the elaſticity of the vapours in action, and prevents their con-denſing into a liquor, it is of great conſequence in diſtillation that the receiver be kept as cool as poſſible.   With this view a thick plank ſhould be placed between the receiver and the body of the furnace, to intercept the heat of the latter, and prevent its reaching the former.   As the vapours themſelves riſe very hot from the diſtilling veſſel, they ſoon communicate their heat to the receiver, and eſpecially to its upper part, againſt which they ſtrike firſt.   For this reaſon it is proper that linen cloths dipt in very cold water be laid over the re-ceiver, and frequently ſhifted.   By this means the vapours will be conſiderably cooled, their elaſticity weakened, and their condenſation promoted.

By what hath been ſaid in this firſt part, con-cerning the properties of the principal agents in Chymiſtry, the conſtruction of the moſt neceſſary veſſels and furnaces, and the manner of uſing them, we are ſufficiently prepared for proceeding directly to the operations, without being obliged to make frequent and long ſtops, in order to give the ne-ceſſary explanations on thoſe heads.

Neverthelefs we ſhall take every proper occaſion to extend the theory here laid down, and to im-prove it by the addition of ſeveral particulars, which will find their places in our Treatiſe of Chymical Operations.

E L E-

ELEMENTS

OF THE

PRACTICE of CHYMISTRY;

WHEREIN

The Fundamental Operations are defcribed, and illuftrated by Obfervations on each Procefs.

ELEMENTS

OF THE

PRACTICE of CHYMISTRY.

INTRODUCTION.

AS the Elements of the Theory of Chymiftry, delivered in the former part of this work, were intended for the ufe of perfons fuppofed to be altogether unacquainted with the art, they could not properly admit of any thing more than fundamental principles, fo difpofed as conftantly to lead from the fimple to the compound, from things known to things unknown : for which reafon I could not therein obferve the ufual order of Chymical Decompofition, which is not fufceptible of fuch a method. I therefore fuppofed all the analyfes made, and bodies reduced to their fimpleft principles; to the end that, by obferving the chief properties of thofe primary elements, we might be enabled to trace them through their feveral combinations, and to form fome fort of judgment *a priori* of the qualities of fuch compounds as may refult from their junctions.

But this latter part is of a different nature. It is a Practical Treatife, intended to contain the manner of performing the principal Operations of Chymiftry; the operations which ferve as ftandards for regulating all the reft, and which confirm the fundamental truths laid down in the Theory.

As

As thefe operations confift almoft wholly of analyfes and decompofitions, there can be no doubt
concerning the order proper to be obferved in giving an account of them : it evidently coincides with
that of the analyfis itfelf.

But as all bodies, which are the fubjects of Chymical operations, are divided by nature into three
claffes or kingdoms, the mineral, the vegetable,
and the animal, the analyfis thereof may naturally
be divided into three branches : fome difference may
alfo arife from the different order in which thefe
three may be treated of.

As the reafons affigned for beginning with one
kingdom rather than with another have never been
thoroughly canvaffed, and may perhaps feem e-
qually good when viewed in a particular light, Chymical writers differ in their opinions on this point.
For my part, without entering into a difcuffion of
the motives which have determined others to follow
a different order, I fhall only produce the reafons
that led me to begin with the mineral kingdom,
to examine the vegetable in the fecond place, and
to conclude with the animal.

· Firft then, feeing vegetables draw their nourifhment'from minerals, and animals derive theirs from
vegetables, the bodies which conftitute thefe three
kingdoms feem to be generated the one by the other,
in a manner that determines their natural rank.

Secondly, this difpofition procures us the advantage of tracing the principles, from their fource in
the mineral kingdom, down to the laft combinations
into which they are capable of entering, that is, into animal matters ; and of obferving the fucceffive
alterations they undergo in paffing out of one kingdom into another.

Thirdly and laftly, I look upon the analyfis of
minerals to be the eafieft of all ; not only becaufe
they confift of fewer principles than vegetables and
animals, but alfo becaufe almoft all of them are capable

of

of enduring the moſt violent action of fire, when that is neceſſary to their decompoſition, without any conſiderable change or diminution of their principles, to which thoſe of other ſubſtances are frequently liable.

Beſides, I am not ſingular in this diſtribution of the three claſſes of bodies, which are the ſubjects of the chymical analyſis : as it is the moſt natural, it has been adopted by ſeveral authors, or rather by moſt who have publiſhed Treatiſes of Chymiſtry. But there is ſomething peculiarly my own in the manner wherein I have treated the analyſis of each kingdom. In the mineral kingdom, for inſtance, will be found a conſiderable number of operations not to be met with in other Treatiſes of Chymiſtry; the authors having probably conſidered them as uſeleſs, or in ſome meaſure foreign, to the purpoſe of Elementary Books, and as conſtituting together a diſtinct art. I mean the proceſſes for extracting ſaline and metallic ſubſtances from the minerals containing them.

Yet, if it be conſidered that ſalts, metals, and ſemi-metals are far from being produced by nature in a ſtate of perfection, or in that degree of purity, which they are commonly ſuppoſed to have when they are firſt treated of in Books of Chymiſtry; but that, on the contrary, theſe ſubſtances are originally blended with each other, and adulterated with mixtures of heterogeneous matters, wherewith they form compound minerals ; I imagine it will be allowed that the operations by which theſe minerals are decompoſed, in order to extract the metals, ſemi-metals, and other ſimpler ſubſtances, eſpecially as they are founded on the moſt curious properties of theſe ſubſtances, are ſo far from being uſeleſs or foreign to the purpoſes of an Elementary Treatiſe, that they are, on the contrary, abſolutely neceſſary thereto.

After I had made theſe reflections, I could not help thinking that an analyſis of minerals, which ſhould treat of ſaline and metallic ſubſtances, without

out taking any notice of the manner in which their matrices muſt be analyſed, in order to extract them, would be no leſs defective, than a treatiſe of the analyſis of vegetables, in which Oils, eſſential Salts, fixed and volatile Alkalis, ſhould be amply treated of, without ſaying one word of the manner of analyſing the plants from which theſe ſeveral ſubſtances are obtained. I therefore thought myſelf indiſpenſably obliged to deſcribe the manner of decompoſing every ore or mineral, before I attempted to treat of the ſaline or metallic ſubſtance which it yields.

For example: as the Vitriolic Acid, with the conſideration of which I begin my Mineral Analyſis, is originally contained in Vitriol, Sulphur, and Alum; and as theſe ſubſtances again derive their origin from the ſulphureous and ferruginous Pyrites, the firſt operations I deſcribe under this head are the proceſſes for decompoſing the Pyrites in order to extract its Vitriol, Sulphur, and Alum. I then proceed to the particular analyſis of each of theſe ſubſtances, with a view to extract their Vitriolic Acid; and afterwards deliver, in their order, the other operations uſually performed on this Acid. Thus it appears that this ſaline ſubſtance occaſions my deſcribing the analyſes of the Pyrites, Vitriol, Sulphur, and Alum. The whole of the Treatiſe on Minerals proceeds on the ſame plan.

The operations by which we decompoſe ores and minerals are of two ſorts: thoſe employed in working by the great, and thoſe for trying in ſmall the yield of any ore. Theſe two manners of operating are ſometimes a little different; yet in the main they are the ſame, becauſe they are founded on the ſame principles, and produce the ſame effects.

As my chief deſign was to deſcribe the operations that may be conveniently performed in a laboratory, I have preferred the proceſſes for ſmall aſſays; eſpecially as they are uſually performed with more

care

care and accuracy than the operations in great works : and here I muſt acknowledge that I am obliged to M. Cramer's *Docimaſia*, or Art of Aſſaying, for all the operations of this kind in my analyſis of minerals. As M. Hellot's work on that ſubject did not appear till after I had finiſhed this, M. Cramer's *Docimaſia*, in which ſound Theory is joined with accurate practice, was the beſt book of the kind I could at that time conſult. I therefore preferred it to all others; and as I have not quoted it in my analyſis of minerals, becauſe the quotations would have been too frequent, let what I ſay here ſerve for a general quotation. I have been careful to name, as often as occaſion required, the other authors whoſe proceſſes I have borrowed : it is a tribute juſtly due to thoſe who have communicated their diſcoveries to the publick.

Though I have told the reader that in my analyſis of minerals he will find the proceſſes for extracting out of each the ſaline or metallic ſubſtances contained in it, yet he muſt not expect that this book will inſtruct him in all that it is neceſſary he ſhould know to be able to determine, by an accurate aſſay, the contents of every mineral. My intention was not to compoſe a Treatiſe of Aſſaying; and I have taken in no more than was abſolutely neceſſary to make the analyſis of minerals perfectly underſtood, and to render it as complete as it ought to be in an Elementary Treatiſe. I have therefore deſcribed only the principal operations relating thereto ; the operations which are fundamental, and which, as I ſaid before, are to ſerve as ſtandards for the reſt, abſtracted from ſuch additional circumſtances as are of conſequence only to the Art of Aſſaying, properly ſo called.

Such therefore as are deſirous of being fully inſtructed in that Art, muſt have recourſe to thoſe works which treat profeſſedly of the ſubject ; and

particularly to that publifhed by M. Hellot: a per-
formance moft efteemed by fuch as are beft-fkilled
in Chymiftry, and rendered fo complete by the numerous and valuable obfervations and difcoveries
of the Author, that nothing better of the kind can
be wifhed for. I thought it proper to give thefe
notices in relation to my analyfis of minerals ; and
fhall now proceed to fhew the plan of my analyfes
of vegetables and of animals.

Seeing all vegetable matters are fufceptible of fermentation, and when analyfed after fermentation,
yield principles different from thofe we obtain from
them before they are fermented, I have divided
them into two claffes ; the former including vegetables in their natural ftate, before they have undergone fermentation; and the latter thofe only
which have been fermented. This analyfis opens
with the proceffes by which we extract from vegetables all the principles they will yield without the
help of fire : and then follow the operations for decompofing plants by degrees of heat, from the
gentleft to the moft violent, both in clofe veffels
and in the open air.

I have not made the fame divifion in the animal
kingdom, becaufe the fubftances that compofe it
are fufceptible only of the laft degree of fermentation, or putrefaction; and moreover the principles
they yield, whether putrefied or unputrefied, are the
very fame, and differ only with regard to their proportions, and the order in which they are extricated
during the analyfis.

I begin this analyfis with an examination of
the milk of animals that feed wholly on vegetables ; becaufe though this fubftance be elaborated
in the body of the animal, and by that means
brought nearer to the nature of animal matters, yet
it ftill retains a great fimilitude to the vegetables
from which it derives its origin, and is a fort of
inter-

intermediate fubftance between the vegetable and animal. Then.I proceed to the analyfis of animal matters properly fo called, thofe which actually make a part of the animal body. I next examine the excrementitious fubftances, that are thrown out of the animal body as fuperfluous and ufelefs. And then I conclude this latter part with operations on the Volatile Alkali ; a faline fubftance of principal confideration in the decompofition of animal matters.

Though, in the general view here given of the order obferved in this Treatife of Practical Chymiftry, I have mentioned only fuch proceffes as ferve for analyfing bodies, yet I have alfo inferted fome other operations of different kinds. The book would be very defective if it contained no more : for the defign of Chymiftry is not only to analyfe the mixts produced by nature, in order to obtain the fimpleft fubftances of which they are compofed, but moreover to difcover by fundry experiments the properties of thofe elementary principles, and to recombine them in various manners, either with each other, or with different bodies, fo as to reproduce the original mixts with all their properties, or even form new compounds which never exifted in nature. In this book therefore the reader will find proceffes for combining and recompounding, as well as for refolving and decompofing bodies. I have placed them next to the proceffes for decompofition, taking all poffible care not to interrupt their order, or break the connection between them.

PRACTICE of CHYMISTRY.

❖❖❖❖❖❖❖❖❖❖❖❖❖❖❖❖❖❖❖❖❖❖❖❖

PART I.
OF MINERALS.

SECTION I.
Operations performed on Saline Mineral Subſtances.

CHAP. I.
Of the VITRIOLIC ACID.

PROCESS I.
To extraƈt Vitriol from the Pyrites.

TAKE any quantity you pleaſe of Iron-Pyrites; leave them for ſome time expoſed to
the air; they will crack, ſplit, loſe their
brightneſs, and fall into powder. Put this powder into a glaſs cucurbit, and pour upon it twice its weight
of hot water; ſtir the whole with a ſtick, and the
liquor will grow turbid. Pour it, while it is yet
warm, into a glaſs funnel lined with brown filtering
paper; and having placed your funnel over another
glaſs cucurbit, let the liquor drain into it. Pour

more

more hot water on the powdered Pyrites, filter as
before, and so go on, every time lessening the quantity of water, till that which comes off the Pyrites
appears to have no astringent vitriolic taste.

Put all these waters together into a glass vessel
that widens upwards ; set it on a sand bath, and
heat the liquor till a confiderable smoke arises; but
take care not to make it boil. Continue the same
degree of fire till the surface of the liquor begins
to look dim, as if some dust had fallen into it ;
then cease evaporating, and remove the vessel into
a cool place : in the space of four-and-twenty hours
will be formed therein a quantity of cryftals, of a
green colour and a rhomboidal figure : these are
Vitriol of Mars, or Copperas. Decant the remaining liquor; add thereto twice its weight of water ;
filter, evaporate, and cryftallize as before ; repeat
these operations till the liquor will yield no more
cryftals, and keep by themselves the cryftals obtained at each cryftallization.

OBSERVATIONS.

The Pyrites are minerals which, by their weight
and shining colours, frequently impose on such as
are not well acquainted with ores. At first sight
they may be taken for very rich ones ; and yet they
consist only of a small quantity of metal combined with
much sulphur or arsenic, and sometimes with both.

They strike fire with a steel as flints do, and emit
a sulphureous smell : so that they may be known by
this extemporaneous proof. The metal most commonly and most abundantly found in the Pyrites is
iron ; the quantity whereof sometimes equals, or
even exceeds, that of the sulphur. Besides metallic and sulphureous matters, the Pyrites contain also
some unmetallic earth.

There are several sorts of Pyrites : some of them
contain only iron and arsenic. They have not all
the property of efflorescing spontaneously in the air,

and

and turning into vitriol: none do fo but fuch as confift only of iron and fulphur, or at leaft contain but a very fmall portion of copper, or of arfenic : and even amongft thofe that are compofed of iron and fulphur alone, there are fome that will continue for years together expofed to the air without fhooting, and indeed without fuffering the leaft fenfible alteration.

The efflorefcence of the Iron-Pyrites, and the changes they undergo, are phenomena well worth our notice. They depend on the fingular property which iron poffeffes of decompofing fulphur by the help of moifture. If very fine iron-filings be accurately mingled with flowers of fulphur, this mixture, being moiftened with water, grows very hot, fwells up, emits fulphureous vapours, and even takes fire : what remains is found converted into Vitriol of Mars. On this occafion, therefore, the fulphur is decompofed; its inflammable part is diffipated or confumed; its acid combines with the iron, and a Vitriol arifes from that conjunction.

This is the very cafe with the Pyrites that confift only of iron and fulphur; yet fome of them, as we faid before, do not efflorefce fpontaneoufly and turn to Vitriol. The reafon probably is that, in fuch minerals, the particles of iron and fulphur are not intimately mixed together, but feparated by fome earthy particles.

In order to procure Vitriol from Pyrites of this kind, they muft be for fome time expofed to the action of fire, which by confuming part of their fulphur, and rendering their texture lefs compact, makes way for the air and moifture, to which they muft be afterwards expofed, to penetrate their fubftance, and produce in them the changes with which thofe others are affected that germinate fpontaneoufly.

The Pyrites which contain copper and arfenic, and for that reafon do not efflorefce, muft likewife

under-

undergo the action of fire; which befides the effects it produces on Pyrites that confift of iron and fulphur only, diffipates alfo the greateft part of the arfenic. Thefe Pyrites being firft roafted, and then expofed to the air for a year or two, do alfo yield Vitriol; but then it is not a pure Vitriol of Iron, but is combined with a portion of blue Vitriol, the bafis of which is Copper.

Sometimes alfo there is Alum in the vitriolic waters drawn off the Pyrites. It was on account of this mixture of different falts that we recommended the keeping apart the cryftals obtained from each different cryftallization: for by this means they may be examined feparately, and the fpecies to which they belong difcovered.

When Vitriol of iron is adulterated with a mixture of the Vitriol of copper only, it is eafy to purify it and bring it to be entirely martial, by diffolving it in water, and fetting plates of iron in the folution: for iron having a greater affinity than copper with the vitriolic acid, feparates the latter from it, and affuming its place produces a pure Vitriol of Mars.

In large works for extracting Vitriol from the Pyrites they proceed thus. They collect a great quantity of Pyrites on a piece of ground expofed to the air, and pile them up in heaps of about three feet high. There they leave them expofed to the action of the air, fun, and rain, for three years together; taking care to turn them every fix months, in order to facilitate the efflorefcence of thofe which at firft lay undermoft. The rain-water which has wafhed thofe Pyrites is conveyed by proper channels into a ciftern; and when a fufficient quantity thereof is gathered, they evaporate it to a pellicle in large leaden boilers, having firft put into it a quantity of iron, fome part of which is diffolved by the liquor, becaufe it contains a vitriolic acid that is not fully faturated therewith. When it is fufficiently evaporated, they draw it off into large leaden or wooden coolers, and there leave it to fhoot into cryftals. In thefe laft veffels feveral fticks are placed, croffing each other in all manner of directions, in order to multiply the furfaces on which the cryftals may faften.

The Pyrites are not the only minerals from which Vitriol may be procured. All the ores of iron and copper that contain fulphur may alfo be made to yield green or blue Vitriol, according to the nature of each, by torrefying them, and leaving them long expofed to the air: but this ufe is feldom made of them, as there is more profit to be got by extracting the metals they contain. Befides, it is eafier to obtain Vitriol from the Pyrites, than from thofe other mineral fubftances.

PROCESS II.

To extract Sulphur from the Pyrites, and other ful-phureous Minerals.

REDUCE to a coarfe powder any quantity of yellow Pyrites, or other Mineral containing Sulphur. Put this powder into an earthen or glafs retort, having a long wide neck, and fo large a body that the matter may fill but two thirds of it. Set the retort in a fand-bath fixed over a reverberating furnace: fit to it a receiver half full of water, and fo placed that the nofe of the retort may be about an inch under the water: give a gradual fire, taking care you do not make it fo ftrong as to melt the matter. Keep the retort moderately red for one hour, or an hour and half, and then let the veffels cool.

Almoft all the Sulphur feparated by this operation from its matrix will be found at the extremity of the neck of the retort, being fixed there by

the

the water. You may get it out either by melting
it with such a gentle heat as will not set it on fire,
or by breaking the neck of the retort.

OBSERVATIONS.

OF all minerals the Pyrites contain the moſt Sulphur; thoſe eſpecially which have the colour of fine
braſs, a regular form, ſuch as round, cubical, hexagonal, and being broken preſent a number of ſhining needles, all radiating, as it were, from a center.

A very moderate heat is ſufficient to ſeparate the
Sulphur they contain. We directed that the retort
employed ſhould have a long and wide neck, with
a view to procure a free paſſage for the Sulphur: the
water ſet in the receiver detains the Sulphur, fixes it,
and prevents it from flying off; ſo that it is unneceſ-
ſary to cloſe the joints of the veſſels. But it is proper to take notice that when ever you uſe an apparatus for diſtilling, which requires the beak of the
retort to be under water, it is of very great conſequence that the fire be conſtantly ſo regulated that
the retort may not cool in the leaſt; for in that caſe,
as the rarefied air contained therein would be condenſed, the water in the receiver would riſe into
the retort and break it.

If in diſtilling Sulphur, according to the preſent
proceſs, the matter contained in the retort ſhould
happen to melt, the operation would be thereby
conſiderably protracted, and it would require a
great deal more time to extract all the Sulphur;
becauſe all evaporation is from the ſurface only,
and the matter, while it remains in a coarſe powder, preſents a much more extenſive ſurface than
when it is melted.

This remark holds with regard to all other diſtillations. Any quantity of liquor, ſet to diſtil in its
fluid ſtate, will take much more time to riſe in vapours, and paſs from the retort into the receiver,

than

than if it be incorporated with fome folid body reduced to minute parts, fo that the whole fhall make
a moift powder; and this though the very fame degree of fire be applied in both cafes.

If the matter from which it is propofed to extract
Sulphur be fuch as will melt with the degree of fire
neceffary to this operation; that is, with a heat which
will make the retort but faintly red, it muft be mixed with fome fubftance that is not fo fufible. Very
pure coarfe fand, or clean gravel, may be ufed with
fuccefs : but abforbent earths are altogether improper for this purpofe, becaufe they will unite with
the Sulphur,

The fulphureous minerals which are moft apt to
fufe are the cupreous Pyrites, or yellow copper
ores: common lead ores are alfo very fufible.

The Pyrites are by this operation deprived of almoft all the Sulphur they contain; and confequently
little is left behind, but the particles of iron and
copper, together with a portion of unmetallic earth,
which we fhall fhew how to feparate from thefe metals, when we come to treat of them. I fay that by
this operation the Pyrites are deprived of almoft all,
and not entirely of all their Sulphur; becaufe, this
feparation being made in clofe veffels only, there
always remains a certain quantity of Sulphur, which
adheres fo obftinately to the metals, that it would
be almoft impoffible to get it all out, even though
a much ftronger fire than that directed in the procefs were applied for this purpofe, and though choice
had been, as it ought to be, made of fuch Pyrites,
or other fulphureous Minerals, as part moft eafily
with their Sulphur. Nothing but a very ftrong
fire in the open air is capable of carrying it wholly
off, or confuming it entirely.

In feveral places are found great quantities of native Sulphur. The Volcanos abound with it, and
people gather it at the foot of thofe burning moun-
tains,

tains. Several springs of mineral waters alfo yield
Sulphur, and it is fometimes found fublimed to
the vaulted roofs of certain wells, and among
others in one at Aix-la-Chapelle.

The Germans and Italians have large works,
for extracting Sulphur in quantities out of Pyrites, and other minerals which abound therewith.
The procefs they work by is the fame with that
here delivered; but with this difference only, that
Sulphur being but of fmall value they do not ufe
fo many precautions. They content themfelves
with putting the fulphureous minerals into large
crucibles, or rather earthen cucurbits, which they
place in the furnace in fuch a manner that, when
the fulphureous part melts, it runs into veffels filled with water, and is thereby fixed.

The Sulphur obtained, either by diftillation or
by fimple fufion, is not always pure.

When it is obtained by diftillation, if the matters
from which you extract it contain moreover fome
other minerals of nearly the fame volatility, fuch,
for inftance, as Arfenic, or Mercury, thefe minerals will come over with it. This is eafily perceived: for pure fublimed Sulphur is always of a beautiful yellow, inclining to a lemon colour. If it
look red, or have a reddifh caft, it is a fign that
fome Arfenic hath rifen along with it.

Mercury fublimed with Sulphur likewife gives it
a red colour; but Sulphur is very feldom adulterated with this metallic fubftance: for Arfenic is
frequently found combined with the Pyrites, and
other fulphureous minerals; whereas on the contrary
we very rarely meet with any Mercury in them.

But if Mercury fhould happen to rife with the
Sulphur in diftillation, it may be difcovered by examining the fublimate; which, in that cafe, will
have the properties of Cinabar: on being broken its
infide will appear to confift of needles adhering laterally

terally to each other; its weight will be very considerable; and laftly, the great heat of the place
where it is collected will furnifh another mark to
know it by; for as Cinabar is lefs volatile than
Arfenic or Sulphur, it faftens on places too hot for
either Sulphur or Arfenic to bear.

Sulphur may alfo be adulterated with fuch fixed
matters, either metallic or earthy, as it may have
carried up along with it in the diftillation, or as may
have been fublimed by the Arfenic, which has a ftill
greater power than Sulphur to volatilize fixed bodies.

If you defire to free the Sulphur from moft of
thefe heterogeneous matters, it muft be put into
an earthen cucurbit, and fet in a fand bath. To
the cucurbit muft be fitted one or more aludels,
and fuch a degree of heat applied as fhall but juft
melt the Sulphur; which is much lefs than that
neceffary to feparate the Sulphur from its matrix.
As foon as the Sulphur is melted it will fublime in
lemon-coloured flowers, that will ftick to the in-
fides of the aludels.

When nothing more appears to rife with this degree of heat, the veffels muft be fuffered to cool.
At the bottom of the cucurbit will be found a fulphureous mafs, containing the greateft part of the
adventitious matters that were mixed with the Sulphur, and more or lefs red or dark-coloured, according to the nature of thofe matters.

When we come to treat of Arfenic and Mercury,
we fhall give the methods of feparating Sulphur
entirely from thofe metallic fubftances. -

PROCESS III.

To extraɛt Alum from aluminous Minerals.

TAKE fuch materials as are known or fufpeɛted
to contain Alum. Expofe them to the air,
that they may efflorefce. If they remain there a
year

year without any fenfible change, calcine them, and
then leave them expofed to the air, till a bit thereof
being put on the tongue imparts an aftringent aluminous tafte.

When your matters are thus prepared, put
them into a leaden or glafs veffel; pour upon them
thrice their weight of hot water; boil the liquor;
filter it; and repeat thefe operations till the earth
be fo edulcorated that the water which comes off it
hath no tafte. Mix all thefe folutions together,
and let them ftand four-and-twenty hours, that the
grofs and earthy parts may fettle to the bottom;
or elfe filter the liquor: then evaporate till it will
bear a new-laid egg. Now let it cool, and ftand
quiet four-and-twenty hours: in that time fome
cryftals will fhoot, which are moft commonly vitriolic; for Alum is rarely obtained by the firft cryftallization. Remove thefe vitriolic cryftals: if
any cryftals of Alum be found amongft them,
thefe muft be diffolved anew, and fet to cryftallize
a fecond time in order to their purification; becaufe
they partake of the nature as well as of the colour
of vitriol. By this method extract all the Alum
that the liquor will yield.

If you get no cryftals of Alum by this means,
boil your liquor again, and add to it a twentieth
part of its weight of a ftrong alkaline lixivium, or
a third part of its weight of putrefied urine, or a
fmall quantity of quick-lime. Experience and repeated trials muft teach you which of thefe three
fubftances is to be preferred, according to the particular nature of the mineral on which you are to
operate. Keep your liquor boiling, and if there be
any Alum in it, there will appear a white precipitate: in that cafe let it cool and fettle. When the
white precipitate is entirely fallen, decant the clear,
and leave the cryftals of Alum to fhoot at leifure,

till

till the liquor will yield no more : it will then be exceeding thick.

O B S E R V A T I O N S.

A L U M is obtained from several sorts of Minerals. In some parts of Italy, and in sundry other places, it effloresces naturally on the surface of the earth. There it is swept together with brooms, and thrown into pits full of water. This water is impregnated therewith till it can dissolve no more. Then it is filtered, and set to evaporate in large leaden vessels ; and when it is sufficiently evaporated, and ready to shoot into cryftals, it is drawn off into wooden coolers, and there left for the salt to cryftallize.

In aluminous soils there are often found springs strongly impregnated with Alum ; so that to obtain it the water need only be evaporated.

In the country about Rome there is a very hard stone, which is hewn out of the quarry juft like other stones for building : this stone yields a great deal of Alum. In order to extract it, the stones are calcined for twelve or fourteen hours ; after which they are exposed to the air in heaps, and carefully watered three or four times a day for forty days together. In that time they begin to effloresce, and to throw out a reddish matter on their surface. Then they are boiled in water, which dissolves all the Alum they contain, and being duly evaporated gives it back in cryftals. This is the Alum called *Roman Alum.*

Several sorts of Pyrites also yield a great deal of Alum. The Englifh have a stone of this kind, which in colour is very like a flate. This stone contains much Sulphur, which they get rid of by roafting it. After this they fteep the calcined stone in water, which dissolves the Alum it contains, and to this solution they add a certain quantity of a lye made of the afhes of sea-weeds.

The

The Swedes have a Pyrites of a bright, golden colour, variegated with filver fpots, from which they procure Sulphur, Vitriol, and Alum. They feparate from it the Sulphur and the Vitriol by the methods above prefcribed. When the liquor which hath yielded Vitriol is become thick, and no more vitriolic cryftals fhoot in it, they add an eighth part of its weight of putrefied urine, mixed with a lye made of the afhes of green wood. Upon this there appears and falls to the bottom a copious red fediment. They decant the liquor from this precipitate, and when it is duly evaporated find it fhoot into beautiful cryftals of Alum.

What hath been faid, concerning the feveral matrices from which Alum is obtained, fufficiently fhews that it is feldom folitary in the waters with which aluminous fubjects have been lixiviated. It is almoft always accompanied with a certain quantity of Vitriol, or other faline mineral matters, which obftruct its cryftallization, and prevent its being pure. 'Tis with a view to free it from thefe matters, that the waters impregnated with Alum are mixed with a certain quantity of the lye of fome fixed Alkali, or with putrefied urine, which contains much volatile Alkali. Thefe Alkalis have the property of decompounding all the Neutral falts which have for their bafis either an abforbent earth or a metallic fubftance; and fuch as have a metallic fubftance for their bafis more readily than thofe whofe bafis is an earth. Confequently, if they are mixed with a liquor in which both thefe forts of falts are diffolved, they muft decompound that fort whofe bafis is metallic fooner than the other whofe bafis is an earth. This is what comes to pafs in a folution of Alum and Vitriol. The metallic part of the latter is feparated from its acid by the Alkalis when mixed with that folution; and 'tis this metallic part, which is generally iron, that appears in the form of a reddifh precipitate, as above-mentioned.

But

But becaufe Alkalis decompound alfo thofe Neutral falts which have an earth for their bafrs, care muft be taken that too much thereof be not added; elfe what you put in, more than is neceffary to decompound the vitriolic falts in your liquor, will attack the Alum, and decompound it likewife.

The Alkali made ufe of to promote the cryftallization of the Alum joins with the Vitriolic Acid, which had diffolved the fubftances now precipitated, and therewith forms different Neutral falts according to its particular nature.   If the Alkali be a lixivium of common wood-afhes, the Neutral falt will be a vitriolated Tartar; if a lixivium of the afhes of a maritime plant like Soda, the Neutral falt will be a Glauber's falt; if putrefied urine, the Neutral falt will be a vitriolic Ammoniacal falt. Some of thefe falts incorporate with the Alum, which in large works cryftallizes in vaft lumps : and hence it comes that fome forts of Alum when mixed with a fixed Alkali fmell like a volatile Alkali.

The cryftals of Alum are octaedral, that is, they are folids with eight fides.   Thefe octaedral folids are triangular pyramids, having their angles cut away, fo that four of their furfaces are hexagons, and the other four triangles.

Sulphur, Vitriol, and Alum are the three principal fubjects in which we certainly know that the univerfal or Vitriolic Acid particularly refides, and from which we extract it when we want to have it pure.   For this reafon we thought it proper, before we treated of the extraction of this Acid, to fhew the method of feparating thofe matters themfelves from the other minerals out of which we obtain them.

Moreover, all the other matrices, in which the Vitriolic Acid is moft commonly lodged, may be referred-to one or other of the matters which ferve as bafes to thefe three minerals.

To

To Sulphur we may refer all combinations of the Vitriolic Acid with an inflammable matter: but we muſt take care not to confound Sulphur with thoſe Bitumens in which the Vitriolic Acid may be found: for the baſis of thoſe bitumens is a real Oil; whereas the baſis of Sulphur is the pure Phlogiſton. Yet as Oils themſelves contain the Phlogiſton, which in union with the Vitriolic Acid forms a true Sulphur, it follows that ſuch bitumens may in a certain reſpect be claſſed with Sulphur.

The ſame is to be ſaid of Vitriol. The name is uſually given to ſuch combinations only as are formed of the Vitriolic Acid with Iron or Copper, which make the green and blue Vitriol; and to a third ſpecies of Vitriol, which is white, and has Zinc for its baſis: but as the Vitriolic Acid may, by particular combinations, be united with many other metallic ſubſtances, all ſuch Metallic Salts muſt be referred to the claſs of Vitriols.

The ſame may alſo be ſaid of Alum, which is no other than a combination of the Vitriolic Acid with a particular kind of abſorbent earth; ſo that all combinations of this Acid with any earth whatever may be placed in the ſame claſs.

This laſt claſs of mixts is the moſt extenſive of all that contain the Vitriolic Acid; becauſe there are a vaſt many earths, all differing from one another, with which that Acid may be united. Alum properly ſo called, the Gypſums, Talcs, Selenites, Boles, and all the other compounds of this kind, differ from each other only in their particular earths.

The different properties of theſe earthy ſalts depend on the nature of their baſes. Thoſe which are of the aluminous kind retain much water in cryſtallizing, which makes them very ſoluble in water, and gives them the property of acquiring readily the aqueous fluor when expoſed to the fire. Thoſe which are of the nature of the Selenites admit but very little water in their cryſtals, and conſequently

are

are almoſt inſoluble in water; nor does the fire give
them an aqueous fluor.   Laſtly, the Gypſums and
Talcs are ſtill more deſtitute of theſe properties.
The natures of the earths in theſe ſeveral compounds are hitherto but very imperfectly known,
and may give the Chymiſts occaſion for enquiries
equally curious and uſeful.

The Vitriolic Acid is ſometimes found complicated with a fixed alkaline baſis.   This is almoſt always the Alkali of Sea-ſalt; ſo that the compound is
a Glauber's Salt.   Some mineral waters are impregnated therewith; which happens when theſe waters
contain Vitriol or Alum, together with Sea-ſalt.

From the principles laid down, in our Elements of
the Theory, it appears that the Vitriolic Acid hath
not ſo great an affinity with earthy and metallic ſubſtances as with fixed Alkalis; and alſo that it is
ſtronger than the Marine Acid, and hath a greater
affinity with fixed Alkalis.   This being allowed,
the generation of native Glauber's Salts is eaſily
accounted for.   The Acid of aluminous or Vitriolic
ſalts quits the earth or the metal with which it was
combined, and expelling the Acid of ſea-ſalt unites
with its baſis.   Warmth greatly promotes theſe
decompoſitions.

If the common foſſil ſalt, uſually called *Sal Gem*, or
any other kind of Sea-ſalt, ſhould happen to be near
a Vulcano, when it diſcharges flaming Sulphur, as
is frequently the caſe, and if this Sulphur ſhould
run among the Sea-ſalt, a Glauber's Salt would in-
ſtantly be formed in that place; becauſe when Sulphur burns, its Acid is ſeparated and ſet at liberty.

Laſtly, if aluminous or vitriolic matters, or burning Sulphur, ſhould meet with the aſhes of plants or
trees conſumed by fire, a vitriolated Tartar would
be formed, becauſe theſe aſhes contain a fixed Alkali of the ſame nature with that of Tartar.

The Vitriolic Acid when combined with an earthy
baſis adheres ſtrongly thereto; ſo that the force of
fire

fire is able to expel very little or none of it. There is no way of feparating it from fuch a bafis, but by prefenting to it an Alkaline Salt, with which it will unite: nor is it ever extracted from fuch matters when it is required pure. It does not adhere fo firmly to metallic fubftances; but is feparated from them by the force of fire: fo that it may be obtained from the feveral forts of Vitriol. It is ufually drawn from Green Vitriol; that being the commoneft fort.

As to Sulphur, the Phlogifton which is its bafis being the fubftance wherewith the Vitriolic Acid hath the greateft affinity, it would be altogether impoffible to decompofe it, and to feparate its Acid, if it were not inflammable; but by burning it the Phlogifton is deftroyed, and leaves the Acid at liberty. By this means therefore it may be feparated. We fhall now give the proceffes for extracting the Acid from Vitriol and Sulphur.

PROCESS IV.

To extract the Vitriolic Acid from Green Vitriol.

TAKE any quantity of Green Vitriol: put it in an unglazed earthen veffel, and heat it gradually. Vapours will foon begin to rife. Encreafe the fire a little, and it will liquefy by means of the water contained in it, and acquire what we called an *aqueous* fluor. Continue the calcination, and it will become lefs and lefs fluid, grow thick, and turn of a greyifh colour. Now raife your fire, and keep it up till the falt recover its folidity, acquire an orange colour, and begin to grow red where it immediately touches the fides of the veffel. Then take it out, and reduce it to powder.

Put the Vitriol thus calcined and pulverized into a good earthern retort, of which one half at leaft

muſt remain empty. Set the retort in a reverberatory furnace: fit thereto a large glaſs receiver, and,
having luted the joint well, give fire by degrees.
You will ſoon ſee white clouds riſe into the receiver,
which will render it opaque, and heat it. Continue
the ſame degree of fire till theſe clouds diſappear:
they will be ſucceeded by a liquor which will trickle
down the ſides of the receiver in veins. Still keep
up the fire to the ſame degree as long as theſe veins
appear. When they begin to abate, encreaſe the
fire, and puſh it to the utmoſt extremity: upon this,
there will come over a black, thick liquor: it will
even be found congealed, and prove the Icy Oil of
Vitriol, if care hath been taken to change the receiver, keep the veſſels perfectly cloſe, and give a
ſufficient degree of heat. Proceed thus till nothing
more comes over, or at leaſt very little. Let the
veſſels cool, unlute them, pour the contents of the
receiver into a bottle, and ſeal it hermetically.

OBSERVATIONS.

GREEN Vitriol retains much water in cryſtallizing; and in order to free it from that ſuperfluous
phlegm, it muſt be calcined before you diſtill it.
Without this precaution the operation will be exceedingly protracted, and a great deal of time waſted
in diſtilling ſuch a quantity of water; which will
moreover greatly weaken the Acid by commixing
with it, unleſs care be taken to change the recipient as ſoon as the water is all come over.

But there is alſo another advantage in calcining
the Vitriol before you put it into the retort: for
otherwiſe this ſalt would melt on the firſt application
of heat, and run into a maſs; which would prove
a great hindrance to its diſtillation. This inconvenience is avoided by a previous calcination, in conſequence whereof the Vitriol is eaſily reduced to a
powder which never becomes fluid.

Vitriol

Vitriol calcined as directed in the procefs grows fo hard, and adheres fo firmly to the veffel in which the calcination is performed, that it requires no fmall pains to feparate and pulverize it. Care muft be taken to put it into the retort as foon as it is pulverized, and to ftop that veffel very clofe if you do not begin the diftillation immediately: for otherwife it will naturally attract from the air almoft all the moifture it hath loft.

The Acid which Vitriol yields by diftillation is fulphureous; probably becaufe it ftill retains fome of the Phlogifton, with which it was united when under the form of fulphur in the Pyrites; or elfe hath laid hold on a portion of that belonging to the iron which ferved for its bafis in Vitriol. But this ful-phureous part is volatile, and flies off in time.

This decompofition of Vitriol in clofe veffels is a difficult and laborious procefs. To carry the opera-tion to its utmoft perfection requires a fire of ex-treme violence, kept up without intermiffion during four or five days; fuch in fhort as few veffels are able to bear. Of courfe this operation is feldom performed in laboratories. The French Chymifts fetch their Oil of Vitriol from Holland, where it is extracted from Vitriol in large quantities, by means of furnaces erected for the purpofe, in which many retorts are employed at once.

In the Memoirs of the Academy of Sciences M. Hellot hath given us the moft material circumftances of a very fine experiment of this kind, in which he pufhed the diftillation of Green Vitriol to the utmoft. Into a German retort * he put fix pounds of Green Englifh Vitriol calcined to rednefs, which he ex-pofed to a fire of the extremeft violence, conftantly kept up during four days and four nights. At the expiration of that time he found in the veffels em-

* They are much the beft, and bear a very fierce heat.

ployed

ployed as receivers an Icy Oil of Vitriol, which was altogether in a cryftalline form and black. The precautions neceſſary to make this experiment fuc-ceed he reprefents in the following terms.

" The fuccefs of this operation, which produces
" an Oil of Vitriol perfectly Icy and without any
" liquor, depends on the care taken to prevent the
" acid vapours, driven by the fire out of Vitriol
" calcined to rednefs, from having any communi-
" cation with the external air while they are diftil-
" ling: for otherwife they will attract from it a
" moifture which will keep them fluid in the re-
" ceiver. The receiver muſt be at ſuch a diftance
" from the furnace that it may remain cool enough
" for the vapours to condenfe in it. There muſt
" alfo be fufficient room for thofe vapours to circu-
" late in, and to prevent the fulphureous explofions,
" which are every now and then difcharged out of
" the retort, from burfting the veffels: for though
" the previous calcination of the Vitriol hath car-
" ried off the moft volatile, yet there ftill remains
" enough of the inflammable principle, even in the
" iron itfelf, to form a Sulphur with the Acid as
" it is extricated, or at leaſt a mixt that would be
" as apt to take fire as common Sulphur, if it were
" not over-dofed with the Acid.

" As the beſt means of gaining thefe ends M.
" Hellot contrived to adapt to the neck of his re-
" tort a receiver with two necks, the lowermoſt of
" which was inferted into a large ballon. Receivers
" applied to each other in this manner are called
" Adopters.

" It is no eafy matter to get this Icy Oil out of
" the ballon: for as foon as the air touches it fuch
" a thick cloud of fulphureous fumes arifes, that
" it is abfolutely neceffary to place the veffel on
" fome fhelf over head, becaufe a man cannot ftand

5   " expofed

" expofed thereto for a fingle minute without be-
" ing fuffocated." .

This Icy Acid muft be fhut up with all poffible
expedition in a cryftal bottle accurately clofed with
a glafs ftopple, which fhould be ground with emery
in its neck fo as to fit it exactly: for it attracts
moifture fo powerfully, that, unlefs exceeding great
care be taken to prevent all communication with
the external air, it will foon diffolve into a fluid.

" The Icy Oil is black; becaufe the acid vapours
" carry over with them fomething of a greafy mat-
" ter, from which Vitriol is feldom free, and
" which always appears, after repeated folutions
" and cryftallizations of this Salt, in the mother-
" water which will fhoot no more. Now the
" fmalleft portion of inflammable matter prefently
" blackens the moft highly rectified Oil of Vitriol,
" which is perfectly clear.

" The Vitriolic Acid, when forced over by a
" violent heat, carries along with it fome ferrugi-
" nous particles alfo, that want nothing but to be
" united with a Phlogifton to become true iron.
" They are eafily difcovered, either in the common
" black Oil of Vitriol, or in the blackifh cryftals
" of the Icy Oil, by only diffolving them in a large
" quantity of diftilled water: for after feven or
" eight days digeftion a light powder or downy
" fediment precipitates, which being calcined in a
" violent fire is partly attracted by the magnet;
" and being again calcined with bees-wax becomes
" almoft entirely iron."

The *Caput mortuum* of this diftillation of Vitriol
is the ferruginous earth of this Salt, and is called
Colcothar. When this Colcothar hath undergone a
violent fire, as in the experiment now related, fcarce
any Acid remains therein. Out of fix pounds of
Vitriol that M. Hellot ufed, he could recover no
more, by lixiviating what was left in the retort,

than

than two ounces of a Vitriolic Salt; and even that was very earthy.

If Vitriol be expofed to a fire neither fo violent nor fo long continued, its Colcothar will yield a greater quantity of Vitriol that hath not been decompofed. A white cryftalline falt is alfo obtained from it, and called *Salt of Colcothar*; which is no other than the fmall portion of Alum ufually contained in Vitriol, and not fo eafily decompofed by the action of fire.

PROCESS V.

To decompofe Sulphur, and extract its Acid, by burning it.

TAKE any quantity of the pureft Sulphur: fill therewith a crucible or other earthen difh: heat it till it melts: then fet it on fire; and when its whole furface is lighted place it under a large glafs head, taking care that the flame of the Sulphur do not touch either its fides or bottom; that the air have free accefs, in order to make the Sulphur burn clear; and that the head incline a little toward the fide on which its beak is, that as the vapours condenfe therein the liquor may run off with eafe. To the beak of this veffel fit a receiver: the fumes of the lighted Sulphur will be condenfed, and gather into drops in the head, out of which they will run into the receiver. There, when the Sulphur has done burning, you will find an Acid liquor, which is the Spirit of Sulphur.

OBSERVATIONS.

In the burning of Sulphur, the Phlogifton which ferves for its bafis is diffipated, and feparated from the Acid which is left at liberty. The acid fumes which rife from the lighted fulphur ftrike againft the

5   infide

infide.of the head placed over it, are there condenfed, and appear in the form of a liquor. But as Sulphur, like all other inflammable bodies, Nitre excepted,. will not burn in clofe veffels, it is neceffary that the air be freely admitted here; which occafions the lofs of a great deal of the Acid of the Sulphur, as is evident from the pungent fuffocating fmell perceived in the laboratory during the operation.

This Acid, while combined with the Phlogifton, is incapable of contracting any union with water; but when alone is very apt to mix therewith: it is even proper to put fome in its way, that it may incorporate therewith as foon as it is difcharged from the Sulphur; for it is then very free from phlegm, very volatile, and confequently very little difpofed to condenfe into a liquor, but on the contrary very apt to fly off in vapours. The water, which it imbibes with a kind of avidity, fixes and detains it; fo that by this means a much greater quantity thereof is obtained from Sulphur, than if it were diftilled without this precaution.

It is proper therefore now and then to introduce a difh full of hot water under the head which receives the fumes of the Sulphur. The vapours that exhale from the water bedew the infide of the head, and procure the advantage we are fpeaking of.

The fame thing may be effected feveral other ways: thus, the crucible containing the Sulphur may be fet on a foot placed in an earthen difh with fome water in it; which however muft not rife above the foot; for if it fhould reach the crucible, it might cool and fix the fulphur. The difh thus prepared muft be placed on a fand-bath hot enough to make the water fmoke continually; and over all is to be placed the head as directed in the procefs.

The fize and form of the veffel which immediately receives the fulphureous fumes may alfo contri-

bute

bute to increafe the quantity of the Acid Spirit. A very large veffel, with a hole at bottom no wider than is juft fufficient to admit the vapours, is the propereft for this operation.

After the Sulphur has burnt for fome time, it often happens that a fort of fkin or cruft forms on its furface, which is not inflammable, but gradually leffens the quantity and vigour of the flame as it increafes in thicknefs, and at laft puts it quite out. This cruft proceeds from the impurities, and hete-rogeneous uninflammable particles contained in the fulphur. Care muft be taken to remove it with an iron wire as faft as it forms.

Two quantities of Sulphur may alfo be kept in two crucibles, and heated alternately. That in which the Sulphur is hot and melted may be fub-ftituted for the other in which the Sulphur is grown cold and fixed; becaufe cold Sulphur does not burn well.

The Spirit of Sulphur is at firft pungent and vo-latile, becaufe it ftill retains a fmall portion of the Phlogifton: but that fulphureous part flies off, efpecially if the bottle in which the Spirit is kept · be left for fome time unftopped.

The Acid obtained from Sulphur appears by all chymical proofs perfectly like that obtained from Vitriol: they differ in this only that the former is the pureft; for the Acid obtained from Vitriol car-ries over with it fome metallic parts, as we ob-ferved before, which can never happen to that ob-tained from Sulphur.

If linen rags dipped in a folution of Fixed Al-kali be expofed to the fumes of burning brimftone, the Spirit of Sulphur joins with the Alkali, and therewith forms a Vitriolated Tartar. This Salt is known to be formed when the rags grow ftiff, and appear fpangled with a vaft many glittering points, which

which are nothing but little cryftals of the Salt we
are fpeaking of.

When the Sulphur burns very gently and flowly
the Spirit that exhales from it is fo much the more
fulphureous and volatile; and hence the Salt formed
by the combination of this Spirit with the Alkali
expofed to it in linen rags, as in the above-mentioned experiment, is not at firft a Vitriolated Tartar;
but a Neutral Salt of a particular kind, which is
capable of being decompofed by any other Mineral
Acid, the fulphureous Acid having lefs affinity than
any of the reft with Alkalis. Neverthelefs this Salt
becomes in time a true Vitriolated Tartar, becaufe
the fulphureous part which weakened its. Acid
eafily quits it and flies off.

PROCESS VI.

To concentrate the vitriolic Acid.

TAKE the Vitriolic Acid you intend to concentrate, that is, to dephlegmate and make
ftronger: pour it into a good glafs retort, of fuch a
fize that your quantity of Acid may but half fill it:
fet this retort in the fand-bath of a reverberating
furnace : fit to it a receiver; lute it on, and give a
gradual fire. There will come over into the receiver a clear liquor, the firft drops of which will be
but faintly Acid : this is the moft aqueous part.

When the drops begin to follow one another
much more flowly, raife your fire, till the liquor begin to bubble a little in the middle. Keep it thus
gently boiling, till one half or two thirds thereof
be come over into the receiver. Then let your
veffels cool; unlute them ; what remains in the retort pour into a cryftal bottle, and ftop it exactly
with a glafs ftopple rubbed with emery.

OBSER-

O B S E R V A T I O N S.

THE Acid obtained from Sulphur is generally
very aqueous; either becaufe in preparing it water
muft neceffarily be adminiftered, that it may unite
therewith as it feparates from the Sulphur; or becaufe it is fo greedy of moifture as to attract a great
deal from the air, which muft needs be admitted to
make the Sulphur burn.,

The Acid obtained from Vitriol, excepting that
which rifes laft, is alfo mixed with a pretty confiderable quantity of phlegm; becaufe the Vitriol,
though calcined, ftill retains a great deal thereof,
which rifes with the Acid in diftillation.   Now, as
there are many chymical experiments that will not
fucceed without Acids exceedingly dephlegmated,
it is proper to have in a laboratory all the Acids
thus conditioned: becaufe if they happen to be
too ftrong for particular operations, as is fometimes
the cafe, it is very eafy to lower them to the defired
degree, by adding a fufficient quantity of water.

The Vitriolic Acid is much heavier and much
lefs volatile than water.   If therefore a mixture of
thefe two liquors be expofed to the fire, the aqueous
part will rife with a degree of heat which is not able
to carry up the Acid : by this means they may be
feparated from each other ; and thus is the Vitriolic
Acid concentrated.

Neverthelefs, as this Acid combines moft clofely
with water, and is in a manner ftrongly connected
with it, the water carries up fome portion thereof
along with it; and hence it comes that the liquor
which rifes into the receiver is Acid : it is called
Spirit of Vitriol.

As the fire carries off the moft aqueous part,
the other which remains in the retort increafes in
fpecific gravity.   The Acid particles are brought
nearer together, retain the aqueous particles more

obfti-

obftinately; and therefore to feparate them the degree of heat muft be increafed.

It is ufual to draw off one half or two thirds of the liquor that was put into the retort: but this depends on the degree of ftrength the Acid was of before concentration, and the degree of concentration intended to be given it.

If the Acid to be concentrated be Oil of Vitriol, from being brown or black it grows clearer as the operation advances, and at laft becomes perfectly colourlefs and tranfparent; becaufe the fat matter, which tinged it black, is diffipated during the procefs. Some of it depofites a white cryftalline earth.

A fulphureous fmell is generally perceived about the veffels in this operation. This arifes from a fmall portion of the Phlogifton from which the Acid is not free; and 'tis this inflammable matter which gives the Oil of Vitriol its black colour: for the cleareft and beft rectified Oil of Vitriol will become brown, and even black, in a fhort time, if any inflammable matter, though in a very fmall quantity, be diffolved therein.

The veffels are luted in this operation, to prevent any lofs of the Spirit of Vitriol, which being very acid is of ufe in many chymical experiments, and may itfelf alfo be again concentrated.

We obferved that in this operation it is neceffary the retort fhould be of very good glafs. Indeed the Acid is fo active and fo ftrong, that if the glafs be tender and have a little too much falt in its compofition, it will be fo corroded thereby that it will fall to pieces.

Though we directed the retort to be fet in a fand-bath for this operation, it does not follow that it may not alfo be placed in a naked fire: on the contrary, when the heat is not conveyed through a bath the operation advances fafter, and is much lefs tedious. But then great caution muft be ufed, and the

the clofeft attention given to the management of the fire, which muft be raifed by almoft imperceptible degrees, efpecially at the beginning of the operation; otherwife it is next to a certainty that the veffels will break. In general, a naked fire may be employed in almoft all diftillations which require a greater degree of heat than that of boiling water, or the *balneum mariæ* : the operation will be fooner finifhed; but it requires an experienced hand, that has by practice acquired a habit of governing the fire with judgment.

There is moreover another advantage in not ufing the fand-bath; which is, that if in the time of the operation you perceive the fire too fierce, you can quickly check it, either by ftopping clofe all the apertures of the furnace, or by drawing out all or part of the lighted coals. This inconvenience is not near fo eafily remedied when you ufe the fand-bath; becaufe when once heated it retains its heat very long after the fire is quite extinguifhed.

PROCESS VII.

To decompofe Vitriolated Tartar by means of the Phlogifton; or to compofe Sulphur by combining the Vitriolic Acid with the Phlogifton.

TAKE equal parts of Vitriolated Tartar, and very dry Salt of Tartar, feparately reduced to powder; add an eighth part of their weight of charcoal-duft; and mix the whole together very accurately. Throw this mixture into a red hot crucible, placed in a furnace filled with burning coals. Cover it very clofe, and keep it very hot, till the mixture melt, which may be known by uncovering the crucible from time to time. There

will

will then appear a bluifh flame, accompanied with a pungent fmell of Sulphur.

Take the crucible out of the fire: diffolve its contents in hot water: filter the folution through brown paper fupported by a glafs funnel: drop into the filtered liquor by little and little any Acid whatever. As you add the Acid the liquor will grow more and more turbid, and let fall a grey precipitate. Continue dropping in more Acid till the liquor will yield no more precipitate. Filter it a fecond time, to feparate it from the precipitate: what remains on the filter is a true inflammable Sulphur, which you may either melt or fublime into flowers.

OBSERVATIONS.

ALL bodies that contain the Vitriolic Acid may contribute, as well as Vitriolated Tartar, to the generation of Sulphur: fo that all the neutral Salts in which this Acid is a principle, the Alums, Selenites, Gypfums, Vitriols, may be fubftituted for it in this experiment. All thefe matters, with the addition of charcoal-duft only, being fufed in a crucible, conftantly produce Sulphur; becaufe the Vitriolic Acid having a greater affinity with the Phlogifton than with any thing elfe, will quit its bafis, whatever it be, to join with the Phlogifton of the charcoal, and therewith form a Sulphur.

The fixed Alkali added thereto helps to promote the fufion of the ingredients, which is neceffary for effecting the defired combination. It alfo ferves to unite with the Sulphur, when formed; and thus makes the combination called *Liver of Sulphur*, which prevents the Sulphur from being confumed as foon as formed: for the fixed Alkalis, which are incombuftible, hinder Sulphur from burning fo eafily as it would do if they were not joined with it. They may afterwards be feparated from each other, by the means of any Acid whatever.

This

This procefs, in which Sulphur is regenerated, by re-combining together the principles of which it was originally compofed, is one of the moft beautiful experiments that modern Chymiftry hath produced. We are indebted for it to M. Stahl ; and Dr. Geoffroy hath given a particular account of it in the Memoirs of the Academy of Sciences.

Before thefe gentlemen Glauber and Boyle had indeed publifhed methods of producing Sulphur, Glauber made ufe of his *Sal mirabile* and powdered charcoal : Boyle employed the Vitriolic Acid and Oil of Turpentine. But neither of thofe Chymifts underftood the true theory of their operations : they did not thoroughly know the principles of Sulphur : they did not imagine they had compofed Sulphur; they thought they only extracted what they fuppofed to exift previoufly in the matters they employed in their experiments.

M. Stahl was the firft who difcovered and explained the nature of Sulphur, and proved that in Glauber's and Boyle's experiments Sulphur was actually produced, by uniting together the principles of which it is conftituted. This beautiful experiment gives the ftrongeft luftre of evidence to the theory of the compofition of that mixt, which acts fuch a capital part in Chymiftry ; and it can no longer be doubted that Sulphur is actually a combination of the Vitriolic Acid with the Phlogifton.

Befides this important truth, our procefs for compofing Sulphur by art proves feveral others that are equally effential and fundamental.

The firft is that the Vitriolic Acid hath a greater affinity with the Phlogifton than with any other thing, feeing it quits metallic and earthy fubftances, as well as Alkaline falts, in order to combine therewith.

The fecond is that Sulphur combines with fixed Alkalis without fuffering any decompofition; feeing
it

it may be feparated from them entire and unaltered; and feeing that very Sulphur, which is naturally in-diffoluble in water, is rendered foluble therein by the union it hath contracted with the fixed Alkali.

The third is that the Vitriolic Acid, which when it is pure hath the greateft affinity with Alkalis of any Acid whatever, lofes a great deal of that affinity by contracting an union with the Phlogifton; feeing the weakeft Acids are capable of decompofing the Liver of Sulphur, and feparating the Sulphur from the Alkali. And this alfo confirms one of the general propofitions concerning affinities advanced in our theory; to wit, that the affinities of compound or mixed fubftances are weaker than thofe of the fame fubftances in a purer or more fimple ftate.

CHAP. II.

Of the NITROUS ACID.

PROCESS I.

To extract Nitre out of nitrous Earths and Stones. The Purification of Salt Petre. Mother of Nitre. Magnefia.

TAKE any quantity of nitrous earths or ftones; reduce them to powder; and therewith mix a third part of the afhes of green-wood and quick-lime. Put this mixture into a barrel or vat, and pour on it hot water to about twice the weight of the whole mafs. Let it ftand thus for twenty-four hours, ftirring it from time to time with a ftick. Then filter the liquor through brown paper, or pafs it through a flannel bag, till it come clear: it will then have a yellowifh colour. Boil this liquor, and evaporate till you perceive that a drop of it let fall

on

on any cold body coagulates. Then stop the evaporation, and set your liquor in a cool place. In the space of four-and-twenty hours crystals will be formed in it, the figure of which is that of an hexagonal prism, having its opposite planes generally equal, and terminated at each extremity by a pyramid of the same number of sides. These crystals will be of a brownish colour, and deflagrate on a live coal.

Decant the liquor from these crystals; mix it with twice its weight of hot water; evaporate and crystallize as before. Repeat the same operation till the liquor will yield no more crystals: it will then be very thick, and goes by the name of *Mother of Nitre.*

OBSERVATIONS.

. EARTHS and stones that have been impregnated with animal or vegetable juices susceptible of putrefaction, and have been long exposed to the air, but sheltered from the sun and rain, are those which yield the greatest quantity of Nitre. But all sorts of earths and stones are not equally fit to produce it. None is ever found in flints or sands of a crystalline nature.

Some earths and stones abound so with Nitre, that it effloresces spontaneously on their surface, in the form of a crystalline down. This Nitre may be collected with brooms, and accordingly has the name of *Salt-Petre Sweepings.* Some of this sort is brought from India.

Hitherto we are much in the dark as to the origin and generation of Nitre. Some Chymists pretend that the Nitrous Acid is diffused through the air, and gradually deposited in such earths and stones as are qualified to receive it.

Others, considering that none of it is ever obtained but from earths that have been impregnated with vegetable or animal juices, have from thence
concluded

concluded thofe two kingdoms to be the general repofitories of the Nitrous Acid; that if we do not perceive it to exift in fuch matters at all, or at leaft in any great quantity, till they have undergone putrefaction, and are in fome meafure incorporated with fuitable earths and ftones, 'tis becaufe the Acid is fo entangled with heterogeneous particles that it requires the affiftance of putrefaction, and much more of filtration through an earth, to difengage it; and enable it to appear in its proper nature.

Laftly, others are of opinion that this Acid is no other than the univerfal or Vitriolic Acid; difguifed indeed by a portion of the Phlogifton, which is combined with it in a peculiar manner by the means of putrefaction. They ground this opinion chiefly on the analogy or refemblance which they find between the Nitrous Acid and the Volatile Sulphureous Spirit. Its volatility, its pungent fmell, its properties of taking fire, and of deftroying the blue and violet colours of vegetables, ferve them as fo many proofs.

Their opinion is the more probable on this account, that even though the Nitrous Acid fhould actually be produced by vegetable and animal fubftances, yet as thefe fubftances themfelves draw all their component principles from the earth; and as the Vitriolic Acid is diffufed through all the foils which afford them nourifhment, there is great reafon to think that the Nitrous Acid is no other than the Vitriolic Acid altered by the changes and combinations it hath undergone in its paffage into and through thofe fubftances. In 1750 the Royal Academy of Sciences at Berlin propofed an account of the generation of Nitre as the fubject for their prize, which was conferred on a Memoir wherein this laft opinion was fupported by fome new and very judicious experiments.

The proceſs by which our Salt-petre makers extract Nitre in quantities, out of rubbiſh and nitrous
earths, is very nearly the ſame with that here ſet
down : ſo that I ſhall not enter into a particular
account of it. I ſhall only take notice of one thing,
which it is of ſome confequence to know ; namely,
that there is no nitrous earth which does not contain ſea-ſalt alſo. The greateſt quantities of this
ſalt are to be found in thoſe earths which have
been drenched with urine, or other animal excre-
ments. Now as the rubbiſh of old houſes in great
cities is in this claſs, it comes to paſs that when the
Salt-petre workers evaporate a nitrous lixivium
drawn from that rubbiſh, as ſoon as the evaporation is brought to a certain pitch, a great many
little cryſtals of ſea-ſalt form in the liquor, and fall
to the bottom of the veſſel.

The Salt-petre workers in France call theſe ſaline
particles *the Grain*, and take great care to ſeparate
them from the liquor, (which as long as it continues
hot keeps the ſalt-petre diſſolved) before they ſet it
to cryſtallize. This fact ſeems a little ſingular, confidering that ſea-ſalt diſſolves in water more eaſily
than ſalt-petre, and cryſtallizes with more difficulty.

In order to diſcover the cauſe of this phenomenon,
we muſt recollect ſome truths delivered in our theoretical Elements. The firſt is, that water can keep
but a determinate quantity of any ſalt in ſolution,
and that if water fully ſaturated with a ſalt be evaporated, a quantity of ſalt will cryſtallize in proportion
to the quantity of water evaporated. The ſecond is,
that thoſe ſalts which are the moſt ſoluble in water,
particularly thoſe which run in the air, will diſſolve
in cold and in boiling water equally ; whereas much
greater quantities of the other ſalts will diſſolve in
hot and boiling water than in cold water. Theſe
things being admitted, when we know that ſea-ſalt
is one of the firſt ſort, and ſalt-petre of the ſecond,

I   the

the reafon why fea-falt precipitates in the preparation of falt-petre appears at once. For

When the folution of falt-petre and fea-falt comes
to be evaporated to fuch a degree that it contains as
much fea-falt as it poffibly can, this falt muft begin
to cryftallize, and continue to do fo gradually as
the evaporation advances. But becaufe at the fame
time it does not contain as much falt-petre as it can
hold, feeing it is capable of diffolving a much greater
quantity thereof when it is boiling hot than when
it is cold, this laft named falt will not cryftallize fo
foon. If the evaporation were continued till the
cafe of the falt-petre came to be the fame with that
of the fea-falt, then the falt-petre alfo would begin
to cryftallize gradually in proportion to the water
evaporated, and the two falts would continue cryftallizing promifcuoufly together: but it is never
carried fo far; nor is it ever neceffary; for as the
water cools it becomes more and more incapable
of holding in folution the fame quantity of falt-
petre as when it was boiling hot.

And then comes the very reverfe, with regard to
the cryftallizing of the two falts; for then the Salt-
petre fhoots, and not the Sea-falt. The reafon of
this fact alfo is founded on what has juft been faid.
The Sea-falt, of which cold water will diffolve as
much as boiling water, and which owed its cryftallizing before only to the evaporation, now ceafes to
cryftallize as foon as the evaporation ceafes; while
the Salt-petre, which the water kept diffolved only
becaufe it was boiling hot, is forced to cryftallize
merely by the cooling of the water.

When the folution of Salt-petre has yielded as
many cryftals of that Salt as it can yield by cooling,
it is again evaporated, and being then fuffered to
cool yields more cryftals. And thus they continue
evaporating and cryftallizing, till the liquor will
afford no more cryftals. It is plain that as the Salt-

petre cryftallizes, the proportion of Sea-falt tó
the diffolving liquor increafes; and as a certairi
quantity of water evaporates alfo during the time
employed in cryftallizing the Salt-petre, a quantity
of Sea-falt, proportioned to the watér- fo évaporat-
ing, muft cryftallize in that time: and this is the
reafon why Salt-petre is adulterated with a mixture
of Sea-falt. It likewife follows that the laft cryftals
of Nitre, obtained from a folution of Salt-petre
and Sea-falt, contain much more Sea-falt than the
firft.

From all that has been faid concerning the cryftallization of Salt-petre and Sea-falt, it is eafy to
deduce the proper way of purifying the former of
thefe two Salts from a mixture of the latter. For
this purpofe the Salt-petre to be refined need only
be diffolved in fair water. The proportion betweeri
the two falts in this fecond folution is very different
from what it was in the former; for it contains no
more Sea-falt than what had cryftallized along witH
the Salt-petre under favour of the evaporation, the
reft having been left diffolved in the liquor that re-
fufed to yield any more nitrous cryftals.

As there is therefore a much greater quantity of
Salt-petre than of Sea-falt in this fecond folution, it
is eafy to evaporate it to fuch a degree that a great
deal of Salt-petre fhall cryftallize, while much more
of the water muft neceffarily be evaporated before
any of the Sea-falt will cryftallize.

However, the Salt-petre is not yet entirely freed
from all mixture of Sea-falt by this firft purification;
for the cryftals obtained from this liquor, in which
Sea-falt is diffolved, are ftill encrufted, and, as it
were, infected therewith : hence it comes, that, to
refine the Salt-petre thoroughly, thefe cryftallizations muft be repeated four or five times.

The Salt-petre men commonly content themfelves
with cryftallizing it thrice, and call the produce Salt-

petre

petre of the firft, fecond, or third fhoot, according to the number of cryftallizations it has undergone. But their beft refined Salt-petre, even that of the third fhooting, is not yet fufficiently pure for Chymical experiments that require much accuracy: fo that it muft be further purified; but ftill by the fame method.

. The Nitrous Acid is not pure in the earths and ftones from which it is extracted. It is combined partly with the very earth in which it is formed, and partly with the Volatile Alkali produced by the putrefaction of the vegetable or animal matters that concurred to its generation. A Fixed Alkali and Quick-lime are added to the lixivium of a nitrous earth, in order to decompofe the nitrous Salt formed in that earth, and to feparate the Acid from the Volatile Alkali and the abforbent earth with which it is united: thence comes that copious fediment which appears in the lye at the beginning of the evaporation. Thefe matters form with that Acid a true Nitre, much more capable than the original Nitrous Salts of cryftallization, detonation, and the other properties which are effential thereto. The bafis of Nitre is therefore a Fixed Alkali mixed with a little lime.

The Mother of Nitre, which will yield no more cryftals, is brown and thick: by evaporation over a fire it is further infpiffated, and becomes a dry, folid body; which however being left to itfelf foon gives, and runs into a liquor. This water ftill contains a good deal of Nitre, Sea-falt, and the Acids of thefe Salts united with an abforbent earth. It contains moreover a great deal of a fat, vifcid matter, which prevents its cryftallizing.

All faline folutions in general, after having yielded a certain quantity of cryftals, grow thick, and refufe to part with any more, though they ftill contain much Salt. They are all called *Mother-waters*, as

well

well as that which hath yielded Nitre. The Mother-waters of different Salts may prove the subjects of curious and useful enquiries.

If a Fixed Alkali be mixed with the Mother of Nitre, a copious white precipitate immediately falls, which being collected and dried is called *Magnesia*. This precipitate is nothing but the absorbent earth that was united with the Nitrous Acid, together with a good deal of the lime that was added, and was also united with that Acid, from which they are now separated by the Fixed Alkali, according to the usual laws of affinities.

The Vitriolic Acid poured upon Mother of Nitre causes many Acid vapours to rise, which are a compound of the nitrous and marine Acids, that is, an *Aqua regia*. On this occasion also there falls a large quantity of a white powder, which is still called *Magnesia*; yet it differs from the former in that it is not, like it, a pure absorbent earth, but combined with the Vitriolic Acid.

An *Aqua regis* may also be drawn from nitrous earths by the force of fire only, without the help of any additament.

PROCESS II.

To decompose Nitre by means of the Phlogiston. Nitre fixed by Charcoal. Clyssus *of Nitre.* Sal Polychrestum.

TAKE the purest Salt-petre in powder; put it into a large crucible, which it may but half fill; set the crucible in a common furnace, and surround it with coals. When it is red-hot the Nitre will melt, and become as fluid as water. Then throw into the crucible a small quantity of charcoal-dust: the Nitre and the charcoal will immediately deflagrate with violence; and a great commotion will be raised, raifed, accompanied with a confiderable hiffing, and abundance of black fmoke. As the charcoal waftes, the detonation will abate, and ceafe entirely as foon as the coal is quite confumed.

Then throw into the crucible the fame quantity of charcoal-duft as before, and the fame phenomena will be repeated. Let this coal alfo be confumed : then add more, and go on in the fame manner till you can excite no further deflagration ; always obferving to let the burning coal be entirely confumed before you add any frefh. When no deflagration enfues, the matter contained in the crucible will have loft much of its fluidity.

OBSERVATIONS.

NITRE will not take fire, unlefs the inflammable matter added to it be actually burning, or the Nitre itfelf red-hot, and fo thoroughly ignited as immediately to kindle it. Therefore, if you would procure the detonation of Nitre with charcoal, and make ufe of cold charcoal, as in the procefs, the Nitre in the crucible muft be red-hot, and in perfect fufion : but you may alfo ufe live coals, and then the Nitre need not be red-hot.

It is proper that the crucible ufed in this experiment fhould be only half full ; for during the detonation its contents fwell, and might run over without this precaution. For the fame reafon the charcoal-duft is to be thrown in by little and little ; and that firft put in muft be entirely confumed before any frefh be added.

The matter remaining in the crucible after the operation is a very ftrong Fixed Alkali. Being expofed to the air it quickly attracts the moifture thereof, and runs into a liquor. It is called *Alkalizated Nitre*, or to diftinguifh it from Nitre alkalizated by other inflammable matters, *Nitre fixed by charcoal*.

R 4　　　　　　　However,

However, this Alkali is not abfolutely pure. It ftill contains a portion of the Nitre that hath not been decompofed : for when there remains but a little of this Salt mixed with a great quantity of Alkali, which is not inflammable, the Alkali in fome meafure fhelters it, coats it over, and obftructs that immediate contact with the inflammable matters applied, which is neceffary to make it deflagrate.

If the Fixed Alkali be defired perfectly free from any mixture of undecompofed Nitre, the fire about the crucible muft be confiderably increafed as foon as the detonation is entirely over ; the matter muft be made to flow, which requires a much ftronger heat than would melt Nitre, and kept thus in fufion for about an hour. After this no perfect Nitre will be found therein : for the little that was left, being unable to abide the force of the fire, as not being extremely fixed, either is entirely diffipated, or lofes its Acid, which is carried off by the violence of the heat.

Fixed Nitre contains alfo a portion of the earth that conftituted the bafis of the Nitre, which is no other than the lime employed in its cryftallization, or elfe fome of the earth with which its Acid was originally combined, and which it retained in cryftallizing. When Nitre is deflagrated with fuch matters as produce afhes, thefe afhes likewife furnifh a certain quantity of earth, which mixes with the Fixed Alkali. To feparate thefe feveral earths from the Alkali, nothing more is requifite than to let it run *per deliquium*, or to diffolve it in water, and filter the folution through brown paper. Whatever is faline will pafs through the filtre with the water, and the earthy part will be left upon it.

The Nitrous Acid is not only diffipated during the deflagration of the Nitre, but is even deftroyed, and perfectly decompofed. The fmoke that rifes during the operation has not the leaft odour of an Acid. Its nature may be accurately examined by catching it in proper veffels, and condenfing it into a liquor.

Nitre differs from Sulphur, and from all other inflammable bodies whatever, in this, that the free accefs of the air is indifpenfably neceffary to make any of the others burn; whereas Nitre, and Nitre only, is capable of burning in clofe veffels: and this property furnifhes us with the means of collecting the vapours which it difcharges in deflagration.

For this purpofe to a tubulated earthen retort you muft fit two or three large adopters: fet the retort in a furnace; and under it make a fire fufficient to keep its bottom moderately red. Then take a fmall quantity, two or three pinches for example, of a mixture of three parts of Nitre with one of charcoal-duft, and drop it into the retort through its tube, which muft be uppermoft, and immediately. ftopped clofe. A detonation inftantly enfues, and the vapours that rife from the inflamed mixture of Nitre and charcoal, paffing out through the neck of the retort into the adopters, circulate therein for a while, and at laft condenfe into a liquor.

When the detonation is over, and the vapours condenfed, or nearly fo, drop into the retort another equal quantity of the mixture; and repeat this till you find there is liquor enough in the recipients to be examined with eafe and accuracy. This liquor is almoft infipid, and fhews no tokens of acidity; or at moft but very flight ones. It is called *Clyffus* of Nitre.

It is eafy to perceive why feveral adopters are required in this experiment, and why a very fmall quantity of the mixture muft be introduced into the retort at once. The explofion, and the quantity of air and vapours difcharged on this occafion, would quickly burft the veffels, if all thefe precautions were not attended to. This plainly appears from
the

the terrible effects of gun-powder, which is nothing but a compofition of Nitre, Sulphur, and Charcoal.

Nitre is alfo decompofed and takes fire by the means of Sulphur; but the circumftances and the refult differ widely from thofe produced therewith by charcoal, or any other inflammable body.

Nitre deflagrates with Sulphur on account of the Phlogifton which the latter contains. If one part of Sulphur be mixed with two or three parts of Nitre, and the mixture thrown by little and little into a red-hot crucible, upon every projection there arifes a detonation accompanied with a vivid flame.

The vapours difcharged on this occafion have the mingled fmell of a Sulphureous Spirit and Spirit of Nitre; and if they be collected by means of a tubulated retort, and fuch an apparatus of veffels as was ufed in the preceding experiment, the liquor contained in the recipients is found to be an actual mixture of the Acid of Sulphur, the Sulphureous Spirit, and the Acid of Nitre; the firft being in greater quantity than the other two, and the fecond greater than the laft.

Nor is the remainder after detonation a Fixed Alkali, as in the former experiments; but a Neutral Salt, confifting of the Acid of Sulphur combined with the Alkali of Nitre; a fort of Vitriolated Tartar known in medicine by the name of *Sal Polychreftum.*

There are evidently two effential differences between this laft experiment and the preceding one. What remains after the deflagration of Nitre with Sulphur is not a Fixed Alkali: and moreover, the vapours emitted in the operation are impregnated with a quantity of the Nitrous Acid; which is not the cafe when Nitre is decompofed by any other inflammable matter which contains no Vitriolic Acid.

The reafon of thefe differences is naturally deducible from what hath been already faid concerning the
properties

properties of the Vitriolic and Nitrous Acids. We
have seen that by burning Sulphur its Acid is not
decompofed, but only feparated from its Phlogifton.
We alfo know that its Acid has a great affinity with
Fixed Alkalis. Thefe things being granted, it follows that, as foon as the Nitrous Acid quits its Alkaline bafis, by deflagrating with the Phlogifton of the
Sulphur, the Acid of this very Sulphur, being fet
at liberty by that deflagration, muft unite with the
Alkaline bafis deferted by the Acid of Nitre, and
therewith form a Neutral Salt. Hence, inftead of
a Fixed Alkali, we find at the end of the operation
a fort of Vitriolated Tartar; the Acids of Sulphur
and of Vitriol being the fame, as is evident from
what hath been above faid concerning them.

In order to difcover the caufe of the other phenomenon, we muft recollect two things advanced in
our Elements of the Theory; to wit, that the affinity of the Vitriolic Acid with Fixed Alkalis is
greater than that of the Nitrous Acid; and again,
that the Nitrous Acid is not capable of combining and
taking fire with the Phlogifton, but when it is in the
form of a Neutral Salt, that is, when it is united
with fome alkaline, earthy, or metallic bafis. If
thofe two principles be applied to the effect in queftion, the folution is eafy and natural. For, in the
deflagration of Nitre with Sulphur, the Phlogifton
is not the only fubftance capable of feparating the
Nitrous Acid from its bafis: the Acid of the Sulphur,
more and more of which is fet at liberty as the Phlogifton is confumed, is alfo capable of producing
the fame effect; but with this difference, that the
portion of the Nitrous Acid which is detached from
its Alkali by the Phlogifton is at the fame inftant
fet on fire and decompofed by that union; whereas the portion thereof which is feparated by the
Vitriolic Acid, being when fo feparated incapable
of uniting with the Phlogifton, and of confuming
there-

therewith, is preferved entire, and rifes in vapours, together with that portion of the Vitriolic Acid which could not unite with the bafis of the Nitre.

PROCESS III.

To decompofe Nitre by means of the Vitriolic Acid. The Smoking Spirit of Nitre. Sal de duobus. *The Purification of Spirit of Nitre.*

TAKE equal parts of well purified Nitre and Green Vitriol: dry.the Nitre thoroughly, and bruife it to a fine powder. Calcine the Vitriol to rednefs : reduce it likewife to a very fine powder; and mingle thefe two fubftances well together. Put the mixture into an earthen long-neck, or a good glafs retort coated, of fuch a fize that it may be but half full.

Set this veffel in a reverberating furnace covered with its dome; apply a large glafs receiver, having a fmall hole in its body, ftopped with a little lute. Let this receiver be accurately luted to the retort with the fat lute, and the joint covered with a flip of canvas fmeared with lute made of quicklime and the white of an egg. Heat the veffels very gradually. The receiver will foon be filled with very denfe red vapours, and drops will begin to diftill from the nofe of the retort.

Continue the diftillation, encreafing the fire a little when you obferve the drops to follow each other but flowly, fo that above two thirds of a minute paffes between them; and, in order to let out the redundant vapours, open the fmall hole in the receiver from time to time. Towards the end of the operation raife the fire fo as to make the retort red. When you find that, even when the retort is red-hot, nothing more comes over, unlute the receiver, and without delay pour the liquor it con-

tains

tains into a cryſtal bottle, and cloſe it with a cryſtal ſtopple ground in its neck with emery. This liquor will be of a reddiſh yellow colour, ſmøking exceedingly, and the bottle containing it will be conſtantly filled with red fumes like thoſe obſerved in the receiver.

OBSERVATIONS:

THE Vitriolic Acid having a greater affinity with Fixed Alkalis than with any other ſubſtance, the Phlogiſton excepted, and being in the Vitriol united with a ferruginous baſis, will naturally quit that baſis to join with the Fixed Alkali of the Nitre; the Acid whereof being weaker than the Vitriolic, as we have already obſerved on ſeveral occaſions, muſt needs be thereby expelled from its baſis. The Nitre therefore is decompoſed by the Vitriol, and its Acid being ſet at liberty, is carried up by the force of the fire.

Indeed the Nitrous Acid, being thus ſeparated from its alkaline baſis, might be expected to combine with the ferruginous baſis of the Vitriol : but as it has, like all other Acids, much leſs affinity with Metallic ſubſtances than with Alkalis, even a moderate degree of fire is ſufficient to ſeparate it from them. Moreover, this Acid hath either no effect, or very little, upon iron that has loſt much of its Phlogiſton by contracting an union with any Acid ; which is the caſe of the ferruginous baſis of Vitriol.

By the proceſs here delivered a very ſtrong, perfectly dephlegmated, and vaſtly ſmoking Spirit of Nitre is obtained. If the precautions of drying the Nitre and calcining the Vitriol be neglected, the Acid that comes over, greedily attracting the water contained in theſe ſalts, will be very aqueous, will not ſmoke, and will be almoſt colourleſs, with a very ſlight tinge of lemon.

The

The fumes of highly concentrated Spirit of Nitre, such as that obtained by the above procefs, are light, corrofive, and very dangerous to the lungs; being no other than the moft dephlegmated part of the Nitrous Acid. The perfon therefore who unlutes the veffels, or pours the liquor out of the receiver into the bottle, ought with the greateft caution to avoid drawing them in with his breath; and for that reafon ought to place himfelf fo that a current of air, either natural or artificial, may carry them off another way. It is alfo neceffary that care be taken, during the operation, to give the vapours a little vent every now and then, by opening the fmall hole in the recipient; for they are fo elaftic, that, if too clofely confined, they will burft the veffels.

When the operation is over, you will find a red mafs at the bottom of the retort, caft as it were in a mould. This is a Neutral Salt of the nature of Vitriolated Tartar, refulting from the union of the Acid of the Vitriol with the Alkaline bafis of the Nitre.

The ferruginous bafis of the Vitriol, which is mixed with this falt, gives it the red colour. To feparate it therefrom, you muft pulverife it, diffolve it in boiling water, and filter the folution feveral times through brown paper; becaufe the ferruginous earth of the vitriol is fo fine, that fome of it will pafs through the firft time. When the folution is very clear, and depofites no fediment, let it be fet to fhoot, and it will yield cryftals of Vitriolated Tartar; to which Chymifts have given the peculiar title of *Sal de duobus.*

In this *Caput mortuum* we frequently find, befides the ferruginous earth of Vitriol, a portion of Nitre and Vitriol not decompofed; either becaufe the two falts were not thoroughly mingled, or becaufe the fire

fire was not raifed high enough towards the end of the operation.

Nitre may alfo be decompofed, and its Acid obtained, by the interpofition of any of the other Vitriols, Alums, Gypfums, Boles, Clays ; in fhort, by means of any compound in which the Vitriolic Acid is found, provided it have not a Fixed Alkali for its bafis.

The diftillers of *Aqua fortis*, who make large quantities at a time, and who ufe the leaft chargeable methods, do their bufinefs by the means of earths impregnated with the Vitriolic Acid ; fuch as Clays and Boles. With thefe earths they accurately mix the Nitre from which they intend to draw their Spirit : this mixture they put into large oblong earthen pots, having a very fhort curved neck, which enters a recipient of the fame matter and form. Thefe veffels they place in two rows oppofite to each other in long furnaces, and cover them over with bricks cemented with Windfor-loam, which ferves for a reverberatory : then they light the fire in the furnace, making it at firft very fmall, only to warm the veffels ; after which they throw in wood, and raife the fire till the pots grow quite red-hot, in which degree they keep it up till the diftillation is entirely finifhed.

The Acid of Nitre may alfo be feparated from its bafis by means of the pure Vitriolic Acid. For this purpofe the Nitre from which you mean to extract the Acid muft be finely pulverized, put into a glafs retort, and a third of its weight of concentrated Oil of Vitriol poured on it : the retort muft be placed in a reverberating furnace, and a receiver, like that ufed in the preceding operation, expeditioufly applied.

As foon as the Oil of Vitriol touches the Nitre the mixture grows hot, and copious red fumes
begin

begin to appear : fome drops of the Acid come over even before the fire is kindled in the furnace.

On this occafion the fire muft be moderate ; becaufe the Vitriolic Acid, being clogged by no bafis; acts upon the Nitre much more brifkly, and with much greater effect, than when it is not pure.

This operation may be performed by a fand heat; which is a fpeedy and commodious way of obtaining the Nitrous Acid. In other refpects the precautions recommended in the preceding experiment muft be carefully obferved here, both in diftilling the Acid and in taking it out of the receiver.

The Spirit of Nitre extracted by this method is as ftrong, and fmokes as much, as that obtained by calcined Vitriol, provided the oil of Vitriol made ufe of be well concentrated; but it is generally tainted by the admixture of a fmall portion of the Vitriolic Acid, which, having no bafis of its own to reftrain it, is carried up by the heat before it can lay hold of the bafis of the Nitre.

There are feveral experiments in Chymiftry that fucceed equally well whether the Nitrous Acid be or be not thus adulterated with a mixture of the Vitriolic Acid ; but there are fome, as we fhall fee, that will not fucceed without a Spirit of Nitre fo mixed: If the Acid be diftilled with a view to fuch experiments, it muft be kept as it is. But moft experiments require the Spirit of Nitre to be abfolutely pure; and if it be intended for fuch, it muft be perfectly cleanfed from the Vitriolic taint.

This is eafily effected by mixing your Spirit with very pure Nitre, and diftilling it a fecond time. The Vitriolic Acid, with which this Spirit of Nitre is adulterated, coming in contact with a great quantity of undecompofed Nitre, unites with its Alkaline bafis, and expells a proportionable quantity of the Nitrous Acid.

In

In the retort made ufe of to diftill the Nitrous Acid, by means of the pure Vitriolic Acid, is found a *Caput mortuum* differing from that left after the diftillation of the fame Acid by the interpofition of Vitriol, in as much as it contains no red ferruginous earth. This is a very white faline mafs, moulded in the bottom of the retort : if you pound it, diffolve it in boiling water, and evaporate the folution, it will fhoot into cryftals of Vitriolated Tartar : fometimes alfo it contains a portion of undecompofed Nitre, which fhoots after the Vitriolated Tartar, becaufe it is much more foluble in water.

CHAP. III.

Of the MARINE ACID.

PROCESS I.

To extract Sea-Salt from Sea-Water, and from Brine-Springs. Epfom Salt.

FILTER the falt water from which you intend to extract the Salt : evaporate it by boiling till you fee on its furface a dark pellicle : this confifts wholly of little cryftals of falt juft beginning to fhoot : now flacken the fire, that the brine may evaporate more flowly, and without any agitation. The cryftals, which at firft were very fmall, will become larger, and form hollow truncated pyramids, the apices whereof will point downwards, and their bafes be even with the furface of the liquor.

Thefe pyramidal cryftals are only collections of fmall cubical cryftals concreted into this form.

When they have acquired a certain magnitude
they fall to the bottom of the liquor. When they
come to be in fuch heaps as almoft to reach the furface of the liquor, decant it from them, and continue the evaporation till no more cryftals of Sea-
falt will fhoot.

O B S E R V A T I O N S. :

THE Acid of Sea-falt is fcarce ever found, either
in Sea-water or in the earth, otherwife than united
with a fixed Alkali of a particular kind, which is
its natural bafis; and confequently it is in the form
of a Neutral Salt. This Salt is plentifully diffolved
in the waters of the ocean, and when obtained
therefrom bears the name of *Sea-Salt*. It is alfo
found in the earth in vaft cryftalline maffes, and is
then called *Sal-Gem:* fo that Sea-falt and Sal-Gem
are but one and the fame fort of Salt, differing
very little from each other, except as to the places
where they are found.

In the earth are alfo found fprings and foun-
tains, whofe waters are ftrong brines, a great deal
of Sea-Salt being diffolved in them. Thefe fprings
either rife directly from the Sea, or run through
fome mines of Sal-Gem, of which they take up a
quantity in their paffage.

As the fame, or at leaft nearly the fame, quantity
of Sea-falt will continue diffolved in cold water as
boiling water will take up, it cannot fhoot, as Nitre
does, by the mere cooling of the water in which it
is diffolved: it cryftallizes only by the means of
evaporation, which continually leffens the proportion of the water to the Salt; fo that it is always
capable of containing juft fo much the lefs Sea-falt
the more there is cryftallized.

The brine fhould not boil after you perceive the
pellicle of little cryftals beginning to form on its furface; for the calmnefs of the liquor allows them to

form

form more regularly, and become larger. Nor after this fhould the evaporation be hurried on too faft; for a faline cruft would form on the liquor, which, by preventing the vapours from being carried off, would obftruct the cryftallization.

If the evaporation be continued after the liquor ceafes to yield any cryftals of Sea-falt, other cryftals will be obtained of an oblong four-fided form, which have a bitter tafte, and are almoft always moift. This fort of Salt is known by the name of *Epfom Salt*, which it owes to a falt fpring in England, from the water of which it was firft extracted. This Salt, or rather faline compound, is a congeries of Glauber's Salt and Sea-falt, in a manner confounded together, and mixed with fome of the Mother of Sea-falt, in which is contained a kind of bituminous matter. Thefe two Neutral Salts, which conftitute the Epfom Salt, may be eafily feparated from each other, by means of cryftallization only. Epfom Salt is purgative and bitter; and therefore named *Sal Catharticum Amarum*, or bitter purging Salts.

There are different methods ufed in great works for obtaining Sea-falt out of water in which it is diffolved. The fimpleft and eafieft is that practifed in France, and in all thofe countries which are not colder. On the fea fhore they lay out a fort of broad fhallow pits, pans, or rather ponds, which the fea fills with the tide of flood. When the ponds are thus filled, they ftop their communication with the fea, and leave the water to evaporate by the heat of the fun; by which means all the Salt contained in it neceffarily cryftallizes. Thefe pits are called *Salt Ponds*. Salt can be made in this way in the fummer-time only; at leaft in France, and other countries of the fame temperature: for during the winter, when the fun has lefs power and rains are frequent, this method is not practicable.

For

For this reafon, as it often rains in the province of Normandy, the inhabitants take another way to extract Salt from fea-water. The labourers employed for this purpofe raife heaps of fand on the fhore, fo that the tide waters and drenches them when it flows, and leaves the fand dry when it ebbs. During the interval between two tides of flood the fun and the air eafily carry off the moifture that was left, and fo the fand remains impregnated with all the Salt that was contained in the evaporated water. Thus they let it acquire as much falt as it can by feveral returns of flood, and then wafh it out with frefh water, which they evaporate over a fire in leaden boilers.

To obtain the Salt from brine-fprings, the water need only be evaporated : but as feveral of thefe fprings contain too little Salt to pay the charges that would be incurred, if the evaporation were effected by the force of fire only, the manufacturers have fallen upon a lefs expenfive method of getting rid of the greateft part of the water, and preparing the brine for cryftallization, in much lefs time, and with much lefs fire, than would otherwife have been neceffary.

The method confifts in making the water fall from a certain heighth on a great many fmall fpars of wood, which divide it into particles like rain. This is performed under fheds open to all the winds, which pafs freely through this artificial fhower. By this means the water prefents to the air a great extent of furface, being indeed reduced almoft entirely to furface, and the evaporation is carried on with great eafe and expedition. The water is raifed by pumps to the heighth from which it is intended to fall *.

P R O-

---

* The Marquis de Montalembert, in a Memoir read before the Academy of Sciences, propofes a new method of effecting thefe evaporations, together with fome confiderable improvements

·PROCESS II.

Experiments concerning the decompofition of Sea-falt,
by means of the Phlogifton. Kunckel's *Phofphorus.*

" O F pure urine that has fermented five or fix
" days take a quantity in proportion to
" the quantity of Phofphorus you intend to make :
" it requires about one third part of a hogfhead to
" make a dram of Phofphorus. Evaporate it in
" iron pans, till it become clotted, hard, black,
" and nearly like chimney-foot; at which time it
" will be reduced to about a fixtieth part of its
" original weight before evaporation.

" When the urine is brought to this condition
" put it in feveral portions into fo many iron pots,
" under which you muft keep a pretty brifk fire
" fo as to make their bottoms red, and ftir it in-
" ceffantly till the volatile falt and the fetid oil be
" almoft wholly diffipated, till the matter ceafe to
" emit any fmoke, and till it fmell like peach-blof-
" foms. Then put out the fire, and pour on the
" matter, which will now be reduced to a powder,
" fomewhat more than twice its weight of warm wa-
" ·ter. Stir it about in this water, and leave it to
" foak therein for twenty-four hours. Pour off the
" water by inclination; dry the drenched matter,
" and pulverize it. The previous calcination carries
" off from the matter about a third of its weight,
" and the lixiviation wafhes out half the remainder.

" With what remains thus calcined, wafhed,
" and dried, mix half its weight of gravel, or yel-

ments in the ftructure and difpofition of the buildings neceffary
for that purpofe. They are called by the French *Batiments de*
Graduation; which may properly enough be rendered *Brine-*
houfes.

" low

" low freeftone rafped, having fifted out and
" thrown away all the fineft particles. River-fand
" is not proper on this occafion, becaufe it flies in
" a hot fire. Then add to this mixture a fixteenth
" part of its weight of charcoal, made of beech, or
" of any other wood except oak, becaufe that alfo
" flies. Moiften the whole with as much water as
" will bring it to a ftiff pafte, by working and
" kneading it with your hands: now introduce it
" into your retort, taking care not to daub its
" neck. The retort muft be of the beft earth,
" and of fuch a fize, that when your matter is in
" it, a full third thereof fhall ftill be empty.

" Place your retort, thus charged, in a reverbe-
" rating furnace, fo proportioned, that there may
" be an interval of two inches all round between the
" fides of the furnace and the bowl of the retort,
" even where it contracts to form the neck, which
" fhould ftand inclined at an angle of fixty degrees.
" Stop all the apertures of the furnace, except the
" doors of the fire-place and afh-hole.

" Fit on to the retort a large glafs ballon two
" thirds full of water, and lute them together, as
" in diftilling the Smoaking Spirit of Nitre. In
" the hinder part of this ballon, a little above the
" furface of the water, a fmall hole muft be bored,
" This hole is to be ftopped with a fmall peg of
" birch-wood, which muft flip in and out very ea-
" fily, and have a fmall knob to prevent its falling
" into the ballon. This peg is to be pulled out
" from time to time, that by applying the hand
" to the hole it may be known whether the air,
" rarefied by the heat of the retort, iffues out with
" too much or too little force.

" If the air rufhes out with too much rapidity,
" and with a hiffing noife, the door of the afh-hole
" muft be entirely fhut, in order to flacken the fire.
" If it do not ftrike pretty fmartly againft the hand,

" that

" that door muſt be opened wider, and large coals
" thrown into the fire-place to quicken the fire im-
" mediately:

" The operation uſually laſts four and twenty
" hours ; and the following ſigns ſhew that it will
" ſucceed, provided the retort reſiſt the fire.

" You muſt begin the operation with putting
" ſome unlighted charcoal in the aſh-hole, and a
" little lighted charcoal at the door thereof, in or-
" der to warm the retort very ſlowly. When the
" whole is kindled, puſh it into the aſh-hole, and
" cloſe the door thereof with a tile. This moderate
" heat brings over the phlegm of the mixture.
" The ſame degree of fire muſt be kept up four
" hours, after which ſome coals may be laid on
" the grate of the fire-place, which the fire under-
" neath will kindle by degrees. With this ſecond
" heat brought nearer the retort, the ballon grows
" warm, and is filled with white vapours, which
" have the ſmell of fetid oil. In four hours after,
" this veſſel will grow cool and clear ; and then
" you muſt open the door of the aſh-hole one inch,
" throw freſh coals into the fire-place every three
" minutes, and every time ſhut the door of it, left
" the cold air from without ſhould ſtrike againſt '
" the bottom of the retort and crack it.

" When the fire has been kept up to this degree
" for about two hours, the inſide of the ballon be-
" gins to be netted over with a volatile ſalt of a
" ſingular nature, which cannot be driven up but
" by a very violent fire, and which ſmells pretty
" ſtrong of peach-kernels. Care muſt be taken
" that this concrete ſalt do not ſtop the little hole in
" the ballon : for in that caſe it would burſt, the
" retort being then red-hot, and the air exceedingly
" rarefied. The water in the ballon, being heated
" by the vicinity of the furnace, exhales vapours

" which

" which diffolve this fprigged falt, and the ballon
" clears up in half an hour after it has ceafed rifing.

" In about three hours from the firft appearance
" of this falt, the ballon is again filled with new va-
" pours, which fmell like Sal Ammoniac thrown
" upon burning coals. They condenfe on the fides
" of the receiver into a falt which is not branched
" like the former, but appears in long perpendi-
" cular ftreaks, which the vapours of the water do
" not diffolve. Thefe white vapours are the fore-
" runners of the Phofphorus, and a little before
" they ceafe to rife they lofe their firft fmell of Sal
" Ammoniac, and acquire the odour of garlick. ·

" As they afcend with great rapidity, the little
" hole muft be frequently opened, to obferve whe-
" ther the hiffing be not too ftrong ;· for in that
" cafe it would be neceffary to fhut the door of the
" afh-hole quite clofe. Thefe white vapours con-
" tinue two hours. When you find they ceafe
" rifing, make a fmall paffage through the dome,
" by opening fome of its regifters, that the flame
" may juft begin to draw. Keep up the fire in
" this mean ftate till the firft volatile Phofphorus
" begins to appear.

" This appears in about three hours after the
" white vapours firft begin to rife. In order to dif-
" cover it, pull out the little birchen peg once
" every minute, and rub it againft fome hot part
" of the furnace, where it will leave a trail of
" light, if there be any Phofphorus upon it.

" Soon after you obferve this fign, there will
" iffue out through the little hole of the ballon a
" ftream of bluifh light, which continues of a
" greater or fhorter extent to the end of the opera-
" tion. This ftream or fpout of light does not burn.
" If you hold your finger againft it for twenty or
" thirty feconds, the light will adhere to it ; and if

" you rub that finger over your hand, the light
" will befmear it, and render it luminous.

" But from time to time this ftreamer darts out
" to the length of feven or eight inches, fnapping
" and emitting fparks of fire ; and then it burns
" all combuftible bodies that come in its way.
" When you obferve this, you muft manage the
" fire very warily, and fhut the door of the afh-
" hole quite clofe, yet without ceafing to throw
" coals into the fire-place every two minutes.

" The Volatile Phofphorus continues two hours;
" after which the little fpout of light contracts to
" the length of a line or two : and now is the time
" for pufhing your fire to the utmoft : immediately
" fet the door of the afh-hole wide open, throw bil-
" lets of wood into it, unftop all the regifters of the
" reverberatory, fupply the fire-place with large
" coals every minute : in fhort, for fix or feven
" hours all the infide of the furnace muft be kept
" of a white heat, fo that the retort fhall not be
" diftinguifhable.

" In this fierce extremity of heat the true Phof-
" phorus diftills like an oil, or like melted wax :
" one part thereof floats on the water in the reci-
" pient, the other falls to the bottom. At laft,
" the operation is known to be quite over when the
" upper part of the ballon, in which the volatile
" Phofphorus appears condenfed in a blackifh film,
" begins to grow red : for this fhews that the
" Phofphorus is burnt where the red fpot appears.
" You muft now ftop all the regifters, and fhut all
" the doors of the furnace, in order to fmother the
" fire ; and then clofe up the little hole in the bal-
" lon with fat lute or bees-wax. In this condition
" the whole muft be left for two days ; becaufe
" the veffels muft not be feparated till they are
" perfectly cold, left the Phofphorus fhould take
" fire.

" As

" As foon as the fire is out, the ballon, which
" is then in the dark, prefents a moſt agreeable
" object: all the empty part thereof above the wa-
" ter ſeems filled with a beautiful blue light:
" which continues for feven or eight hours, or as
" long as the ballon keeps warm, never diſappear-
" ing till it is cooled.

" When the furnace is quite cold take out the
" veſſels, and feparate them from each other as
" neatly as poſſible. With a linen cloth wipe away
" all the black ſtuff you find in the mouth of the
" ballon; for if that filth ſhould mix with the Phoſ-
" phorus, it would hinder it from being tranſpa-
" rent when moulded. This muſt be done with
" great expedition: after which pour into the bal-
" lon two or three quarts of cold water, to accele-
" rate the precipitation of the Phoſphorus that
" ſwims at top. Then agitate the water in the bal-
" lon, to rinſe out all the Phoſphorus that may
" ſtick to the ſides; pour out all the water thus
" ſhaken and turbid, into a very clean earthen pan,
" and let it ſtand till it grows clear. Then decant
" this firſt uſeleſs water, and on the blackiſh ſedi-
" ment, left at the bottom of the pan, pour ſome
" boiling water to melt the Phoſphorus; which
" thereupon unites with the fuliginous matter, or
" volatile Phoſphorus, that precipitated with it, both
" together forming a maſs of the colour of ſlate.
" When this water, in which you have melted the
" Phoſphorus, is cool enough, take out the Phoſ-
" phorus, throw it into cold water, and therein
" break it into little bits in order to mould it.

" Then take a matras, having a long neck
" ſomewhat wider next the body than at its
" mouth: cut off half the body, ſo as to make a
" funnel of the neck-part, the ſmaller end of which
" muſt be ſtopped with a cork. The firſt mould
" being thus prepared, plunge it endwiſe, with its
" mouth

" mouth uppermoſt, in a veſſel full of boiling wa-
" ter, and fill it with that water. Into this funnel
" throw the little bits of your ſlate-like maſs,
" which will melt again in this hot water, and fall
" ſo melted to the bottom of the tube. Stir this
" melted matter with an iron wire, to promote the
" ſeparation of the Phoſphorus from the fuliginous
" matter with which it is fouled, and which, being
" leſs ponderous than the Phoſphorus, will gra-
" dually riſe above it towards the upper part of
" the cylinder.

" Keep the water in the veſſel as hot as at firſt,
" till on taking out the tube you ſee the Phoſpho-
" rus clean and tranſparent. Let the clear tube
" cool a little, and then ſet it in cold water, where
" the Phoſphorus will congeal as it cools. When
" it is perfectly congealed, pull out the cork, and
" with a ſmall rod, near as big as the tube, puſh
" the cylinder of Phoſphorus towards the mouth
" of the funnel, where the feculency lies. Cut
" off the black part of the cylinder, and keep it
" apart: for when you have got a quantity thereof,
" you may melt it over again in the ſame manner,
" and ſeparate the clean Phoſphorus which it ſtill
" contains. As to the reſt of the cylinder which
" is clean and tranſparent, if you intend to mould
" it into ſmaller cylinders, you may cut it in
" ſlices, and melt it again by the help of boiling
" water in glaſs tubes of ſmaller dimenſions."

OBSERVATIONS.

THIS proceſs for making Phoſphorus is copied
from the Memoirs of the Academy of Sciences for
the Year 1737; where it is deſcribed by M. Hellot,
with ſo much accuracy, clearneſs, and preciſion,
that I thought I could not do better than tranſcribe
it, without departing from the author's own expreſ-
ſions, for the ſake of ſuch as may not have thoſe
Memoirs.

Memoirs. We fhall take occafion, in thefe obfervations, to point out fome effential circumftances which
I have omitted in the defcription of the Procefs, that
I might not break the connection between the phenomena that happen in the courfe of this experiment.

It is proper to obferve, in the firft place, that one
of the moft ufual caufes of mifcarriage in this operation is a defect of the requifite qualities in the retort employed. It is abfolutely neceffary to have
that veffel made of the beft earth, and fo well made
that it fhall be capable of refifting the utmoft violence of fire, continued for a very long time ; as
appears by the defcription of the procefs. The
retorts commonly fold by potters, and other earthen
ware-men ; are not fit for this operation ; and Mr.
Hellot was obliged to fend to Heffe-Caffel for fuch
as he wanted.

We fhall, in the fecond place, obferve with
M. Hellot that, " before you fet your retort in
" the furnace, it is proper to make an effay of your
" matter, to fee if there be reafon to hope for fuc-
" cefs. For this purpofe put about an ounce there-
" of into a fmall crucible, and heat it till the veffel
" be red. The mixture, after having fmoked,
" ought to chop or crack without puffing up, or
" even rifing in the leaft. From thefe cracks will
" iffue undulating flames, white and bluifh, dart-
" ing upwards with rapidity. This is the firft vo-
" latile Phofphorus, which occafions all the dan-
" ger of the operation. When thefe firft flafhes
" are over, increafe the heat of your matter by
" laying a large live coal upon the crucible. You
" will then fee the fecond Phofphorus, like a lu-
" minous, fteady vapour, of a colour inclining to
" violet, covering the whole furface of the matter :
" it continues for a very long time, and diffufes a
" fmell of garlick, which is the diftinguifhing
" odour of the Phofphorus you are feeking.

" When

" When this luminous vapour is entirely gone,
" pour the red hot matter out of the crucible upon
" an iron plate. If you do not find one drop of falt
" in fufion, but that, on the contrary, the whole
" falls readily into powder, 'tis a proof that your
" matter was fufficiently lixiviated, and that it
" contains no more fixed Salt, or Sea-falt, if you
" will, than is requifite. If you find on the plate
" a drop of falt coagulated, it fhews that there is
" too much left in, and that there is danger of
" your mifcarrying in the operation; becaufe the
" redundant falt would corrode, and eat through
" the retort. In this cafe your matter muft be
" wafhed again, and then fufficiently dried."

Our third obfervation fhall be concerning the
furnace proper to be employed in this operation.
This furnace muft be fo conftructed, that within
a narrow compafs it may give a heat at leaft equal
to that of a glafs-houfe furnace, or rather greater,
efpecially during the laft feven or eight hours of the
operation. M. Hellot in his Memoir gives an exact defcription of fuch a furnace.

" As certain accidents may happen in the courfe
" of the operation, fome precautions are to be taken
" againft them. For inftance, if the ballon fhould
" break while the Phofphorus is diftilling, and any
" of it fhould fall on combuftible bodies, it would
" fet them on fire, and probably burn the laborato-
" ry, becaufe it is not to be extinguifhed without the
" greateft difficulty. The furnace muft therefore be
" erected under fome vault, or upon a bed of brick-
" work raifed under fome chimney that draws well:
" nor muft any furniture or utenfil of wood be left
" near it. If a little flaming Phofphorus fhould fall
" on a man's legs or hands, in lefs than three mi-
" nutes it would burn its way to the very bone. In
" fuch a cafe nothing but urine will ftop its progrefs.

I " If

" If the retort crack while the Phofphorus is
" diftilling, there is an unfuccefsful end of your
" operation. It is eafy to perceive this by the ftink
" of garlick which you will fmell about the fur-
" nace; and moreover, the flame that iffues thro'
" the apertures of the reverberatory will be of a
" beautiful violet colour. The Acid of Sea-falt
" always gives this colour to the flame of fuch
" matters as are burnt along with it. But if the
" retort break before the Phofphorus hath made
" its appearance, its contents may be faved by
" throwing a number of cold bricks into the fire-
" place, and upon them a little water to quench
" the fire at once." All thefe ufeful obfervations
we owe alfo to M. Hellot.

. The Phofphorus here defcribed was firft difcovered by a citizen of Hamburgh named Brandt, who
worked upon urine in fearch of the Philofopher's
ftone. Afterwards two other fkilful Chymifts, who
knew nothing more of the procefs than that Phofphorus was obtained from urine, or in general from
the human body, likewife endeavoured to difcover
it; and each of them feparately did actually make
the difcovery. Thefe two Chymifts were Kunckel
and Boyle.

The former perfected the difcovery, and found
out a method of making it in confiderable quantities
at a time; which occafioned it to be called *Kunckel's
Phofphorus*. The other, who was an Englifh gen-
tleman, had not time to bring his difcovery to perfection, and contented himfelf with lodging a vou-
cher of his having difcovered it in the hands of the
Secretary of the Royal Society of London, who
gave him a certificate thereof.

" Though Brandt, fays M. Hellot, who had be-
" fore this fold his fecret to a Chymift named Krafft,
" fold it afterwards to feveral other perfons, and
" even at a very low rate; and though Mr. Boyle
I " publifh-

" publifhed the procefs for making it; yet it is ex-
" tremely probable that both of them kept in their
" own hands the mafter-key; I mean, *the particu-*
" *lar management neceffary to make the operation fuc-*
" *ceed:* for till Kunckel found it out, no other
" Chymift ever made any confiderable quantity
" thereof, except Mr. Godfrey Hankwitz, an
" Englifh Chymift, to whom Mr. Boyle revealed
" the whole myftery.

" Neverthelefs, continues he, we are very far
" from alledging that all thofe who have defcribed
" this operation meaned to impofe upon the world:
" but we conceive that moft of them having ob-
" ferved luminous vapours in the ballon, and fome
" fparks about the juncture of the veffels, were
" contented with thofe appearances. And thus it
" came to pafs, that, after Kunckel and Boyle
" died, Mr. Godfrey Hankwitz was the only Chy-
" mift that could fupply Europe therewith; on
" which account it is likewife very well known by
" the name of *Englifh Phofphorus.*"

Almoft all the Chymifts confider Phofphorus as a
fubftance confifting of the Acid of Sea-falt combined with the Phlogifton, in the fame manner as
Sulphur confifts of the Vitriolic Acid combined
with the Phlogifton. This opinion is founded on
the following principles.

Firft, Urine abounds with Sea-falt, and contains
alfo a great deal of Phlogifton: now thefe are the
ingredients of which they conjecture Phofphorus to
be compofed.

Secondly, Phofphorus has many of the properties
of Sulphur; fuch as being foluble in oils; melting
with a gentle heat; being very combuftible; burning without any foot; giving a vivid and bluifh
flame; and laftly, leaving an acid liquor when
burnt: fenfible proofs that it differs from Sulphur
in nothing but the nature of its Acid.

Thirdly,

Thirdly, this Acid of Phofphorus, being mixed with a folution of filver in Spirit of Nitre, precipi-tates the filver, and this precipitate is a true *Luna cornea*, which appears to be more volatile even than the common fort; as M. Hellot tells us, who made the experiment. This fact proves inconteftably that the Acid of Phofphorus is of the fame nature with that of Sea-falt: for all Chymifts know that the property of precipitating filver in a *Luna cornea* belongs to the Marine Acid only.

Fourthly, M. Stahl obferves that, if Sea-falt be caft on live coals, they inftantly burn with great acti-vity; then they emit a very vivid flame, and are much fooner confumed than if none of this Salt had touched them; that Sea-falt in fubftance, which will bear the violence of fire a confiderable time when fufed in a crucible, without fuftaining any fen-fible diminution, yet evaporates very quickly, and is reduced to white flowers, by the immediate con-tact of burning coals; and laftly, that the flame which rifes on this occafion is of a blue colour in-clining to violet, efpecially if it be not thrown di-rectly on the coals themfelves, but kept in fufion amidft burning coals, in a crucible fo placed that the vapour of the Salt may join with the enflamed Phlogifton as it rifes from the coals.

Thefe experiments of Mr. Stahl's prove that the Phlogifton acts upon the Acid of Sea-falt, even while it is combined with its alkaline bafis. The flame that appears on this occafion may be confi-dered as an imperfect Phofphorus: and indeed its colour is exactly like that of Phofphorus.

All the facts above related evince that the Acid of Phofphorus is akin to that of Sea-falt; or rather that it is the very fame. But there are other facts which prove that this Acid undergoes fome change at leaft, fome peculiar preparation, before it enters into the compofition of a true Phofphorus, and that,

when

when extricated therefrom by burning, it is not a pure Acid of Sea-falt; but is ftill adulterated with a mixture of fome other fubftance; which makes it confiderably different from that Acid. For thefe obfervations we are obliged to M. Marggraff, of the Academy of Sciences at Berlin, a celebrated Chymift. I fhall prefently give an account of his principal experiments as fuccinctly as poffible.

M. Marggraff hath alfo publifhed a procefs for making Phofphorus, and affures us that by means thereof we may obtain in lefs time, with lefs heat; lefs trouble, and lefs expence, a greater quantity of Phofphorus than by any other method. His operation is this:

He takes two pounds of Sal Ammoniac in powder, which he mixes accurately with four pounds of Minium. This mixture he puts into a glafs retort; and with a graduated fire draws off a very fharp; volatile, urinous fpirit.

We obferved in our theoretical Elements that fome metallic fubftances have the property of decompofing Sal Ammoniac; and feparating its volatile Alkali; concerning which phenomenon we there gave our opinion. Minium, which is a calx of lead, is one of thofe metallic fubftances. In this experiment it decompofes the Sal Ammoniac, and feparates its volatile Alkali; what remains in the retort is a combination of the Minium with the Acid of the Sal Ammoniac; which is well known to be the fame with the Marine Acid; and confequently the refidue of this operation is a fort of *Plumbum corneum*.

The quantity thereof is four pounds eight ounces. Of this he mixes three pounds with nine or ten pounds of urine; that has ftood putrefying for two months, evaporated to the confiftence of honey. Thefe he mixes by little and little in an iron pan over the fire, ftirring the mixture from time to time. Then he adds half a pound of charcoal duft, and evaporates the matter, kept continually ſtirring, till the whole be brought to a black powder. He next diſtills the mixture in a glaſs retort with degrees of fire, which he raiſes towards the end ſo as to make the retort red-hot, in order to expel all the urinous ſpirit, ſuperfluous oil, and ammoniacal ſalt. The diſtillation being finiſhed, there remains nothing in the retort but a very friable *caput mortuum.*

This remainder he pulveriſes again, and throws a pinch thereof on live coals, thereby to diſcover whether or no the matter be rightly prepared, and in order, for yielding Phoſphorus. If it be ſo, it preſently emits an arſenical odour, and a blue un-dulating flame, which paſſes over the ſurface of the coals like a wave.

Being thus aſſured of the ſucceſs of his operation, he puts one half of his matter, in three equal parts, in-to three ſmall earthen German retorts, capable of holding about eighteen ounces of water a-piece. Theſe three retorts, none of which is above three quarters full, he places together in one reverberatory furnace, built much like thoſe we have deſcribed, ex-cept that it is ſo conſtructed as to hold the three re-torts diſpoſed in one line. To each retort he lutes a recipient ſomething more than half full of water, or-dering the whole in ſuch a manner, that the noſes of his retorts almoſt touch the ſurface of the water.

He begins the diſtillation with warming the retorts ſlowly, for about an hour, by a gentle heat. When that time is elapſed he raiſes the fire gradually, ſo that in half an hour more the coals begin to touch the bottoms of the retorts. He continues throwing coals into the furnace by little and little, till they riſe half way the heighth of the retorts; and in this he employs another half hour. Laſtly, in the next half hour he raiſes the coals above the bowls of the retorts.

Then the Phoſphorus begins to aſcend in clouds: on this he inſtantly increaſes the heat of the fire as much

much as possible, filling the furnace quite up with
coals, and making the retorts very red. This degree of fire causes the Phosphorus to distill in drops
which fall to the bottom of the water. He keeps
up this intense heat for an hour and half, at the
end of which the operation is finished; so that it
lasts but four hours and a half in all: nay, he further assures us that an artist, well versed in managing
the fire, may perform it in four hours only. In
the same manner he distills the second moiety of
his mixture in three other such retorts.

The advantage he finds in making use of several
small retorts, instead of a single large one, is that
the heat penetrates them with more ease, and the
operation is performed with less fire, and in less
time. He purifies and moulds his Phosphorus much
in the same manner as M. Hellot does. From the
quantity of ingredients above-mentioned, he obtains
two ounces and a half of fine crystalline moulded
Phosphorus,

M. Marggraff considering, as a consequence of
the experiments above related, that a highly concentrated Acid of Sea-salt contributes greatly towards the formation of Phosphorus, proceeded to
try several other experiments, in which he employed
that Acid in a state of combination with other
bases. He mixed, for instance, an ounce of *Luna
cornea* with an ounce and half of putrefied and in-
spissated urine, and from the mixture obtained a
very beautiful Phosphorus.

In short, the several experiments mentioned having thoroughly persuaded him that the Acid of Sea-
salt, provided it were highly concentrated, would
combine with the Phlogiston as readily as the Vitriolic Acid does, he resolved to try whether he could
not make Phosphorus with matters containing that
Acid and the Phlogiston, without making use of
any urine.

With

. With this view he made a great number of different trials, wherein he employed Sea-falt in fubftance, Sal Ammoniac, Plumbum corneum, Luna cornea, fixed Sal Ammoniac, otherwife called *Oil of Lime*. Thefe feveral fubftances, all of which contain the Acid of Sea-falt, he mixed with fundry matters abounding in Phlogifton, different vegetable coals, and even animal matters, fuch as the oil of hartfhorn, human blood, &c. varying the proportions of thefe fubftances many different ways, without ever being able to produce a fingle atom of Phofphorus: which gave this able Chymift juft caufe to fufpect that the Marine Acid, while pure and crude, is not capable of combining with the Phlogifton in the manner requifite to form a Phofphorus; that for this purpofe it is neceffary the Acid fhould have contracted a previous union with fome other matter; and that the Acid found in urine hath probably undergone the neceffary change. M. Marggraff is of opinion that the matter, which by its union renders the Marine Acid capable of entering into the compofition of Phofphorus, is a fort of exceedingly fubtile vitrifiable earth. The experiments he made upon the Acid of Phofphorus will fhew that his notion is not altogether groundlefs. M. Marggraff having let fome urine, evaporated to the confiftence of honey, ftand quiet in a cool place, obtained from it, by cryftallization, a Salt of a fingular nature. By diftilling this urine afterwards, he fatisfied himfelf that it yielded him much lefs Phofphorus than urine from which no Salt had been extracted; and as it cannot be entirely deprived of this Salt, he thinks that the fmall quantity of Phofphorus, which this urine yielded him, came from the Salt that was ftill left in it.

Further, he diftilled this Salt feparately with lampblack, and obtained from it a confiderable quantity of very fine Phofphorus. He even mixed *Luna cor-*

nea

nea with this Salt, in order to fee whether it would not increafe the quantity of his Phofphorus; but without fuccefs: whence he concluded that in this Saline matter refides the true Acid that is fit to enter into the compofition of Phofphorus. This opinion is confirmed by feveral experiments on the Acid of Phofphorus, which he found to have fome properties refembling thofe of this Salt.

The Acid of Phofphorus feems to be more fixed than any other: and therefore if you would feparate it, by burning, from the Phlogifton with which it is united, there is no occafion for fuch an apparatus of veffels as is employed for obtaining the Spirit of Sulphur. For this Acid will remain at the bottom of the veffel in which you burn your Phofphorus: indeed, if it be urged by the force of fire, its moft fubtile part evaporates, and the remainder appears in the form of a vitrified matter.

This Acid effervefces with fixed and volatile Alkalis, and therewith forms Neutral Salts; but very different from Sea-falt, and from Sal-Ammoniac. That which has a fixed Alkali for its bafis does not crackle when thrown on burning coals; but fwells and vitrifies like Borax. That which has a volatile Alkali for its bafis fhoots into long pointed cryftals; and, being urged by fire in a retort, lets go its volatile alkali, a vitrified matter remaining behind. This Salt is like that abovementioned, as obtained from urine and yielding Phofphorus.

It appears from the experiments adduced, that the Acid of Phofphorus tends always to vitrification; which proves that it is not pure, and gave M. Marggraff caufe to think that it is altered by the admixture of a very fubtile vitrifiable earth.

M. Marggraff alfo obtained Phofphorus from feveral vegetable fubftances which we ufe every day for food. This gives him occafion to conjecture that

the

the Salt requifite to the formation of Phofphorus exifts in vegetables, and paffes from thence into the animals that feed upon them.

Laftly, he concludes his differtation by informing us of a very important truth, viz. That the Acid obtained from Phofphorus, by burning it, will ferve to form Phofphorus anew; for which purpofe it need only be combined with fome charred coal, fuch as lamp-black, and diftilled.

From what hath been faid on this fubject it is plain the Chymifts have a great many curious and interefting enquiries to make concerning Phofphorus, and particularly concerning its Acid.

I fhall conclude this article with an account of certain properties of Phofphorus which I have not yet mentioned.

Phofphorus diffolves by lying expofed to the air. What water cannot effect, fays M. Hellot, or at leaft requires eight or ten years to bring about, the moifture of the air accomplifhes in ten or twelve days; whether it be that the Phofphorus takes fire in the air, and the inflammable part evaporating, almoft entirely, leaves the Acid of the Phofphorus naked, which like all other Acids, when exceedingly concentrated, is very greedy of moifture; or elfe that the moifture of the air, being water divided into infinitely fine particles, is fo fubtile as to find its way through the pores of the Phofphorus, into which the groffer particles of common water can by no means infinuate themfelves.

Phofphorus heated by the vicinity of fire, or by being any way rubbed, foon takes fire and burns fiercely. It is foluble in all Oils and in Ether, giving to thofe liquors the property of appearing luminous when the bottle containing the folution is opened. Being boiled in water, it likewife communicates thereto this luminous quality. M. Morin, Profeffor at Chartres, is the author of this obfervation.

The

The late Mr. Grosse, a celebrated Chymist, of the Academy of Sciences, observed that Phospho-rus being dissolved in essential oils crystallizes therein. These crystals take fire in the air, either when thrown into a dry vessel, or wrapt up in a piece of paper. If they be dipped in Spirit of Wine, and taken out immediately, they do not afterwards take fire in the air: they smoke a little, and for a very short time, but hardly waste at all. Though some of them were left in a spoon for a fortnight, they did not seem to have lost any thing of their bulk : but when the spoon was warmed a little they took fire, just like common Phosphorus that had never been dis-solved and crystallized in an essential oil.

M. Marggraff, having put a dram of Phosphorus with an ounce of highly concentrated Spirit of Nitre into a glass retort, observed that, without the help of fire, the Acid dissolved the Phosphorus ; that part of the Acid came over into the recipient which was luted to the retort ; that at the same time the Phosphorus took fire, burnt furiously, and burst the vessels with explosion. Nothing of this kind happens when any of the other Acids, though concentrated, are applied to Phosphorus.

PROCESS III.

To decompose Sea-Salt by means of the Vitriolic Acid. Glauber's Salt. The Purification and Concentration of Spirit of Salt.

PUT the Sea-salt from which you mean to ex-tract the Acid into an unglazed earthen pipkin, and set it amidst live coals. The Salt will decrepi-tate, grow dry, and fall into a powder. Put this decrepitated Salt into a tubulated glass retort, leav-ing two thirds thereof empty. Set the retort in a reverberating furnace ; apply a receiver like that

used

ufed in diftilling the fmoking Spirit of Nitre, and
lute it on in the fame manner, or rather more exactly if poffible. Then through the hole, in the upper convexity of the retort, pour a quantity of
highly concentrated Oil of Vitriol, equal in weight
to about a third part of your Salt, and immediately
fhut the hole very clofe with a glafs ftopple, firft
ground therein with emery fo as to fit it exactly.

As foon as the Oil of Vitriol touches the Salt, the
retort and receiver will be filled with abundance of
white vapours ; and foon after, without lighting
any fire in the furnace, drops of a yellow liquor will
diftill from the nofe of the retort. Let the diftillation proceed in this manner without fire, as long as
you perceive any drops come : afterwards kindle a
very fmall fire under the retort, and continue diftilling and raifing the fire by very flow degrees, and
with great caution, to the end of the diftillation ;
which will be finifhed before you have occafion to
make the retort red-hot. Unlute the veffels, and
without delay pour the liquor, which is a very fmoking Spirit of Salt, out of the receiver into a cryftal
bottle, like that directed for the fmoking Spirit of
Nitre.

OBSERVATIONS.

SEA-SALT, as hath been already faid, is a Neutral
Salt compofed of an Acid, which differs from thofe
of Vitriol and Nitre, combined with a Fixed Alkali
that has fome peculiar properties ; but does not vary
from the others in its affinities. This Salt therefore,
as well as Nitre, muft be decompofed by the Vitriolic Acid ; which accordingly is the cafe in the procefs here defcribed. The Vitriolic Acid unites with
the Alkaline bafis of the Sea-falt, and feparates its
Acid ; and that with much greater facility than it expels the Nitrous Acid from its Alkaline bafis, becaufe

çaufe the Acid of Sea-falt has not fo great an affinity
as the Nitrous Acid with Fixed Alkalis.

As a highly concentrated Oil of Vitriol is ufed
on this occafion, and as the Sea-falt is previoufly
dried and decrepitated, the Acid obtained from it
by diftillation is very free from phlegm, and always
fmokes, even more violently than the ftrongeft
Acid of Nitre. The vapours of this Acid are alfo
much more elaftic and more penetrating than thofe
of the Nitrous Acid : on which account this diftillation of the fmoking Spirit of Salt is one of the
moft difficult, moft laborious, and moft dangerous
operations in Chymiftry.

This procefs requires a tubulated retort, that the
Oil of Vitriol may be mixed with the Sea-falt after
the receiver is well luted to the retort, and not before : for, as foon as thefe two matters come together, the Spirit of Salt rufhes out with fo much impetuofity, that, if the veffels were not luted at the
time, the copious vapours that would iffue through
the neck of the ballon would fo moiften it, as well
as the neck of the retort, that it would be impracti-
cable to apply the lute and fecure the joint as the
operation requires. Moreover, the operator would
be expofed to thofe dangerous fumes, which, on this
occafion, rufh out, and enter the lungs, with fuch incredible activity as to threaten inftant fuffocation.

Having faid fo much of the elafticity and activity
of the fumes of Spirit of Salt, it is needlefs to in-
fift upon the neceffity of giving vent to the veffels
from time to time, by opening the little hole of the
ballon : indeed the beft way to prevent the lofs of a
great many vapours, on this occafion, is to employ
adopters, and cover them with wet canvas, which
will cool and condenfe the vapours they contain.

When the operation is finifhed, we find a white,
faline mafs at the bottom of the retort as in a mould.
If this mafs be diffolved in water, and the folution
cryftal-

cryftallized, it yields a confiderable quantity of Sea-falt that hath not been decompofed, and a Neutral Salt confifting of the Vitriolic Acid united with the Alkaline bafis of that part which hath been decompofed... This Neutral Salt, which bears the name of *Glauber* its inventor, differs from Vitriolated Tartar, or the *Sal de duobus*, which remains after diftilling the Nitrous Acid, efpecially in that it is more fufible, more foluble in water, and hath its cryftals differently figured. But as in thefe two Salts the Acid is the fame, the differences that appear between them muft be attributed to the peculiar nature of the bafis of Sea-falt.

Spirit of Salt drawn by the procefs above defcribed is tainted with a fmall mixture of the Vitriolic Acid, carried up by the force of fire before it had time to combine with the Alkali of the Sea-falt; which happens likewife to the Nitrous Acid procured in the fame manner. If you defire to have it pure, and abfolutely free from the Acid of Vitriol, it muft be diftilled a fecond time from Sea-falt, as the Acid of Nitre was before directed to be diftilled again from frefh Nitre, in order to purify it from any Vitriolic taint.

Sea-falt, as well as Nitre, may be decompofed by any combination of the Vitriolic Acid with a metallic or earthy fubftance : but it is proper to obferve, that if you diftill Spirit of Salt by means of Green Vitriol, the operation will not fucceed fo well as when Spirit of Nitre is diftilled in the fame manner : lefs Spirit is obtained, and a much fiercer fire is required.

The caufe of this lies in the property which the Acid of Sea-falt poffeffes of diffolving Iron, even when deprived of a part of its Phlogifton by having contracted an union with another Acid ; fo that it is no fooner diflodged from its own bafis by the Vitriolic Acid, than it unites with the ferruginous

bafis

basis of the Vitriol, from which it cannot be separated but by a most violent fire. This is the consequence more especially when calcined Vitriol is made use of: for moisture, as we shall presently see, greatly facilitates the separation of the Marine Acid from those substances with which it is united.

When you do not desire a highly dephlegmated and smoking Spirit of Salt, you may distill with the additament of any earth containing the Vitriolic Acid; as Clay, for instance, or Bole. To this end one part of Sea-salt, slightly dried and reduced to a fine powder, must be accurately mingled with two parts of the earth you intend to employ likewise pulverized; of this mixture make a stiff paste with a proper quantity of rain water, and having formed little balls thereof about the size of a hazel nut, let them dry in the sun; when dry put them into a stone or coated glass retort, leaving a third part thereof empty; set this vessel in a reverberating furnace, covered with its dome; apply a receiver, which need not be luted on for some time; and heat the vessels very slowly. At first an insipid water will rise; which must be thrown away: afterwards the Spirit of Salt will appear in white clouds. Now lute your vessels, and raise the fire by degrees; which towards the end must be pushed to the utmost extremity. The operation is known to be finished when no drops fall from the nose of the retort, the receiver cools, and the white vapours that filled it are seen no more.

The Spirit of Salt obtained by the process here delivered does not smoke, and contains much more phlegm than that which is distilled by means of the concentrated Oil of Vitriol; because the earth, though dried in the sun, still retains a great deal of moisture, which commixes with the Acid of the Sea-salt. Consequently it is much easier to collect its vapours; so that this operation is attended with

I   much

much lefs trouble than the other. Neverthelefs it is advifeable to proceed gently; to apply but little heat at firft, and to unftop every now and then the fmall hole of the receiver: for a quantity of the vapours of Spirit of Salt, even when weakened by the admixture of water, is very apt to burft the veffels.

A much greater degree of fire is neceffary to raife the Spirit of Salt by this latter procefs, than by that in which the pure Vitriolic Acid is employed: for, as faft as the Spirit of Salt is diflodged from its own bafis, by the Vitriolic Acid contained in the earth made ufe of, part of it joins that earth, and cannot be feparated from it without the moft violent heat.

A Spirit of Salt that fhall not fmoke may alfo be obtained by means of the pure Vitriolic Acid. Spirit of Vitriol, or Oil of Vitriol, lowered with a good deal of water, will do the bufinefs.

Some Chymifts direct a little water to be placed in the receiver, when Spirit of Salt is to be diftilled by the intermedium of concentrated Oil of Vitriol, in order to make the acid vapours condenfe more readily. By this means indeed fome of the inconveniencies attending the diftillation of fmoking Spirit of Salt may be avoided: but on the other hand, the acid vapours being abfolutely fuffocated by the water as faft as they come over, the Spirit of Salt obtained by this method will be no lefs aqueous than that procured by the interpofition of earths: fo that here is an expence to no manner of purpofe. Therefore, when a Spirit of Salt is defired that fhall not fmoke, it is beft to employ an additament of earth; and that fo much the rather as the Marine Acid obtained by this means is purer and freer from any Vitriolic taint, for the reafons already affigned.

Part of the Acid of Sea-falt may be feparated from its Alkaline bafis by the force of fire alone, without the intervention of any other body. With
this

this view the Salt muſt be put into the retort without being dried. At firſt an inſipid water riſes; but it gradually becomes acid, and hath all the properties of Spirit of Salt. When the Salt in the retort is grown perfectly dry, nothing more can be forced over by any degree of heat whatever. If you would obtain more Acid from the ſame Salt, you muſt take it out of the retort, where you will find it in a lump, reduce it to powder, and expoſe it to the air for ſome time, that it may attract the moiſture thereof; or elſe wet it at once with ſome rain water, and diſtill as before. You will again have an inſipid water, and a little Spirit of Salt; which will in like manner ceaſe to riſe when the Salt in the retort becomes dry. This operation may be repeated as often as ſhall be thought proper : and perhaps it may be poſſible to decompoſe Sea-ſalt entirely by means thereof, without the interpoſition of any other body. The Spirit of Salt thus obtained is exceeding weak, in ſmall quantity, and loaded with much water.

This experiment proves that moiſture greatly facilitates the ſeparation of the Acid of Sea-ſalt from the matters with which it is united : and this is the reaſon that, in diſtilling Spirit of Salt with the additament of an earth, the operation requires much leſs fire at the beginning, while the earth and ſalt retain a great deal of humidity, than towards the end, when they begin to grow dry.

After the operation there remains in the retort a ſaline and earthy maſs, which contains, 1. Some entire Sea-ſalt that has ſuffered no decompoſition ; 2. A Glauber's Salt which is, as we ſaid before, a Neutral Salt conſiſting of the Vitriolic Acid united with the Alkaline baſis of the Sea-ſalt, from which it hath expelled its proper Acid; 3. Part of the earth uſed as an intermedium, ſtill retaining a portion of its original Vitriolic Acid, which, happening not to lie near enough to any particles of Sea-ſalt,

could

could not exert its power in decomposing them, and so remains united with its earthy basis; 4. Another part of the same earth, impregnated with some of the Marine Acid, which combined therewith upon being expelled from its Alkaline basis by the Vitriolic Acid, and which the force of fire was unable to separate from it when the matters were grown perfectly dry. In consequence of what remains in this *caput mortuum,* if the whole mass be triturated, moistened with a little water, and distilled a second time, considerably more Spirit of Salt will be obtained from it: and the same is to be said of all distillations of this sort.

Spirit of Salt obtained by the means of any other additament than concentrated Oil of Vitriol is generally very weak: but it may be dephlegmated and concentrated, if required, much in the same manner as Oil of Vitriol. For this purpose you must put it into a glass cucurbit, set it in a *balneum mariæ,* fit thereto a head and a receiver, and with a moderate degree of heat draw off one third or one half of the liquor. What comes over into the receiver will be the most aqueous part, which being the lightest will rise first, impregnated however with a little acid: in the cucurbit will be left a concentrated Spirit of Salt, or the most acid part, which being the heaviest will not rise with the degree of heat that is capable of carrying up the phlegm. Spirit of Salt thus concentrated, called also *Oil of Salt,* does not smoke: it is of a yellow colour inclining to green, and an agreeable smell, not unlike that of saffron.

PROCESS IV.

To decompose Sea-salt by means of the Nitrous Acid.
Aqua regis.   *Quadrangular Nitre.*

TAKE dried Sea-salt: bruise it to powder: put it into a glass-retort, leaving one half of the vessel empty. Pour upon it a third of its weight

of

of good Spirit of Nitre. Place your retort in the fand-bath of a reverberating furnace; put on the dome; lute to the retort a receiver having a fmall hole in it, and heat the veffels very flowly. There will come over into the receiver fome vapours, and an acid liquor. Increafe the fire gradually till nothing more rifes. Then unlute the veffels, and pour the liquor out of the receiver into a cryftal bottle, ftopped like others containing Acid Spirits.

OBSERVATIONS.

THE Nitrous Acid hath a greater affinity than the Marine Acid with Fixed Alkalis. When therefore Spirit of Nitre and Sea-falt are mixed together, the fame confequences, in fome meafure, will follow, as when the Vitriolic Acid is mixed with that Salt; that is, the Nitrous Acid will, like the Vitriolic, decompofe it, by diflodging its Acid from its Alkaline bafis, and affuming its place. But as the Nitrous Acid is confiderably weaker, and much lighter, than the Vitriolic Acid; a good deal of it rifes along with the Acid of Sea-falt during the operation. The liquor found in the receiver is therefore a true *Aqua regis*.

If decrepitated Salt, and a right fmoking Spirit of Nitre, be employed in this procefs, the *Aqua regis* obtained will be very ftrong; and during the operation very elaftic vapours will rufh out and burft the veffels, if thofe precautions be not taken which we pointed out as neceffary in diftilling the Spirit of Nitre, and the fmoking Spirit of Salt.

The operation being finifhed, there is left in the retort a faline mafs, containing Sea-falt not decompofed, and a new fpecies of Nitre, which having for its bafis the Alkali of Sea-falt, that is, as we have feveral times obferved, an Alkali of a peculiar nature, differs from the common Nitre, 1. In the figure of its cryftals; which are folids of four fides, formed

formed like lozenges : 2. In that it cryftallizes with
more difficulty, retains more water in its cryftals,
attracts the moifture of the air, and diffolves in water with the fame circumftances as Sea-falt.

C H A P. IV.

Of B o r a x,

P R O C E S S.

*To decompofe Borax by the means of Acids, and to fe-
parate from it the Sedative Salt by fublimation and
by cryftallization.*

R EDUCE to a fine powder the Borax from
which you intend to extract the Sedative Salt.
Put this powder into a wide-necked glafs retort.
Pour upon it an eighth part of its weight of common water, to moiften the powder ; and then add
concentrated Oil of Vitriol, to the weight of fomewhat more than a fourth part of the weight of the
Borax. Set the retort in a reverberatory, make a
moderate fire at firft, and augment it gradually till
the retort become red-hot.

A little phlegm will firft come over; and then
with the laft moifture that the heat expels the Sedative Salt will rife; by which means fome of it will
be diffolved in this laft phlegm, and pafs therewith
into the receiver; but moft of it will adhere in the form
of faline flowers to the fore-part of the neck of the retort, juft where it is clear of the groove of the furnace.
There they collect into a heap, which the fucceeding
flowers pufh infenfibly forward till they flightly ftop
the paffage. Thofe which rife after the neck is thus
ftopped ftick to the after-part of it which is hot, vitrify

trify in some meafure, and form a circle of fufed
Salt.   In this ftate the flowers of the Sedative Salt
feem to iffue out of the circle, as from their bafis:
they appear like very thin, light, fhining fcales,
and muft be brufhed off with a feather.

At the bottom of the retort will be left a faline
mafs: diffolve this in a fufficient quantity of hot
water; filter the folution in order to free it from a
brown earth which it depofites; fet the liquor to
evaporate, and cryftals of Sedative Salt will form
in it.

OBSERVATIONS.

Though Borax is of great ufe in many chymical
operations, efpecially in the fufion of metals, as we
fhall have occafion to fee, yet till of late years Chymifts were quite ignorant of its nature, as they ftill
are of its origin; concerning which we know nothing with certainty, but that it comes rough from
the Eaft Indies, and is purified by the Dutch.

M. Homberg was one of the firft that attempted
to analyfe this Salt.   He fhewed that on mixing it
with the Vitriolic Acid, and diftilling the mixture, a
falt fublimes in little fine needles.   This product of
Borax he called by the name of *Sedative Salt*, becaufe he found it had the property of moderating
the great tumult and heat of the blood in fevers.

After M. Homberg other Chymifts alfo exer-
cifed themfelves on Borax.   M. Lemery difcovered that the Vitriolic is not the only Acid by means
of which the Sedative Salt may be obtained from
Borax; but that either of the other two Mineral
Acids, the Nitrous or the Marine, may be ufed in
its ftead.

M. Geoffroy hath greatly facilitated the means of
obtaining the Sedative Salt from Borax; having
fhewn that it may be extracted by cryftallization as
well as by fublimation; and that the Sedative Salt

fo obtained is in no refpeƈt inferiour to that which
was procured before by fublimation only. To
him alfo we are indebted for the difcovery that in
the compofition of Borax there is an Alkaline Salt
of the fame nature as the bafis of Sea-falt. This he
found by obferving that he got a Glauber's Salt
from a folution of Borax into which he had poured
fome Vitriolic Acid with a view to obtain its Sedative Salt.

Laftly, M. Baron, whom we mentioned before
on occafion of this Salt, hath proved, by a great number of experiments, that a Sedative Salt may be procured from Borax by the help of Vegetable Acids,
which was never done by any body before him; that
the Sedative Salt is not a combination of an Alkaline matter with the Acid made ufe of in extraƈting
it, as fome of its properties feemed to indicate; but
that it exifts previoufly and completely formed in the
Borax; that the Acid employed to extraƈt it only
helps to difengage it from the Alkali with which it
is united; that this Alkali is aƈtually of the fame
nature as the bafis of Sea-falt, becaufe that after extraƈting the fedative Salt, which by its union therewith forms the Borax, a Neutral Salt is found, of
the fame fort with that which would be produced by
combining the bafis of Sea-falt with the particular
Acid made ufe of; that is, if with the Vitriolic
Acid, a Glauber's falt; if with the Nitrous Acid, a
quadrangular Nitre; and if with the Marine Acid,
a true Sea-falt; and laftly, that the Sedative Salt may
be re-united to its Alkali, and re-produce a Borax.

Nothing therefore now remains, to give us all the
infight we can defire into the nature of Borax, but
to know what the Sedative Salt is. M. Baron hath
already given us certain negative notices concerning
it, by fhewing what it is not; that is, that the Acid
employed in its extraƈtion doth not enter into its
compofition. We have great reafon to hope that he
will

will carry his enquiries ſtill further, and clear up all our doubts on this ſubjeſt.

The Sedative Salt may be extraſted from Borax, not only by the means of pure and ſimple Acids, but alſo by the ſame Acids combined with a metallic ba- ſis. Thus Vitriols, for inſtance, may be employed for this purpoſe with good ſucceſs. It is eaſy to ſee that the Vitriol muſt be decompoſed on this occaſi- on, and that its Acid cannot unite with the Alkali in which the Sedative Salt is lodged, without quitting its metallic baſis, which muſt of courſe precipitate.

The Sedative Salt actually ſublimes, when a li- quid containing it is diſtilled; but it does not there- fore follow that it is naturally volatile. It riſes on- ly by the aid of the water with which it is mixed. The proof of this aſſertion is, that, when all the humidity of the mixture containing this Salt is dif- ſipated, no more Salt will riſe, be the fire ever ſo violent; and that by adding more water to moiſten the dried maſs containing it, more Salt will every time be obtained, through many repeated diſtilla- tions. In the ſame manner, if ſome Sedative Salt be moiſtened, and expoſed to a proper degree of heat, a ſmall quantity thereof will riſe at firſt by the help of the water; but as ſoon as it grows dry it remains exceedingly fixed. This obſervation we owe to M. Rouelle.

The Sedative Salt hath the appearance and the taſte of a Neutral Salt: it does not change the co- lour of the juice of violets; nor does it eaſily diſſolve in water; for it requires a quart of boiling water to diſſolve two ounces of it: yet, with regard to Alka- lis, it has the properties of an Acid; it unites with thoſe ſalts, forms therewith a ſaline compound which cryſtallizes, and even expells the Acids that happen to be combined with them; ſo that it decompoſes the ſame Neutral Salts that are decompoſed by the Vitriolic Acid.

U 2 The

The Sedative Salt, when fuddenly expofed to the violent heat of a naked fire, lofes near half its weight, melts, puts on and retains the appearance of glafs; but its nature ftill remains unchanged. This glafs diffolves in water, and fhoots anew into cryftals of Sedative Salt. This Salt communicates to the Alkaline falt with which it is united, when in the form of Borax, the property of melting with a moderate heat, and forming a kind of glafs; and 'tis this great fufibility that recommends the frequent ufe of Borax as a flux for affaying ores. It is alfo employed fometimes as an ingredient in the compofition of glafs; but, in time, it always communicates thereto the fault which its own glafs hath, namely that of tarnifhing with the air. The Sedative Salt hath moreover the fingular property of diffolving in Spirit of Wine, and of giving to its flame, when fet on fire, a beautiful green colour. All thefe obfervations we owe to Meff. Geoffroy and Baron.

M. Geoffroy prepares the Sedative Salt by cryftallization only in the following manner. " He dif-
" folves four ounces of refined Borax in a fufficient
" quantity of warm water, and then pours into the
" folution one ounce and two drams of highly con-
" centrated Oil of Vitriol, which makes a crackling
" noife as it falls in. When this mixture has ftood
" evaporating for fome time, the Sedative Salt be-
" gins to make its appearance in little, fine, fhining
" plates, floating on the furface of the liquor. The
" evaporation is then to be ftopped, and the plates
" will by little and little encreafe in thicknefs and
" breadth. They unite together into little tufts,
" forming with each other fundry different groups.
" If the veffel be ever fo little ftirred, the regularity
" of the cryftals will be difturbed; fo that it muft
" not be touched till the cryftallization appears to be
" finifhed. The cryftalline clufters, being grown
" too bulky and too heavy, will then fall of them-
" felves

" felves to the bottom of the veffel. This being ob-
" ferved, the faline liquor muft be gently decanted
" from thofe little cryftals, which, as they are not ea-
" fily diffolved, muft be wafhed clean, by pouring
" cold water flowly on the fides of the pan, three or
" four times fucceffively, in order to rinfe out all re-
" mains of the faline liquor, and then fet firft to drain,
" and afterwards to dry in the fun. This Salt, in the
" form of light flakes of fnow, is now foft to the
" touch, cool in the mouth, flightly bitter, crackling
" a little between the teeth, and leaving a fmall im-
" preffion of acidity on the tongue. It will keep
" long without giving or calcining, if managed
" according to the preceding directions ; that is,
" if it be exactly freed from its faline liquor.

- " It differs from the Sedative Salt obtained by
" fublimation in this refpect only, that notwith-
" ftanding its feeming lightnefs it is a little heavier
" than the other. M. Geoffroy fuppofes the caufe
" of this weight to be, that, as feveral of the thin
" plates adhere together in cryftallizing, they re-
" tain between them fome fmall matter of humidity ;
" or, if you will, that, as they form larger cryftals,
" they prefent lefs furface to the air which elevates
" light bodies : whereas, on the contrary, the other
" Sedative Salt, being driven up by the force of
" fire, rifes into the head of the cucurbit in a more
" fubtile form, having its particles much more ex-
" panded and divided.

" M. Geoffroy, having put his Sedative Salt made
" by cryftallization to all the fame trials with that
" made by fublimation, fatisfied himfelf that there
" is no other difference between the two. If the
" Sedative Salt made by cryftallization happens to
" calcine in the fun ; that is, if its luftre tarnifhes,
" and its furface grows mealy, 'tis a fign that it ftill
" contains either a little Borax or fome Glauber's
" Salt: for thefe two Salts are apt to calcine in this

" man-

" manner, and pure Sedative Salt fhould not bo
" fubject to this inconvenience. In order to pu-
" rify it, and free it entirely from thofe Salts, it
" muft be diffolved once more in boiling water.
" As foon as the water cools, the Sedative Salt re-
" appears in light, fhining, cryftalline plates,
" fwimming in the liquor. After ftanding four-
" and-twenty hours, the liquor muft be decanted,
" and the falt wafhed with frefh water; by which
" means it will be very pure and beautiful."

Glauber's Salt and Borax diffolve in water with
vaftly more eafe than the Sedative Salt, and confequently do not cryftallize fo readily by much: fo
that the fmall portion of thofe falts, which may have
been left on the furface of the Sedative Salt, being diffufed through a large quantity of water, continues in a ftate of folution, while the Sedative Salt
cryftallizes; which being alfo wafhed afterwards
with fair water, it is impoffible that the fmalleft
particle of thofe other Salts fhould remain adhering
to it; and confequently this muft be deemed an
excellent way of purifying it,

SECTION II.
Of Operations on METALS.

CHAP. I.
Of GOLD.

PROCESS I.
To separate Gold, by Amalgamation with Mercury, from the Earths and Stones with which it is found mixed.

PULVERIZE the earths and stones containing Gold. Put the powder into a little wooden
tray; dip this tray in water, gently shaking it and
its contents. The water will grow muddy, by taking
up the earthy parts of the ore. Continue washing it
in this manner till the water cease to appear turbid.
Upon the ore thus washed pour strong vinegar, having first dissolved therein, by the help of heat, about
a tenth part of its weight of alum. The powder must
be quite drenched and covered with this liquor, and
so left to stand for twice twenty-four hours.

Decant the vinegar, and wash your powder with
warm water, till the last that comes off hath no taste:
then dry it, and put it into an iron mortar, with
four times its weight of Quick-silver: triturate the
whole with a heavy wooden pestle, till all the powder be of a blackish colour: then pour in a little
water, and continue rubbing for some time longer.
More earthy and heterogeneous particles will be separated from the metalline parts by means of this

U 4   water,

water, which will look dirty: it muft then be decanted, and more fair water added. Repeat this feveral
times; then dry what remains in the mortar with a
fponge, and by the help of a gentle heat: you will
find it an Amalgam of the Mercury with the Gold.

Put this Amalgam into a chamoy bag: tie a knot
on its neck, and fqueeze it hard between your
fingers, over fome wide-mouthed veffel; there will
iffue through the pores of the leather numberlefs
little jets of Mercury, forming a fort of fhower, that
will colleft into large globules in the veffel placed
underneath. When you can force out no more Mercury by this means, open the bag, and in it you
will find the Amalgam freed from the fuperfluous
Mercury; the Gold retaining only about as much
thereof as nearly equals itfelf in weight.

Put this Amalgam into a glafs retort; fet this retort in the fand-bath of a reverberating furnace; cover it quite over with fand; apply a glafs receiver
half full of water, fo that the nofe of the retort may
be under the water. The receiver need not be luted
to the retort. Give a gradual heat, and raife the
fire till drops of the fublimed Mercury appear in
the neck of the retort, and fall into the water with
a hiffing noife. If you hear any noife in the retort,
flacken your fire a little. Laftly, when you obferve
that, though you raife the fire ftill higher than before, nothing more will come over, take out your
retort, break it, and there you will find the Gold,
which muft be melted in a crucible with Borax.

OBSERVATIONS.

Gold is a perfeft metal, which can by no means
be deprived of its Phlogifton, and on which few,
even of the moft powerful chymical folvents, have
any effeft: and therefore it almoft always hath its
metalline form when found in the earth; from which
it may fometimes be feparated by fimple lotion.

3   The

The Gold duft found in the fands of certain rivers is of this kind. When it refides in ftones, or tena-cious earths, it may be extracted by the procefs here delivered; to wit, by Amalgamation, or combina-tion of Mercury with the Gold. Mercury is inca-pable of uniting with any earthy fubftances, not even with the metallic earths, when they are de-prived of their Phlogifton, and confequently have not the metalline form.

Hence it follows that when Mercury is triturated with particles of Gold, of earth, and of ftone, min-gled together, it unites with the Gold, and feparates it from thofe heterogeneous matters. Yet, if there be along with the Gold any other metal, in its me-talline form, except Iron, the Mercury will amalga-mate with that alfo. This often happens to Silver, which being a perfect metal as well as Gold, is for that reafon fometimes dug up in its metalline form, and even incorporated with Gold. When this is the cafe, the mafs that remains in the retort, after abftracting the Mercury of the Amalgama, is a com-pound of Gold and Silver, which are to be feparated from each other by the methods we fhall give for that purpofe. The prefent procefs is therefore ap-plicable to Silver as well as Gold.

Sometimes Gold is intimately combined with fuch mineral matters as hinder the Mercury from acting upon it. In that cafe the mixed mafs muft be roaft-ed before you proceed to Amalgamation: for if the matters be volatile, fuch, for inftance, as antimony or arfenic, the fire will carry them off; fo that after roafting the Amalgamation will fucceed. But fome-times thefe matters are fixed, and require fufion; if fo, recourfe muft be had to fome particular me-thods, which we fhall defcribe when we come to treat of Silver, as thefe two perfect metals are to be treated in the fame manner.

Ores

Ores containing Gold muſt be waſhed before an
Amalgam is attempted; that the metalline parts, being freed from the numerous particles of earth with
which they are encompaſſed, may the more readily
incorporate with the Mercury. Beſides, it is the
property of Mercury to take the form of a dark
unmetallic powder, after being long rubbed·with
other matters, ſo that it cannot be eaſily diſtinguiſhed from the particles of earth. And hence, if
you ſtill continue to grind the matters together,
after the Amalgamation is completed, and waſh
them again and again, the water that comes off will
always look turbid, being impregnated with ſome
particles of the Amalgam. This is eaſily proved :
for if you let the turbid water ſettle, and diſtill the
ſediment, you will obtain Quick-ſilver from it.

The ore is to be ſteeped in vinegar charged with
alum, in order to cleanſe the ſurface of the Gold,
which is often covered with a thin coat of earth
that obſtructs the Amalgamation.

Great care muſt be taken that the Mercury employed in this operation be very pure. If it be
adulterated with any metallic ſubſtance, it muſt be
freed therefrom by the methods which we ſhall
propoſe in their proper place.

The way of ſeparating Mercury from Gold is
founded on the different properties of theſe two metallic ſubſtances; the one being exceedingly fixed,
and the other very volatile. The union which Mercury contracts with the metals is not intimate enough,
to give the new compound which reſults therefrom
all the properties of either of the two united ſubſtances ; at leaſt ſo far as concerns their degrees of
fixity and volatility. Hence it comes that, in our
Amalgam, the Gold communicates but very little
of its fixity to the Mercury, and the Mercury communicates to the Gold but very little of its volatility. Yet if the Mercury be diſtilled off with a

3   much

much greater degree of heat than is neceſſary to elevate it, a pretty conſiderable quantity of Gold will moſt certainly be carried up along with it.

It is alſo of conſequence, on another account, that the fire be duly governed on this occaſion. For if too great a degree of heat be applied, and the fire afterwards lowered, the water in the receiver, which covers the noſe of the retort, will riſe into its body, break it to pieces, and ſpoil the operation.

The cauſe of this phenomenon depends on the property which air poſſeſſes of rarefying with heat and condenſing with cold, joined to its weight. As ſoon as the retort is acted on by a leſs degree of heat than acted on it the inſtant before, the air contained therein is condenſed, and leaves a *vacuum*, which the external air, by virtue of its weight, tends to occupy ; but, the orifice of the retort being under water, the external air can no way gain admittance, but by puſhing in before it the water which intercepts its paſſage. This caution, as we obſerved above, muſt be applied to all diſtillations, where the veſſels are diſpoſed as they are in this.

Care muſt alſo be taken that the noſe of the retort be not placed too deep under water : for as the neck grows very warm during the operation, becauſe the degree of heat required to raiſe mercury is about three times greater than that which raiſes water, it may eaſily be broken by the contact of the cold water in the receiver.

This method of extracting Gold and Silver from their ores, by Amalgamation with Mercury, is not to be abſolutely depended on as a ſure proof of the quantity of thoſe metals that may be contained in the earth aſſayed by this means : for ſome ſmall part of the Amalgam is always loſt in waſhing it ; and moreover, the Mercury, when ſqueezed through chamoy, always carries with it a ſmall portion of Gold. So that if you deſire to know more exactly

actly, by this method, the quantity of Gold or Silver contained in any earth, the Amalgam muſt not be ſqueezed through chamoy, but diſtilled altogether. Much the ſureſt method of making an accurate aſſay is that by fuſion and ſcorification, which we ſhall deſcribe under the head of Silver.

In ſome countries, and eſpecially in America, the method of Amalgamation is uſed, for extracting Gold and Silver in large quantities, from the matrices which contain them in their metalline form. Agricola and other metallurgiſts have deſcribed the machines by means whereof ſuch Amalgamations are managed.

PROCESS II.

To diſſolve Gold in Aqua regis, *and by that means to ſeparate it from Silver.* Aurum Fulminans. Aurum Fulminans *reduced.*

TAKE Gold that is perfectly pure, or alloyed with Silver only. Reduce it to little thin plates, by hammering it on an anvil. If it be not ſufficiently tough, neal it till it be red in a moderate, clear fire, quite free from ſmoking coals, and then let it cool gradually, which will reſtore its ductility.

When the plates are thin enough, make them red hot once more, and cut them into ſmall bits with a pair of ſheers. Put theſe bits into a tall, narrow-mouthed cucurbit, and pour on them twice their weight of good *Aqua regis,* made of one part Sal Ammoniac, or Spirit of Salt, and four parts Spirit of Nitre. Set the cucurbit in a ſand-bath moderately heated, ſtopping its orifice ſlightly with a paper coffin, to prevent any dirt from falling in. The *Aqua regis* will preſently begin to ſmoke. Round the little bits of Gold will be formed an infinite number of ſmall bubbles, which will riſe to the

surface

surface of the liquor. The Gold will totally dissolve, if it be pure, and the solution will be of a beautiful yellow colour: if the Gold be alloyed with a small quantity of Silver, the latter will remain at the bottom of the vessel in the form of a white powder. If the Gold be alloyed with much Silver, when the Gold is dissolved the Silver will retain the form of the little metalline plates put into the vessel.

When the dissolution is completed, gently pour off the liquor into another low, wide-mouthed, glass cucurbit, taking care that none of the Silver, which lies at the bottom in the form of a powder, escape with the liquor. On this powder of Silver pour as much fresh *Aqua regis* as will cover it entirely; and repeat this till you are sure that nothing more can be taken up by it. Lastly, having decanted the *Aqua regis* from the Silver, wash the Silver with a little Spirit of Salt weakened with water, and add this Spirit of Salt to the *Aqua regis* in which your Gold is dissolved. Then to the body containing these liquors fit a head and a receiver, and distil with a gentle heat, till the matter contained in the cucurbit become dry.

OBSERVATIONS.

It is certain that *Aqua regis* is the true solvent of Gold, and that it does not touch Silver: so that if the Gold dissolved in it were alloyed with Silver, which is often the case, the two metals would by this means be pretty accurately separated from each other. But if you desire to obtain from this solution a Gold absolutely pure, you must free it, before you dissolve it, from every other metallic substance but Silver; because *Aqua regis* acts upon most of the other metals and the semi-metals. We shall shew under the head of Silver, as we promised before, how to purify a mass of Gold and Silver from every
other

other metallic alloy. Thither alfo we refer the common Parting Affay performed by means of *Aqua fortis :* becaufe in that operation the Silver is diffolved, and not the Gold.

If the Gold put to diffolve in *Aqua regis* be pure, the diffolution is eafily and readily effected. But if, on the contrary, it be alloyed with Silver, the *Aqua regis* finds more difficulty in diffolving it. Nay, if the Silver exceed the Gold in quantity, the diffolution will not take place at all, for the reafons adduced in our Theoretical Elements; of which we fhall fpeak more fully when we come to treat of the Parting Affay.

In the procefs we directed the Gold to be diffolved in a tall body. This precaution is neceffary to prevent the lofs of fome part thereof : for it is the property of *Aqua regis* to carry off along with it fome of the Gold, efpecially when there is any Sal Ammoniac in its compofition, if the veffel be heated while the diffolution is going on, or if the *Aqua regis* be very ftrong. Yet it is proper to make ufe of *Aqua regis* that is too ftrong rather than too weak : for if it prove too ftrong, and be obferved not to act upon the metal for that reafon, it is eafy to weaken it, by gradually adding fmall quantities of pure water, till you perceive it begin to act with vigour. This is a general rule regarding all metallic diffolutions in Acids.

When the folution of Gold is evaporated to drynefs, if you defire to reduce into a mafs the Gold duft left at the bottom of the cucurbit, you muft put it into a crucible, and cover it with pulverized borax, mixed with a little nitre and calcined winelees; then cover the crucible clofe, heat it with a moderate fire, which muft be afterwards increafed fo as to melt the contents. At the bottom of the crucible you will find a lump of Gold, over which the falts you added will be as it were vitrified. Thefe

falts

falts are added chiefly to promote the fufion of the metal.

The Gold may, if you will, be feparated from its folvent without evaporating the folution as above directed. You need only mix with the folution a fixed or volatile Alkali by little and little, till you fee no more precipitate fall, and then let the liquor ftand to fettle, at the bottom of which you will find a fediment: filter the whole, and dry what is left on the filter.

Both fixed and volatile Alkalis poffeffing, as hath been frequently repeated, a greater affinity with Acids than metallic fubftances have, they precipitate the Gold, and feparate it from the Acids in which it is diffolved: but it is of great confequence to take notice that, if you attempt to melt this precipitated Gold in a crucible, it will fulminate as foon as it feels the heat, with fuch a terrible explofion, that, if the quantity be at all confiderable, it may prove fatal to the operator: even rubbing it a little hard will make it blow up. This preparation is therefore called *Aurum fulminans,*

Hitherto no fatisfactory explanation hath been given of this phenomenon. Some Chymifts confidering that, in the precipitation of the Gold, a Nitre is regenerated by the union of the Alkali with the Nitrous Acid which enters into the compofition of the *Aquaregis,* imagine that fome of this regenerated Nitre, combining with the precipitated Gold, takes fire and detonates, either by means of fome fmall portion of Phlogifton that may be contained in the Alkali, or by means of that which conftitutes the Gold itfelf. But, in the firft place, 'tis well known that Fixed Alkalis do not contain Phlogifton enough to make Nitre detonate. Indeed, if a Volatile Alkali be employed in the precipitation, a Nitrous Ammoniacal Salt will be formed, containing Phlogifton enough

to

tó be capable of detonating without the concourfe of any additional Phlogifton : but this detonation of the Nitrous Ammoniacal Salt is not to be compared, as to the violence of its effects, with the fulmination of Gold. Befides, we do not find that Gold precipitated by a Volatile Alkali explodes with greater force than that precipitated by a Fixed Alkali. As for the Gold, 'tis certain that it fuffers no decompofition at all by fulminating. When fulminated under a glafs bell, in fuch fmall quantities as not to endanger the operator, the Gold is found fcattered about under the bell in very fine particles, without having undergone any alteration.

Others have fancied this fulmination of the Gold to be nothing but the decrepitation of the Sea-falt that is regenerated, in the precipitation of the metal, by the Fixed Alkali uniting with the Acid of Seafalt which makes part of the *Aqua regis.* But to this it may be faid, that Gold precipitated by a Volatile Alkali fulminates as violently as that precipitated by a Fixed Alkali; and yet no Sea-falt can be formed in the liquor by the addition of a Volatile Alkali, but only a Sal Ammoniac which has not the property of decrepitating. Moreover, there is no comparifon, as to the effects, between the decrepitation of Sea-falt and the fulmination of Gold.

Nor, laftly, can this fulmination be attributed, as it is by fome, to the effort made by the Salts to efcape from amidft the particles of Gold, in whicfi they are fuppofed by them to be imprifoned : for then we might deprive this Gold entirely of its fulminating quality by only boiling it in water, and fo wafhing off all the faline particles, which probably adhere to its furface only. It is plain there is great room for very beautiful difcoveries on this fubject. In Walerius's Mineralogy we find fome obfervations that may throw a little light on the point before us.

" The

" The quantity, fays he, of fulminating Gold
" precipitated exceeds that of the Gold diffolved :
" if the *Aqua regis* be made with Sal Ammoniac the
" explofion will be ftronger ; it will alfo be more
" violent if the folution be precipitated with a Vo-
" latile Alkali, than if a Fixed Alkali be ufed for
" that purpofe."

One of the fpeedieft and eafieft methods to deprive
this Gold of its fulminating quality, is to grind in a
mortar twice as much flowers of Sulphur as you have
Gold to reduce, mixing your fulminating Gold
therewith by little and little as you grind them
together ; then to put the mixture into a crucible,
and heat it juft enough to melt the Sulphur. Part
of the Sulphur will be diffipated in vapours, and the
reft will burn away. When it is quite confumed,
encreafe the fire fo as to make the crucible red-hot.
When you perceive no more fmell of Sulphur, pour
on the Gold a little Borax, previoufly melted in another crucible with a Fixed Alkali, as calcined
Wine-lees, or Nitre fixed with Tartar ; and then
raife the fire fufficiently to make the whole flow. After the fufion is completed you will find a button
of Gold at the bottom of the crucible under the falts.

Fulminating Gold may alfo be reduced by pouring on it a fufficient quantity of Fixed Alkali reduced to a liquor, or of oil of Vitriol, evaporating
all the moifture, and gradually throwing what remains, mixed up with fome pinguious matter, into
a crucible kept red-hot in a furnace. The reafon
why thefe fubftances deprive the Gold of its fulminating quality depends on the caufes that produce
the fulmination.

Gold may alfo be feparated from *Aqua regis*, and
precipitated by the means of feveral metallic fubftances that have a greater affinity, either with
Aqua regis, or with one of the two Acids that compofe it. Mercury is one of the fitteft for this pur-

pose. On dropping a solution of Mercury in the Nitrous Acid by little and little into a solution of Gold, the mixture becomes turbid, and a precipitate is formed. Continue dropping in more of the solution of Mercury till no more precipitate falls; then let the liquor stand to settle, and at the bottom of it you will find a sediment, which is the precipitated Gold: pour off the liquor by inclination, and wash the precipitate with fair water.

Mercury hath a greater affinity with the Marine than with the Nitrous Acid. The affinity which Mercury hath with the Marine Acid is also greater than that of Gold with the Marine Acid; for unless this Acid be associated either with the Nitrous Acid, or at least with a certain proportion of Phlogiston, it will not dissolve Gold. Hence it comes that when a solution of Mercury in the Nitrous Acid is dropped into a solution of Gold in *Aqua regis*, the Mercury unites with the Acid of Sea-salt, which is an ingredient in the *Aqua regis:* but the Marine Acid cannot on this occasion join the Mercury, without deserting the Gold and the Nitrous Acid with which it was united; and then the Gold, which cannot be kept in solution by the Nitrous Acid alone, is forced to quit its solvent and precipitate. The liquor therefore, that now floats over the Gold thus precipitated, must contain Mercury united with the Acid of Sea-salt: and in fact it yields a true Corrosive Sublimate, which is known to be a combination of Mercury with the Marine Acid.

Mercury dissolved in Spirit of Nitre is employed to procure the precipitation we are speaking of; because metallic substances, when so comminuted by an Acid, are much fitter for such experiments than when they are in a concrete form.

Gold precipitated in this manner by a metallic substance doth not fulminate.

PROCESS III.

To diſſolve Gold by Liver of Sulphur.

MIX together equal parts of common Brim-
ſtone, and a very ſtrong Fixed Alkali; for
inſtance, Nitre fixed by charcoal. Put them in a crucible, and melt the mixture, ſtirring it from time to time
with a ſmall rod. There is no occaſion to make the
fire very briſk; becauſe the Sulphur facilitates the
fuſion of the Fixed Alkali. Some ſulphureous vapours will riſe from the crucible; the two ſubſtances
will mix intimately together, and form a reddiſh
compound. Then throw into the crucible ſome
little pieces of Gold beat into thin plates, ſo that the
whole do not exceed in weight one third part of the
Liver of Sulphur: raiſe the fire a little. As ſoon
as the Liver of Sulphur is perfectly melted, it will
begin to diſſolve the Gold with ebullition; and will
even emit ſome flaſhes of fire. In the ſpace of a few
minutes the Gold will be entirely diſſolved, eſpecially if it was cut and flatted into ſmall thin leaves.

OBSERVATIONS.

The proceſs here delivered is taken from M.
Stahl. The deſign of that ingenious Chymiſt's enquiry was to diſcover how Moſes could burn the
golden calf, which the Iſraelites had ſet up and wor-
ſhipped while he was on the mount; how he could
afterwards reduce that calf to powder, throw it
into the water which the people uſed, and make all
who had apoſtatized drink thereof, as related in
the Book of Exodus.

M. Stahl, having firſt obſerved that Gold is abſolutely inalterable and indeſtructible by the force of
fire alone, be it ever ſo violent, concludes, that
without a miracle Moſes could not poſſibly perform

the

the above-mentioned operations on the golden calf, any way but by mixing with the Gold-fome matter qualified to alter and diffolve it. He then takes notice that pure Sulphur does not act upon Gold at all, and that many other fubftances, which are thought capable of dividing and diffolving it, cannot however do it fo completely as is neceffary to render that metal fufceptible of the effects related. He then gives the method of diffolving it by Liver of Sulphur defcribed in the procefs.

Liver of Sulphur diffolves likewife all the other metals : but M. Stahl obferves that it attenuates Gold more than any other metallic fubftance, and unites with it much more intimately than with the reft. This appears from what happens, on attempting to diffolve in water any of the mixts refulting from the union of another metal with Liver of Sulphur : for then the metal feparates, and appears in the form of a powder or fine calx ; whereas, when Gold is united with Liver of Sulphur, the whole compound diffolves in water fo perfectly, that the Gold even paffes with the Liver of Sulphur through the pores of filtering paper.

If an Acid be poured into a folution of this combination of Gold with Liver of Sulphur, the Acid unites with the Alkali of the *Hepar*, and the Gold falls to the bottom of the liquor along with the Sulphur, which doth not quit it. The Sulphur thus precipitated with the Gold is eafily carried off by a flight torrefaction, after which the Gold remains exceedingly comminuted. The Sulphur of this compound may alfo be deftroyed by torrefaction, without the trouble of a previous folution and precipitation, and then alfo the Gold remains fo attenuated as to be mifcible with liquors, and floats on them, or fwims in them, in fuch a manner that it may eafily be fwallowed with them in drinking. From all this M. Stahl concludes there is great reafon to believe

believe it was by means of the Liver of Sulphur that Mofes divided, and in a manner calcined, the golden calf, fo that he could mingle it with water, and make the Ifraelites drink it.

PROCESS IV.

To feparate Gold from all other metallic Subftances by means of Antimony.

HAVING put the Gold you intend to purify into a crucible, fet it in a melting furnace, cover it, and make the Gold flow. When the metal is in fufion caft upon it, by a little at a time, twice its weight of pure crude Antimony in powder, and after each projection cover the crucible again immediately: this done keep the matter in fufion for a few minutes. When you perceive that the metallic mixture is perfectly melted, and that its furface begins to fparkle, pour it out into a hollow iron cone, previoufly heated, and fmeared on the infide with tallow. Immediately ftrike with a hammer the floor on which the cone ftands; and when all is cold, or at leaft fufficiently fixed, invert the cone and ftrike it: the whole metallic mafs will fall out, and the under part thereof, which was at the point of the cone, will be a Regulus more or lefs yellow as the Gold was more or lefs pure. On ftriking the metallic mafs the Regulus will freely part from the fulphureous cruft at top.

Return this Regulus into the crucible, and melt it. Lefs fire will do now than was required before. Add the fame quantity of Antimony, and proceed as at firft. Repeat the fame operation a third time, if your Gold be very impure.

Then put your regulus into a good crucible much larger than is neceffary to hold it. Set your crucible in a melting furnace, and heat the matter

X 3 but

but juſt enough to make it flow, with a ſmooth, brilliant ſurface. When you find it thus conditioned,
point towards it the noſe of a long-ſnouted pair of
bellows, and therewith keep gently and conſtantly
blowing. There will ariſe from the crucible a conſiderable ſmoke, which will abate greatly when you
ceaſe to blow, and increaſe as ſoon as you begin
again. You muſt raiſe the fire gradually as you
approach towards the end of the operation. If
the ſurface of the metal loſe its brilliant poliſh, and
ſeem covered with a hard cruſt, 'tis a ſign the
fire is too weak; in which caſe it muſt be increaſed,
till the ſurface recover its ſhining appearance. At
laſt, when no more ſmoke riſes, and the ſurface of
the Gold looks neat and greeniſh, caſt on it, by
little and little, ſome pulverized Nitre, or a mixture of Nitre and Borax. The matter will ſwell
up. Continue thus adding more Nitre gradually,
till no commotion is thereby pf oduced in the crucible; and then let the whole cool. If you find,
when the Gold is cold, that it is not tough enough,
melt it over again; when it begins to melt caſt in
the ſame Salts as before; and repeat this till it be
perfectly ductile.

OBSERVATIONS.

ANTIMONY is a compound, conſiſting of a ſemi-
metallic part united with about a fourth part of
its weight of common Sulphur. It appears, in the
ninth column of the Table of Affinities, that all
the metals, Mercury and Gold excepted, have a
greater affinity than the reguline part of Antimony
with Sulphur. If therefore Gold, adulterated with
a mixture of Copper, Silver, or any other metal,
be melted with Antimony, thoſe metals will unite
with the Sulphur of the Antimony, and ſeparate it
from the reguline part, which being thus ſet free
will combine and be blended with the Gold. Theſe

two

two metallic fubftances, forming a mafs far heavier than the other metals mixed with the fulphur, fall together to the bottom of the crucible in the form of a Regulus, while the others float over them like a fort of fcoria or flag : and thus the Gold is freed from all alloy but the reguline part of the Antimony.

As all the other metals have a great affinity with Sulphur, and Gold is the only one that is capable of refifting its action, one would think Sulphur alone might be fufficient to free it from the metals combined with it, and that it would therefore be better to employ pure Sulphur, in this operation, than to make ufe of Antimony ; the reguline part of which remaining united with the Gold requires another long and laborious operation to get rid of it.

Indeed, ftrictly fpeaking, Sulphur alone would be fufficient to produce the defired feparation : but it is proper to obferve that, as Sulphur alone is very combuftible, moft of it would be confumed in the operation before it could have an opportunity to unite with the metallic fubftances; whereas when it is combined with the Regulus of Antimony, it is thereby enabled to bear the action of the fire much longer without burning, and confequently is much fitter for the purpofe in queftion. Befides, if we were to make ufe of pure Sulphur, a great part of the Gold, which is kept in perfect fufion, and its precipitation facilitated, by the Regulus of Antimony, would remain confounded with the fulphureous fcoria.

Neverthelefs, feeing the metals with which Gold is alloyed cannot be feparated from it by Antimony, but that a quantity of Regulus proportioned to the quantity of the metals fo feparated will unite with the Gold, and that the more Regulus combines with the Gold, the more tedious, chargeable, and laborious will the operation prove, this confideration

X 4 ought

ELEMENTS *of the*

cess. Therefore, if the Gold be very impure, and
worse than sixteen carats, we must not mix it with
crude Antimony alone, but add two drams of pure
Sulphur for every carat the Gold wants of sixteen,
and lessen the quantity of Antimony in proportion
to that of the real Gold.

It is necessary to keep the crucible close covered,
after mixing the Antimony with the Gold, to prevent any coals from falling into it: for, if that
should happen, the melted mass would puff up
considerably, and might perhaps run over.

The inside of the cone, into which you pour the
melted metallic mass, must be greased with tallow,
to prevent its sticking thereto, and that it may
come easily out. Striking the floor, on which the
cone with the melted metal stands, helps the precipitation and descent of the Regulus of Gold and
Antimony to the bottom of the cone.

Less fire is requisite to melt this compound Regulus, in order to add fresh Antimony, than was
necessary before the Gold was mixed with the reguline part of the Antimony; because this metallic
substance, being much more fusible than Gold, promotes its melting. The Antimony is mixed with
the Gold by repeated projections, that the separation
of the metals may be accomplished with the greater
ease and accuracy. Yet the operation might be
successfully performed, by putting in all the Antimony at once, and with one melting only.

The metalline mass found at the bottom of the
cone after all these operations, is a mixture of Gold
with the reguline part of the Antimony. All the rest
of the process consists only in separating this reguline
part from the Gold. As Gold is the most fixed of
all metals, and as the Regulus of Antimony cannot
bear the violence of fire without flying off in vapours, nothing more is necessary for this purpose,
but

but to expose the compound, as directed in the process, to a heat strong enough, and long enough
continued, to dissipate all the Regulus of Antimony. This semi-metal exhales in the form of a very
thick white smoke. It is proper to blow gently
into the crucible during the whole operation ; because the immediate contact of the fresh air incessantly thrown in promotes and considerably en-
creases the evaporation : and this is a general rule
applicable to all evaporations.

The fire must be gradually raised as the Regulus
of Antimony is dissipated, and the operation draws
toward an end ; because the mixed mass of Regulus of Antimony and Gold becomes so much the
less fusible as the proportion of the Regulus is les-
sened. Though the Regulus of Antimony be separated from the Gold in this operation, because
the latter is of such a fixed nature that it cannot be
volatilized by the degree of fire which dissipates
the Regulus ; yet, as the Regulus is very volatile,
it will undoubtedly carry up some of the Gold along
with it, especially if you hurry on the evaporation
too fast, by applying too great a degree of fire, by
blowing too briskly into the crucible, and still
more if you evaporate your mixture in a broad flat
vessel instead of a crucible. All these things must
therefore be avoided, if you would lose no more
Gold than you needs must.

However, unless the evaporation be carried to
the utmost, by the means above pointed out, a small
portion of the Regulus of Antimony will always
remain combined with the Gold, which defends it
from the action of the fire. This small portion of
Regulus hinders the Gold from being perfectly
pure and ductile. In order therefore to consume
and scorify it, we cast Nitre into the crucible when
we perceive it to emit no more white vapours.

We

We know that Nitre has the property of reducing all metallic fubftances to a calx, Gold and Silver excepted ; becaufe it deflagrates with the phlogifton to which their metalline form is owing : but
as this accenfion of the Nitre occafions a tumid effervefcence, care muft be taken to throw it in but by
little and little at a time; for if too much be pro-
jected at once the melted matter will run over.

This operation might be confiderably abridged
by taking advantage of the property which Nitre
poffeffes of thus confuming the phlogifton of metallic fubftances; as by means thereof we might
deftroy all the Regulus of Antimony incorporated
with the Gold, without having recourfe to a long
and tedious evaporation. But then we fhould at
the fame time lofe a much greater quantity of
Gold, by reafon of the tumult and ebullition which
are infeparable from the detonation of Nitre. On
the whole therefore, if Nitre be made ufe of to purify Gold, great care muft be taken to apply but
very little of it at a time.

All the Silver that was mixed with the Gold, and
indeed a little of the Gold itfelf, remains confounded with the fulphureous fcoria, which floats
upon the Golden Regulus after the addition of the
Antimony : we fhall fhew in the Chapter on Silver how thefe two metals are to be feparated from
the Sulphur.

C H A P.

CHAP. II.

Of SILVER.

PROCESS I.

To separate Silver from its Ore, by means of Scorification with Lead.

BEAT to powder in an iron mortar the ore from which you mean to separate the Silver, having first roasted it well in order to free it from all the Sulphur and Arsenic that it may contain. Weigh it exactly: then weigh out by itself eight times as much granulated Lead. Put one half of this Lead into a test, and spread it equally thereon: upon this Lead lay your ore, and cover it quite over with the remaining half of the Lead.

Place the test thus loaded under the further end of the Muffle in a cupelling furnace. Light your fire, and increase it by degrees. If you look through one of the apertures in the door of the furnace you will perceive the ore, covered with calcined Lead, swim upon the melted Lead. Presently afterwards it will grow soft, melt, and be thrown towards the sides of the vessel, the surface of the Lead appearing in the midst thereof bright and shining like a luminous disc: the Lead will then begin to boil, and emit fumes. As soon as this happens, the fire must be a little checked, so that the ebullition of the Lead may almost entirely cease, for about a quarter of an hour. After this it must be excited to the degree it was at before, so that the Lead may begin again to boil and smoke. Its shining surface will gradually lessen, and be covered with *scoriæ*. Stir the whole with an iron hook, and draw it towards
the

the middle what you obferve towards the fides of the veffel; to the end that, if any part of the ore fhould ftill remain undiffolved by the Lead, it may be mixed therewith.

When you perceive that the matter is in perfect fufion, that the greateft part of what fticks to the iron hook, when you dip it in the melted matter, feparates from it again, and drops back into the veffel; and that the extremity of this inftrument when grown cold, appears varnifhed over with a thin, fmooth, fhining cruft; you may look on thefe as marks that the bufinefs is done, and the more uniform and evenly the colour of the cruft is, the more perfect may you judge the fcorification to be.

Matters being brought to this pafs, take the teft with a pair of tongs from under the muffle, and pour its whole contents into an iron cone, firft heated and greafed with tallow. This whole operation lafts about three quarters of an hour. When all is cold, the blow of a hammer will part the Regulus from the fcoria; and as it is not poffible, how perfect foever the fcorification be, to avoid leaving a little Lead containing Silver in the fcoria, it is proper to pulverife this fcoria, and feparate therefrom whatever extends under the hammer, in order to add it to the Regulus.

OBSERVATIONS.

Silver, as well as Gold, is often found quite pure, and under its metalline form, in the bowels of the earth; and in that cafe it may be feparated from the ftones or fand, in which it is lodged, by fimple wafhing, or by Amalgamation with Mercury, in the fame manner as before directed for Gold. But it alfo happens frequently that Silver is combined in the ore with other metallic fubftances and minerals, which will not admit of this procefs, but force us to employ other methods of feparating it from them.

Sulphur

Sulphur and Arfenic are the fubftances to which Silver and the other metals ufually owe their mineral ftate. Thefe two matters are never very clofely united with Silver; but may be pretty eafily fepa-rated from it by the action of fire, and the addition of Lead. If Arfenic be predominant in a Silver ore, it will unite with the Lead by the help of a pretty moderate heat, and quickly convert a confi-derable quantity thereof into a penetrating fufible glafs, which has the property of fcorifying with eafe all fubftances that are capable of fcorification.

When Sulphur predominates, the fcorification proceeds more flowly, and doth not always fucceed; becaufe that mineral combined with Lead leffens its fufibility, and retards its vitrification. In this cafe part of the Sulphur muft be diffipated by roafting: the other part unites with the Lead; and that, be-ing rendered lighter by this union, floats on the reft of the mixture, which chiefly contains the Silver. At laft the joint action of the air and of the fire, diffipates the portion of Sulphur that had united with the Lead: the Lead vitrifies and reduces to a fcoria whatever is not either Silver or Gold: and thus the Silver being difentangled from the hetero-geneous matters with which it was united, one part thereof being diffipated and the other vitrified, combines with the portion of Lead which is not vi-trified, and falls through the fcoria, which to favour its defcent muft be in perfect fufion.

The whole procefs therefore confifts of three dif-tinct operations. The firft is Roafting, which diffi-pates fome of the volatile fubftances found united with the Silver: the fecond is Scorification, or the Vitri-fication of the fixed matters alfo united with the Sil-ver, fuch as fand, ftones, metals, &c. and the third is Precipitation, or the feparation of the Silver from the fcoria. The two firft are, as hath been fhewn, preparatives for the laft, and indeed produce it.

5   As

As every thing we faid concerning Gold, when we treated of the procefs of Amalgamation, is to be applied to Silver, which may be extracted by the fame method, when it is in its metalline form; in the fame manner all we now advance touching the method of extracting Silver by Scorification, when it is depraved with a mixture of heterogeneous matters, is equally applicable to Gold in the fame circumftances: and indeed Silver almoft always contains more or lefs Gold naturally.

In the procefs we directed that the ore fhould be pulverized before it be expofed to the fire, with a view to enlarge its furface, and by that means facilitate the action of the Lead upon it, as well as the evaporation of its volatile parts.

We recommended the precaution of flackening the fire a little at the beginning of the operation, only to prevent the Lead from being too haftily converted into litharge, left it fhould penetrate and corrode the teft before it had wholly diffolved the ore: but if we were perfectly certain of the veffel's being fo good as to be in no danger of penetration by the Lead, this precaution would be needlefs.

It is proper to add eight parts of Lead for one of ore; though fo much is not always abfolutely neceffary, efpecially when the ore is very fufible. The fuccefs of this operation depends chiefly on the completenefs of the Scorification; and therefore the addition of more Lead than enough is attended with no inconvenience: for, as it always promotes the fcorification, it can never do any harm.

If the ore be mixed with fuch earthy and ftony parts as cannot be feparated from it by wafhing, it is the more difficult of fufion, even though the ftones fhould be fuch as are moft difpofed to vitrify; becaufe the moft fufible earths and ftones are always lefs fo than moft metallic fubftances. In that cafe it will be neceffary, for effecting the Scorification, to mix thoroughly with the pulverized ore an equal Quantity of Glafs of Lead, to add twelve times as much granulated Lead, and then to proceed as directed for a fufible ore; expofing the mixture to a degree of fire ftrong enough, and long enough kept up, to give the fcoria all the properties above required as figns of a perfect fcorification.

Silver ore is fometimes mixed with pyrites, and the ore of Arfenic, or cobalt, which alfo make it refractory. As the pyrites contain a large quantity of Sulphur, which is very volatile as well as Arfenic; in this cafe it is proper to begin with freeing the ore from thefe two extraneous fubftances. This is eafily done by roafting: only be fure, when you firft expofe the ore to the heat, to cover the veffel in which you roaft it, for fome minutes, with an inverted veffel of the fame width; becaufe fuch forts of ore are very apt to fly when they firft feel the heat.

After this uncover it, and leave it expofed to the fire till no more fulphureous or arfenical matters rife. Then mix it with the fame quantity of Glafs of Lead as we ordered for ores rendered refractory by the admixture of earths or ftones, and proceed in the fame manner.

It is the more neceffary to roaft Silver ore infected with Sulphur and Arfenic, becaufe, as Sulphur obftructs the fufion of Lead, it cannot but do hurt, and protract the operation; and Arfenic does mifchief, on the other hand, by fcorifying a very great quantity of Lead too haftily.

When the Sulphur and Arfenic are diffipated by roafting, the ore muft be treated like that which is rendered refractory by ftony and earthy matters; for as the pyrites contain much iron, there remains, after the Sulphur is evaporated, a confiderable quantity of martial earth, which is difficult to fcorify. The pyrites as well as the cobalts, contain moreover an unmetallic earth, which is hard to fufe.

The

The general rule therefore is, when the ore is
rendered refractory by any caufe whatever, to mix
it with Glafs of Lead, and to add a larger quantity
of granulated Lead. Yet fome ores are fo refractory
that Lead alone will not do the bufinefs, and recourfe muft be had to fome other flux. That which
is fitteft for the prefent purpofe is the *Black Flux*,
compofed of one part of Nitre and two parts of Tartar deflagrated together. The Phlogifton contained
in this quantity of Tartar is more than fufficient to
alkalizate the Nitre. This Flux therefore is nothing
more than Nitre alkalizated by Tartar, mixed with
fome of the fame Tartar that hath not loft its Phlogifton, and is only reduced to a fort of coal.

The *White Flux* is alfo very fit to promote fufion;
but on this occafion the Black Flux is preferable,
becaufe the Phlogifton of the Black Flux prevents
the Lead from being too foon converted to litharge,
and fo gives it time to diffolve the metallic matters.
The White Flux, which is the refult of equal parts
of Tartar and Nitre alkalizated together, being no
more than an Alkali deftitute of Phlogifton, or containing but very little, doth not poffefs this advantage.

If Silver fhould be combined in the ore with
Iron in its metalline ftate, which however does not
commonly happen, then, in order to feparate them,
the Iron muft be deprived of its Phlogifton, and converted to a *crocus* before the mixed mafs be melted
with Lead; which may be done by diffolving it in
the Vitriolic Acid, and then evaporating the Acid.

We are neceffitated to make ufe of this contrivance, becaufe Iron in its metalline form cannot be
diffolved either by Lead or by the Glafs of Lead;
but when it is reduced to a calx, litharge unites with
it, and fcorifies it.

If you have not at hand the utenfils neceffary for
performing the operation we have been defcribing in
a teft,

a teſt, and under the muffle; or if you have a mind to work on a greater quantity of ore at a time, you may make uſe of a crucible for the purpoſe, and perform the operation in a melting furnace.

In this caſe the ore muſt be prepared, as above directed, according to its nature, and mixed with a proper quantity of Lead and Glaſs of Lead; the whole put into a good crucible, leaving two thirds thereof empty, and covered with a mixture of Sea-ſalt and a little Borax, both very dry, to the thick-neſs of a full half inch.

This being done, ſet the crucible in the midſt of a melting furnace, raiſe the coals quite to the lip of the crucible; light the fire; cover the furnace with its dome; but do not urge the fire more than is neceſſary to bring the mixture to perfect fuſion: leave it thus in fuſion for a good quarter of an hour; ſtir the whole with a bit of ſtrong iron wire; then let it cool; break the crucible, and ſeparate the Regulus from the ſcoria.

The Salts added on this occaſion are fluxes, and their uſe is to procure a perfect fuſion of the ſcoria.

If the melted matters be left expoſed to the fire, either in a teſt, or in a crucible, longer than is above preſcribed, the portion of Lead, that hath united and precipitated with the Silver, will at laſt vitrify, and at the ſame time ſcorify all the alloy with which that metal may be mixed. But as there are no veſſels that can long endure the action of litharge, without being pierced like a ſieve, ſome of the ſilver may eſcape through the holes or fiſ-ſures of the veſſel, and ſo be loſt. It is better therefore to complete the purification of your Sil-ver by the operation of the Cupel, the deſcription of which follows.

PROCESS II.

The refining of Silver by the Cupel.

TAKE a cupel capable of containing one third more matter than you have to put into it : set it under the muffle of a furnace, like that defcribed in our Theoretical Elements, as peculiarly appropriated to this fort of operation. Fill the furnace with charcoal; light it; make the cupel redhot, and keep it fo till all its moifture be evaporated; that is, for about a good quarter of an hour, if the cupel be made wholly of the afhes of burnt bones; and for a whole hour, if there be any wafhed wood-afh in its compofition.

Reduce the Regulus which remained after the preceding operation to little thin plates, flatting them with a fmall hammer, and feparating them carefully from all the adherent fcoria. Wrap thefe in a bit of paper, and with a fmall pair of tongs put them gently into the cupel. When the paper is confumed, the Regulus will foon melt, and the fcoria, which will be gradually produced by the Lead as it turns to litharge, will be driven to the fides of the cupel, and immediately abforbed thereby. At the fame time the cupel will affume a yellow, brown, or blackifh colour, according to the quantity and nature of the fcoria imbibed by it.

When you fee the matter in the cupel in a violent ebullition, and emitting much fmoke, lower the fire by the methods formerly prefcribed. Keep up fuch a degree of heat only that the fmoke which afcends from the matter may not rife very high, and that you may be able to diftinguifh the colour which the cupel acquires from the fcoria.

Increafe the fire by degrees, as more and more litharge is formed and abforbed. If the Regu-

lus

lus examined by this affay contain no Silver, you will fee it turn wholly into fcoria, and at laft difappear. When it contains Silver, and the quantity of Lead is much diminifhed, you will perceive little vivid irifes, or beautiful rain-bow colours, fhooting fwiftly along its furface, and croffing each other in many different directions. At laft, when all the Lead is deftroyed, the thin dark fkin, that is continually protruded by the Lead while it is turning into litharge, and which hitherto covered the Silver, fuddenly difappears; and, if at this moment the fire happen not to be ftrong enough to keep the Silver in fufion, the furface of that metal will at once dart out a dazzling fplendour: but, if the fire be ftrong enough to keep the Silver in fufion, though freed from all mixture of Lead, this change of colour, which is called its *fulguration*, will not be fo perceptible, and the Silver will appear like a bead of fire.

These phenomena fhew that the operation is finifhed. But the cupel muft ftill be left a minute or two under the muffle, and then drawn flowly out with the iron hook towards the door of the furnace. When the Silver is fo cooled as to be but moderately red, you may take the cupel from under the muffle with your little tongs, and in the middle of its cavity you will find an exceeding white bead of Silver, the lower part whereof will be unequal, and full of little pits.

OBSERVATIONS.

THE Regulus obtained by the former procefs confifts altogether of the Silver contained in the ore, alloyed with the other metals that happened to be mixed therewith in its mineral ftate, and a good deal of the Lead that was added to precipitate the Silver. The operation of the cupel may be confidered as the fequel of that procefs, being intended only to reduce into a fcoria whatever is not Gold or

Silver.

Silver. Lead being of all metals that which vitrifies the moſt eaſily, which moſt promotes the vitrification of the reſt, and the only one which, when
vitrified, penetrates the cupel, and carries along
with it the other metals which it hath vitrified, is
conſequently the fitteſt for that purpoſe. We ſhall
ſee in its place that Biſmuth hath the ſame properties with Lead, and may be ſubſtituted for it in
this operation.

Care muſt be taken to chooſe a cupel of a proper
capacity. Indeed it ſhould rather be too big than
too little: becauſe the operation is no way prejudiced
by an exceſs in its ſize; whereas, if it be too ſmall,
it will be over-doſed with Lead, and at laſt the litharge, which deſtroys every thing, will corrode its
cavity, and eat holes through the very body of the
veſſel. Add that the aſhes, of which the cupel is
made, being once glutted with litharge, abſorb it
afterwards but ſlowly, and that the quantity of this
vitrified litharge, becoming too great to be contained
in the ſubſtance of the veſſel, exſudes through it,
and drops on the floor of the muffle, which it corrodes and renders unequal; and moreover ſolders
to it the veſſels ſet thereon. It may be laid down
as a general rule for determining the ſize of a cupel, that it weigh, at leaſt, half as much as the
metallic maſs to be refined in it.

It is alſo of the utmoſt conſequence that the cupel
be well dried before the metal be put into it. In
order to make ſure of this point it muſt be kept
red-hot for a certain time, as is above directed : for
tho' to the ſight and to the touch it may appear very
dry, it nevertheleſs obſtinately retains a ſmall matter of moiſture, ſufficient to occaſion the loſs of ſome
of the metal; which, when it comes to melt, will be
thereby ſpirited up, in the form of little globules, to
the very roof of the muffle. The cupels, that ſtand
moſt in need of an intenſe heat to dry them, are thoſe
chiefly

chiefly in whose composition wood-ashes are employed : for whatever care be taken to lixiviate those ashes before they are used, they will still retain a little alkaline salt ; and that, we know, is very greedy of moisture, will not part entirely with it, but by the means of a violent calcination, and presently re-imbibes it when exposed to the air.

A little Phlogiston also may still be left in the ashes of which the cupels are made ; and that is another reason for calcining them before they are used. By this means the remaining Phlogiston is dissipated; which might otherwise combine with the litharge during the operation, reduce it, and occasion such a ferment in the matter as to make some of it run over : to these inconveniencies, which any remainder of moisture or Phlogiston may produce, we must add the cracks and flaws, which are very incident to cupels not perfectly freed from both those matters.

It is of no less importance to the success of this operation, that a due degree of heat be kept up. In the process we have described the marks which shew the heat to be neither too strong nor too weak ; when it exceeds in either of these respects may be known by the following signs.

If the fume emitted by the Lead rise like a spout to the roof of the muffle ; if the surface of the melted metal be extremely convex, considering the quantity of the mass: if the cupel appear of such a white heat, that the colour communicated thereto by the imbibed scoria cannot be distinguished : all these shew that the heat is too great, and that it ought to be diminished. If, on the contrary, the vapours only hover, as it were, over the surface of the metal; if the melted mass be very flat, considering its quantity; if its ebullition appear but faint ; if the *scoriæ*, that appear like little fiery drops of rain, have but a languid motion ; if the scoria gather in heaps, and do not penetrate the cupel ; if the metal be covered with

it

it as with a glaffy coat; and laftly, if the cupel look dull; thefe are proofs that the heat is too weak, and ought to be increafed.

The defign of this operation being to convert the Lead into litharge, and to give it fufficient time and opportunity to fcorify and carry off with it whatever is not Gold or Silver; the fire muft be kept up to fuch a degree that the Lead may eafily be turned into litharge; and yet that litharge not be abforbed too haftily by the cupel, but that a fmall quantity thereof may all along remain, like a ring, round the melted metal.

The fire is to be gradually increafed as the operation draws nearer to its end : for, as the proportion of the Lead to the Silver is continually leffening, the metallic mafs gradually becomes lefs fufible; while the Silver defends the Lead mixed with it from the action of the fire, and prevents its being eafily converted into litharge.

When the operation is finifhed, the cupel muft ftill be left under the muffle, till it has imbibed all the litharge, to the end that the bead of Silver may be eafily taken out: for, without this precaution, it would ftick fo faft as not to be removed, but by breaking off part of the cupel along with it. Care muft alfo be taken to let this bead of Silver cool gradually, and be perfectly fixed, before you draw it from under the muffle : for if you expofe it at once to the cold air, before it be fixed, it will fwell, fhoot into fprigs, and even dart out feveral little grains to a confiderable diftance, which will be loft.

If the bead appear to have a yellowifh tinge, 'tis a fign that it contains a great deal of Gold, which muft be feparated from it by the methods to be hereafter fhewn.

It is proper to obferve that there is fcarce any Lead that does not contain fome Silver; too little perhaps to defray the charges neceffary to feparate it,

yet

yet confiderable enough to lead us into an error, by mixing with the Silver obtained from an ore, and increafing its weight. And therefore, when the operations above defcribed are applied to the affaying of an ore, in order to know how much Silver it yields, it is previoufly neceffary to examine the Lead to be ufed, and to afcertain the quantity of Silver it contains, which muft be deducted from the total weight of the bead of Silver obtained by purifying it in this manner.

Silver may be feparated from its ore, and at the fame time refined, by the fingle operation of the cupel, without any previous fcorification with Lead. In order to do this, you muft pound the ore; roaft it, to diffipate all its volatile parts; mix it with an equal quantity of litharge, if it be refractory; divide it into five or fix parcels, wrapping each in a bit of paper; weigh out eight parts of granulated Lead for one of ore, if it be fufible, and from twelve to fixteen, if it be refractory; put one half of the Lead into a very large cupel under the muffle, add thereto one of the little parcels of ore, when the Lead begins to fmoke and boil; immediately flacken the fire a little; continue the fame degree of heat till you perceive that the litharge formed round the metal, and on its furface, begins to look bright; then raife the fire; add a frefh parcel of ore; continue proceeding in the fame manner till you have put in all the ore; then add the remaining half of the granulated Lead, and conduct the fucceeding part of the operation in the fame manner as that of cupelling.

In this operation it is neceffary that the fire be not too ftrongly urged, and that it be diminifhed every time you add a frefh parcel of ore; that fo the Lead and the litharge may have time to diffolve, fcorify, and carry off into the pores of the cupel, all the adventitious matters with which your Silver may be mixed. Notwithftanding this precaution, when the

Y 4 ore

ore is refractory, there often gathers in the cupel a great quantity of fcoria, together alfo with fome of the ore that could not be diffolved and fcorified. It is with a view to remedy this inconvenience that the fecond moiety of the Lead is added towards the end, which completes the diffolution and fco-rification of the whole; fo that by means thereof no fcoria, or very little, is left in the cupel at the end of the operation.

The operation of the cupel is chiefly ufed to pu-rify Silver from the alloy of Copper; becaufe this metal, being more fixed and harder to calcine than other metallic fubftances, is the only one that re-mains united with Silver and Lead, after roafting and fcorification with Lead. It requires no lefs than fix-teen parts of Lead to deftroy it in the cupel, and fe-parate it from Silver. It melts into one mafs with the Lead; and the glafs produced by thefe two metals, deprived of their phlogifton, inclines to a brown or a black colour; by which appearance chiefly we know that our Silver was alloyed with Copper.

PROCESS III.

To purify Silver by Nitre.

GRANULATE the Silver you intend to pu-rify, or reduce it to thin plates; put it into a good crucible; add thereto a fourth part in weight of very dry pulverized Nitre, mixed with half the weight of the Nitre of calcined Wine-lees, and about a fixth part of the fame weight of common glafs in powder. Cover this crucible with another crucible inverted; which muft be of fuch a fize that its mouth may enter a little way into that of the lower one, and have its bottom pierced with a hole of about two lines in diameter. Lute the two cru-cibles together with clay and Windfor-loam. When

the

the lute is dry, place the crucibles in a melting furnace. Fill the furnace with Charcoal, taking care however that the fuel do not rife above the upper crucible.

Kindle the fire, and make your veffels of a middling-red heat. When they are fo, take up with the tongs a live coal, and hold it over the hole of the upper crucible. If you immediately perceive a vivid fplendour round the coal, and at the fame time hear a gentle hiffing noife, it is a fign that the fire is of a proper ftrength; and it muft be kept up at the fame degree till this phenomenon ceafe.

Then increafe the fire to the degree requifite to keep pure Silver in fufion; and immediately after take your veffels out of the furnace. You will find the Silver at the bottom of the lower crucible, covered with a mafs of Alkaline fcoria of a greenifh colour. If the metal be not rendered perfectly pure and ductile by this operation, it muft be repeated a fecond time.

O·B S E R V A T I O N S.

The purification of Silver by Nitre, as well as the procefs for refining it on the cupel, is founded on the property which this metal poffeffes of refifting the force of the ftrongeft fire, and the power of the moft active folvents, without lofing its phlogifton. The difference between thefe two operations confifts wholly in the fubftances made ufe of to procure the fcorification of the imperfect metals, or femi-metals, that may be combined with the Silver. In the former this was obtained by Lead, and here it is effected by Nitre. This Salt, as we have fhewn, hath the property of calcining and quickly deftroying all metallic fubftances, by confuming their phlogifton, except the perfect metals, Gold and Silver, which alone are able to refift its force. This method may therefore be employed to purify Gold as well as Silver, or indeed both the two mixed together.

In

In this operation the Nitre is gradually alkalizated, as its Acid is confumed with the phlogifton of
the metallic fubftances. The Alkaline Salt and
pounded glafs are added, with a view to promote
the fufion of the metalline calxes, as faft as they
are formed, and to fix and retain the Nitre, which,
as we fhall prefently fee, is apt to fly off in a certain degree of heat.

The precaution of covering the crucible with another crucible inverted, which hath only a fmall hole
in its bottom, is defigned to prevent any of the Silver
from being loft in the operation : for when the Nitre comes to be acted on by a certain degree of heat,
and efpecially when it deflagrates with any inflammable matter, part of it flies off, and fo rapidly
too as to be capable of carrying off with it a good
deal of the Silver. The little hole left in the covering crucible is neceffary for giving vent to the
vapours, which rife during the deflagration of the
Nitre, as they would otherwife open themfelves a
paffage by burfting the veffels. After the operation this vent-hole is found befet with many little
particles of Silver, which would have been loft if
the crucible had not been covered.

If you fhould obferve, during the detonation of
the Nitre, that a great many vapours iffue through
the vent-hole with a confiderable hiffing noife, even
without applying the coal, you muft take it for a
fign that the fire is too brifk, and accordingly check
it; elfe a great deal of the Nitre will be diffipated,
and with it much Silver.

You muft obferve to take the Silver out of the
fire as foon as it is in fufion : for if you neglect this,
the Nitre being entirely diffipated or alkalizated,
the calxes of the metals deftroyed by it may poffibly
recover a little phlogifton, communicated either by
the vapours of the charcoal, or by little bits of coal
accidentally falling into the crucible; by which

5   means

means fome portion of thofe metals being reduced will mix again with the Silver, prevent its having the defired degree of purity and ductility, and oblige you to begin the operation afrefh.

PROCESS IV.

To diffolve Silver in Aqua Fortis, *and thereby fepa-rate it from every other metalline Subftance. The Purification of* Aqua Fortis. *Silver precipitated by Copper.*

THE Silver you intend to diffolve being beaten into thin plates, put it into a glafs cucurbit; pour on it twice its weight of good precipitated *Aqua Fortis*; cover the cucurbit with a piece of paper, and fet it on a fand-bath moderately heated. The *Aqua Fortis* will begin to diffolve the Silver as foon as it comes to be a little warm. Red vapours will rife; and from the upper furfaces of the Silver there will feem to iffue ftreams of little bubbles, afcending to the top of the liquor, between which and the Silver they will form, as it were, a number of fine chains : this is a fign that the diffolution proceeds duly, and that the degree of heat is fuch as it ought to be. If the liquor appear to boil and be agitated, a great many red vapours rifing at the fame time, it is a fign that the heat is too great, and fhould be leffened till it be reduced to the proper degree indicated above: having obtained that, keep it equally up till no more bubbles or red vapours appear.

If your Silver be alloyed with Gold, the Gold will be found, when the diffolution is finifhed, at the bottom of the veffel in the form of a powder. The folution muft now be decanted while it is yet warm: on the powder pour half as much frefh *Aqua Fortis* as before, and make it boil; again decant this fecond *Aqua Fortis*, and repeat the fame a third time;

time; then with fair water waſh the remaining powder well : it will·be of a brown·colour inclining to red. In the obſervations we ſhall ſhew how the Silver is to be ſeparated from the *Aqua Fortis*.

OBSERVATIONS.

ALL the proceſſes on Silver already delivered, whether for extracting it from its ores, or for refining it, either by the Cupel or by Nitre, are applicable to Gold alſo. And if Silver be alloyed with Gold before it undergo thoſe ſeveral operations, it will ſtill remain alloyed therewith after them, in the ſame manner, and in the ſame quantity ; becauſe both metals bear them equally. All therefore, that can be expected from thoſe ſeveral aſſays, is the ſeparation of every thing that is neither Silver nor Gold from theſe two metals. But in order to ſeparate theſe two from each other, recourſe muſt be had either to the proceſs laid down under the head of Gold, or to that here deſcribed, which is the moſt commodious, the moſt uſual, and known by the names of *Quartation* and the *Parting Aſſay*.

Aqua Fortis is the true Solvent of Silver, and is utterly incapable of diſſolving the leaſt atom of Gold. If therefore a maſs conſiſting of Gold and Silver be expoſed to the action of *Aqua Fortis*, that Acid will diſſolve the Silver contained in the compound, without touching the gold, and the two metals will be ſeparated from each other. This method of parting them is juſt the reverſe of that deſcribed before under the head of Gold, which is effected by the means of *Aqua Regis*.

To the ſucceſs of this ſeparation, by means of *Aqua Fortis*, ſeveral conditions are eſſentially neceſſary. The firſt is, that the Gold and Silver be in due proportion to each other ; that is, there muſt be at leaſt twice as much Silver as Gold in the metalline maſs, otherwiſe the *Aqua Fortis* will not be able to diſſolve

it,

it, for the reason formerly given. If therefore
the mass contain too little Silver, it muft either be
melted down again, and a proper quantity of Silver
added ; or elfe, if the Gold be in a fufficient proportion to the Silver, they may be parted by means
of *Aqua Regis.*

Secondly, it is neceffary that the *Aqua Fortis* employed in this operation be abfolutely pure, and free
from any taint of the Vitriolic or Marine Acid : for,
if it be adulterated with the Vitriolic Acid, the Silver will precipitate as faft as it diffolves, and fo the
precipitated Silver will again mix with the Gold. If
the *Aqua Fortis* contain any of the Marine Acid,
the Silver will be precipitated in that cafe alfo; and
this inconvenience will be attended with another,
namely, that the menftruum, being partly an *Aqua
Regis*, will diffolve fome of the Gold. You muft
therefore be very fure that your *Aqua Fortis* is pure,
before you fet about the operation. In order to difcover its quality, you muft try it by diffolving, in a
fmall portion thereof, as much Silver as it will take
up : if the *Aqua Fortis* grow opaque and milky as
it diffolves the Silver, 'tis a a fign it contains fome
foreign Acid, from which it muft be purified.

In order to effect this, let the portion of *Aqua
Fortis* ufed for the above trial ftand to fettle : the
white milky part will gradually fall to the bottom
of the veffel. When it is all fallen, gently decant
the clear liquor, and pour a few drops of this decanted folution of Silver into the *Aqua Fortis* which
you want to purify. It will inftantly become milky.
Let the white particles precipitate as before, and
then add a few more drops of your folution of Silver,
If the *Aqua Fortis* ftill become milky, let it precipitate again, and repeat this till you find that a drop
of your Solution of Silver, let fall into this *Aqua
Fortis*, does not make it in the leaft turbid. Then
filter

filter it through brown paper, and you will have an *Aqua Fortis* perfectly fit for the Parting Affay.

The white particles that appear and fettle to the bottom, on diffolving Silver in an *Aqua Fortis* adulterated with a mixture of fome foreign Acid, are no other than that very Silver, which is no fooner diffolved by the Nitrous Acid than it deferts that folvent to unite with the Vitriolic or Marine Acid, wherewith it has a greater affinity, and falls to the bottom with them. And this happens as long as there remains in the *Aqua Fortis* a fingle atom of either of thofe two Acids.

When therefore your *Aqua Fortis* hath diffolved as much Silver as it is capable of taking up, and when all the white particles formed during the diffolution are fettled to the bottom, you may be affured that the portion which remains clear and limpid is a folution of Silver in an exceeding pure *Aqua Fortis*. But if the folution of Silver thus depurated be mixed with an *Aqua Fortis* adulterated with the Vitriolic or Marine Acid, a like precipitation will immediately enfue, for the reafons above given, till the very laft particle of the heterogeneous Acid contained in the *Aqua Fortis* be precipitated. .

Aqua Fortis purified by this method contains no extraneous fubftance whatever, except a fmall portion of Silver ; fo that it is very fit for the parting procefs. But if it be intended for other chymical purpofes, it muft be rectified in a glafs retort with a moderate heat, in order to feparate it from the fmall portion of Silver it contains, which will remain at the bottom of the retort.

The third condition neceffary to the fuccefs of this operation is, that your *Aqua Fortis* be neither too aqueous, nor too highly concentrated. If too weak, it will not act upon the Silver: and the confequence will be the fame if it be too ftrong. Both thefe inconveniencies are eafily remedied : for in the former

cafe

cafe part of the fuperfluous phlegm may be drawn off by diftillation; or a fufficient quantity of much ftronger *Aqua Fortis* may be mixed with that which is too weak: and in the latter cafe, very pure rain water, or a weaker *Aqua Fortis*, may be mixed with that which is too ftrong.

You may fatisfy yourfelf whether or no your *Aqua Fortis* hath the requifite degree of ftrength, by diffolving therein a thin plate confifting of one part Gold and two or three parts Silver; which plate muft be rolled up in form of a paper coffin. If, when all the Silver contained in the plate is diffolved, the Gold remains in the form of the coffin, it is a fign that your Solvent has a due degree of ftrength. If, on the contrary, the Gold be reduced to a powder, it is a proof that your *Aqua Fortis* is too ftrong, and ought to be weakened.

The Gold remaining after the diffolution of the Silver muft be melted in a crucible with Nitre and Borax, as hath already been faid under the procefs for parting Gold and Silver by means of *Aqua Regis*. As to the Silver which remains diffolved in the *Aqua Fortis*, there are feveral ways to recover it.

The moft ufual is to precipitate it by the interpofition of Copper, which hath a greater affinity than Silver with the Nitrous Acid*. For this purpofe the folution is weakened by adding twice or thrice as much very pure rain water. The cucurbit containing the folution is fet on a fand-bath gently heated, and very clean plates of copper put into it. The furfaces of thefe plates are foon covered with little white fcales, which gradually fall to the bottom of the veffel, as they come to be collected in quantities. It is even proper to ftrike the cucurbit gently now and then, in order to fhake the fcales of Silver from the copper plates, and fo make room for a new crop.

* See the Table of Affinities, Column IV.

The

The *Aqua Fortis* parts with the Silver by degrees only, as it diffolves the Copper; and therefore the liquor gradually acquires a bluifh green colour as the precipitation advances. This precipitation of the Silver is to be continued as long as any remains diffolved in the *Aqua Fortis:* you may be fure that your liquor contains no more Silver, if the furface of a frefh plate of Copper laid therein remain clean and free from afh-coloured or greyifh particles; or if one drop of a folution of Sea-falt let fall into it produce no white or milky cloud.

The precipitation being finifhed, the liquor is to be gently poured off from the precipitated Silver, which muft be rinfed in feveral waters, and even made to boil therewith, in order to free it wholly from the diffolved Copper. The Silver thus well wafhed muft be thoroughly dried, mixed with a fourth of its weight of a flux compounded of equal parts of Nitre and calcined Borax, and then melted in a crucible. On this occafion care muft be taken to raife the fire gently and gradually, till the Silver be brought to fufion.

With what accuracy foever the precipitated Silver be wafhed, in order to free it from the folution of Copper, yet the Silver will always be found alloyed with a fmall portion of the Copper : but then this Copper is eafily deftroyed by the Nitre, with which the Silver is afterwards melted; fo that the latter metal remains perfectly pure after the operation.

Though the Silver be not previoufly cupelled, but be alloyed with other metallic fubftances at the time it is thus diffolved, yet the diffolving, precipitating, and fufing it with Nitre would be fufficient to feparate it accurately from them all, and refine it to a degree of purity equal to that obtained by the cupel.

The Copper that remains diffolved in the *Aqua Fortis*, after the precipitation of the Silver, may in like manner be precipitated by Iron, and, as it re-

tains

tains a fmall portion of Silver, ought not to be neglected when thefe operations are performed on confiderable quantities.

In the two next procefses we fhall fhew two other
methods of feparating Silver from *Aqua Fortis.*

PROCESS V.

*To feparate Silver from the Nitrous Acid by Diftilla-
tion. Cryftals of Silver. The Infernal Stone.*

INTO a large, low, glafs body put the folution
of Silver, from which you intend to feparate the
Silver by diftillation. To this body fit a tubulated
head provided with its ftopple. Set this alembic
in a fand-bath, fo that the body may be almoft
covered with fand : apply a receiver, and diftill with
a moderate heat, fo that the drops may fucceed
each other at the diftance of fome feconds. If the
receiver grow very hot, check the fire. When red
vapours begin to appear, pour into the alembic,
through the hole in its head, a frefh quantity of
your folution of Silver, firft made very hot. Continue diftilling in this manner, and repeating the
addition of frefh liquor, till all your folution be put
into the alembic. When you have no more frefh
folution to put in, and when, the phlegm being all
come over, red vapours begin again to appear, convey into the alembic half a dram or a dram of tallow, and diftill to drynefs ; which being done, increafe your fire fo as to make the veffel containing
the fand-bath red-hot. In the alembic you will
find a calx of Silver, which muft be melted in a
crucible with fome foap and calcined wine-lees.

OBSERVATIONS.

A low cucurbit is recommended for this opera-tion, to the intent that the particles of the Nitrous Acid, which are ponderous, may the more eafily be carried up and pafs over into the receiver. For the fame reafon the cucurbit is directed to be al-moft wholly covered with fand; left otherwife the acid vapours fhould be condenfed about that part of the cucurbit, which, being out of the fand, would be much cooler than that which is encom-paffed therewith, and from thence fhould fall back again to the bottom; by which means the diftilla-tion would certainly be retarded, and the veffel pro-bably be broken.

Notwithftanding thefe precautions the veffels are liable to break in fuch diftillations; efpecially when they contain a great deal of liquor. With a view therefore to prevent this accident, we or-dered that the whole quantity of the folution of Silver to be diftilled fhould not be put at once into the alembic. The little bit of tallow, added to-wards the end of the operation, is intended to hin-der the metal from adhering clofely to the veffel, as it would otherwife do, when all the moifture is diffipated.

The Soap and Fixed Alkali mixed with the Sil-ver to flux it, after its feparation from the *Aqua Fortis* in this way, ferve to abforb fuch of the moft fixed particles of the Acid as may ftill remain united with the metal.

If the diftillation be ftopped when part of the phlegm is drawn off, and the liquor be then fuffered to cool, many cryftals will fhoot therein, which are a Neutral Salt conftituted of the Nitrous Acid and Silver. If the diftillation be carried further, and ftopped when near its conclufion, the liquor being
then

then fuffered to cool will wholly coagulate into a blackifh mafs called the *Infernal Stone*.

This way of feparating Silver from its folvent is attended with the advantage of faving all the *Aqua Fortis*, which is excellent, and fit to be employed in other operations.

PROCESS VI.

To feparate Silver from the Nitrous Acid by Precipitation. Luna Cornea. Luna Cornea *reduced.*

INTO your folution of Silver pour about a fourth part in weight of Spirit of Salt, folution of Sea-falt, or folution of Sal Ammoniac. The liquor will inftantly become turbid and milky. Add twice or thrice its weight of fair water, and let it ftand fome hours to fettle. It will depofite a white powder. Decant the clear liquor, and on the precipitate pour frefh *Aqua Fortis*, or Spirit of Salt, and warm the whole on a fand-bath with a gentle heat for fome time. Pour off this fecond liquor, and boil your precipitate in pure water, fhifting it feveral times, till the precipitate and the water be both quite infipid. Filter the whole, and dry the precipitate, which will be a *Luna Cornea*, and muft be reduced in the following manner.

Smear the infide of a good crucible well with foap. Put your *Luna Cornea* into it; cover it with half its weight of Salt of Tartar, thoroughly dried and pulverifed; prefs the whole hard down; pour thereon as much oil, or melted tallow, as the powder is capable of imbibing; fet the crucible thus charged, and clofe covered, in a melting furnace, and, for the firft quarter of an hour, make no more fire than is neceffary to make the crucible moderately red : after that raife it fo as to melt the Sil-

ver

ver and the Salt, throwing into the crucible from time to time little bits of tallow. When it ceafes to fmoke, let the whole cool; or pour it into a hollow iron cone, warmed and tallowed.

OBSERVATIONS.

THE procefs here delivered furnifhes us with the means of procuring Silver in a degree of purity which is not to be obtained by any other method of treating it whatever. That which is refined on the cupel always retains a fmall portion of copper, from which it cannot poffibly be feparated in that way: but if it be diffolved in *Aqua Fortis*, and precipitated thence in a *Luna Cornea* by the Marine Acid, the precipitate will be an abfolutely pure Silver, unalloyed with that fmall portion of Copper which it retained on the cupel. The reafon of this effect is, that the Copper remains as perfectly diffolved in Spirit of Salt and in *Aqua Regia* as in *Aqua Fortis*: fo that when the Silver, and the Copper with which it is alloyed, are diffolved together in the Nitrous Acid, if the Acid of Sea-falt be mixed with the folution, part of this latter Acid unites with the Silver, and therewith forms a new compound, which not being foluble in the liquor, falls to the bottom. The other part of the Acid mixing with the Nitrous, forms an *Aqua Regis*, in which the Copper remains diffolved, without feparating from it.

Frefh Acid is poured on the precipitated calx of Silver, in order to complete the folution of the fmall portion of Copper that may have efcaped the action of the firft folvent. It is indifferent whether the Spirit of Salt or the Spirit of Nitre be employed for this purpofe, becaufe they both diffolve Copper alike, and becaufe Silver precipitated by Spirit of Salt is not foluble in either.

After

After this it is neceſſary to waſh the precipitate well with pure water, in order to free it entirely from the particles of *Aqua Fortis* adhering to the Silver; becauſe they may poſſibly contain ſomething of Copper, which would mix with the Silver in melting, and taint its purity.

If this precipitate of Silver be expoſed to the fire, unmixed with any other ſubſtance, it melts as ſoon as it begins to be red; and if the fire be increaſed, part thereof will be diſſipated in vapours, and the reſt will make its way through the crucible. But being poured out as ſoon as melted, it coagulates into a cake of a purpliſh red colour, ſemitranſparent, ponderous, and in ſome degree pliable, eſpecially if it be very thin. It bears ſome reſemblance to horn, which hath occaſioned it to be called *Luna Cornea*.

As *Luna Cornea* is not ſoluble in water, recourſe muſt be had to fuſion, in order to reduce it, by ſeparating from the Silver thoſe acids which give it the abovementioned properties. Fixed Alkalis and fatty matters are very fit to produce that ſeparation.

We directed that the inſide of the crucible, in which the reduction is to be made, ſhould be carefully ſmeared with ſoap, and that the *Luna Cornea* ſhould be quite covered with a Fixed Alkali and fat, to the end that when the heat is ſtrong enough to diſſipate it in vapours, or to attenuate it ſo as to render it capable of penetrating the crucible, it may be forced to paſs through matters qualified to abſorb its Acid, and reduce it.

Luna Cornea may alſo be reduced by being melted with ſuch metalline ſubſtances as have a greater affinity than Silver with the Acids wherewith it is impregnated. Of this kind are Tin, Lead, Regulus of Antimony: but the *Luna Cornea* ruſhes ſo impetuouſly into conjunction with thoſe

Z 3

metal-

metalline fubftances, that a vaft many vapoufs arife, and carry off with them part of the Silver: if therefore you chufe to effect the reduction by the interpofition of fuch metalline fubftances, you muft employ a retort inftead of a crucible.

But this method is attended with another incon-venience; which is, that fome part of thofe metal-line fubftances may unite with the Silver, and adul-terate it; for which reafon it is beft to keep to the method firft propofed.

P R O C E S S VII.

To diſſolve Silver, and ſeparate it from Gold, by Cementation.

MIX thoroughly together fine brick-duft four parts, Vitriol calcined to rednefs one part; and Sea-falt or Nitre one part. Moiften this pow-der with a little water. With this cement cover the bottom of a crucible half an inch thick; on this firft bed lay a thin plate of the mafs of Gold and Silver you intend to cement, and which you muft previ-oufly take care to beat into fuch thin plates. Cover this plate with a fecond layer of cement, of the fame thicknefs as the former; on this fecond bed lay another plate of your metal; cover it in like manner with cement; and fo proceed till the cru-cible be filled to within half an inch of its brim. Fill up the remaining fpace with cement, and clofe the crucible with a cover, luted with a pafte made of Windfor-loam and water: fet your crucible thus charged in a furnace, whofe fire-place is deep enough to let it be entirely furrounded with coals, quite up to its mouth. Light fome coals in the furnace, taking care not to make the fire very brifk at firft; encreafe it by degrees, but only fo far as

to make the crucible moderately red ; keep up the
.fire in this degree for eighteen or twenty hours :
then let the fire go out ; open the crucible when it
is cold, and feparate the cement from your plates
of Gold. Boil the.Gold repeatedly in fair water,
till the water come off quite infipid.

OBSERVATIONS.

IT cannot but feem ftrange that, after having fo
often declared the Acid of Sea-falt to be incapable
of diffolving Silver, we fhould direct either Nitre
or Sea-falt indifferently to be employed in compofing a cement, which is to produce an Acid capable of eating out all the Silver mixed with Gold.
It is eafy to conceive how the Nitrous Acid extricated from.its bafis by means of the Vitriolic Acid
may produce this effect : but if Sea-falt inftead of
Nitre be made an ingredient in the cement, its
Acid, though fet at liberty in the fame manner by
the Vitriolic Acid, muft at firft fight appear unable
to anfwer the end.

In order to remove this difficulty we muft here
obferve that there are two very effential differences
between the Marine Acid collected in a liquor, as
it is when diftilled in the ufual manner, and the
fame Acid feparated from its bafis in a crucible, as
it is in cementation.

The firft of thefe two differences is, that the Acid
being reduced into vapours when it acts on the Silver in cementation, its activity is thereby greatly
encreafed : the fecond is, that in the crucible it
fuftains a vaftly greater degree of heat than it can
ever bear when it is in the form of a liquor. For,
after it is once diftilled and feparated from its bafis,
it cannot fuftain any extraordinary degree of heat
without being volatilized and entirely diffipated :
whereas while it continues united with its bafis it is
much more fixed, and cannot be feparated but by a

very

very intenfe heat. Confequently, if it meet with
any body to diffolve, at the very inftant of its feparation from its bafis, while it is actuated by a much
fiercer heat than can ever be applied to it on any
other occafion, it muft operate upon that body
with fo much the more efficacy : and thus it comes
to pafs that in cementation it has the power of diffolving Silver,· which it would be incapable of
touching if it were not fo circumftanced.

But herein Gold differs from Silver : for, whatever force the Nitrous or the Marine Acid may
exert, when extricated from their bafes in the cementing crucible, this metal obftinately refufes to
yield to either of thofe Acids feparately, and can
never be diffolved by them, unlefs both be united
together.

Our cementation therefore is actually a parting
procefs in the dry way. The Silver is diffolved,
and the Gold remains unaltered. Nay, as the
action of the Acids is much ftronger when they are
applied this way, than when they are ufed for diffolution in the moift way, the Nitrous Acid, which
in the common parting procefs will not diffolve
Silver unlefs its weight be double that of the Gold,
is able in cementation to diffolve a very fmall quantity of Silver diffufed through a large quantity of
Gold.

It fometimes happens that after the operation the
cement proves extremely hard, fo that it is very
troublefome to feparate it entirely from the Gold.
In this cafe it muft be foftened by moiftening it with
hot water. This hardnefs which the cement acquires is occafioned by the fufion of the Salts, which
is the effect of too ftrong a heat. It was in order to
prevent this, and that a due degree of heat might
be applied, without the danger of melting the falts, ·
that we directed the cement to be mixed with
a confiderable quantity of earthy matter incapable

of

of fufion, fuch as brick-duft. A greater inconvenience ftill will enfue, if the fire be made fo ftrong
as to melt the Gold: for then it will partly com-
mix again with the other metalline fubftances diffolved by the cement, and confequently will not
be purified.

The crucible is covered, and its cover luted on,
to prevent the acid vapours from being too foon
diffipated, and to force them to circulate the longer
in the crucible. However, it is neceffary that thofe
vapours fhould find a vent at laft, otherwife they
would burft the veffel : and for this reafon we directed the crucible to be luted only with Windfor-
loam, which does not grow very hard by the action
of fire, and fo is capable of yielding and giving
paffage to the vapours, when a certain quantity of
them is collected in the crucible, and they begin
to ftruggle for an efcape on every fide.

When the operation is finifhed, the Silver diffolved by the Acid of the cement is partly diftributed through the cement, and partly in the Gold
itfelf, which is impregnated therewith. For this
reafon the Gold muft be wafhed feveral times in
boiling water, till the water become abfolutely infipid : for, if the Gold be melted without this precaution, it will mix again with the Silver : the cement alfo may be wafhed in the fame manner to recover the Silver it contains.

Though this cementation be, properly fpeaking,
a purification of Gold, yet we have placed it among
the proceffes on Silver, becaufe it is the Silver that
is diffolved on this occafion, and becaufe this is a
particular way of diffolving that metal. Moreover,
moft of the proceffes hitherto delivered, either on
Gold or Silver, are equally applicable to both thefe
metals.

If the Gold do not appear quite pure after the
cementation, the procefs muft be repeated.

There

There are several ways to know the fineness of Gold, the quantity of Silver with which it is alloyed, and the proportion in which these two metals are mixed in a mass purified by the cupel.

One of the simplest is the trial by the Touchstone; which indeed is hardly any more than judging by the eye only, from the colour of the compound metal, what proportion of Gold and Silver it contains.

The Touch-stone is a sort of black marble, whose surface ought to be half polished. If the metalline mass which you want to try be rubbed on this stone, it leaves thereon a thin coat of metal, the colour of which may be easily observed. Such as are accustomed to see and handle Gold and Silver can at once judge very nearly from this sample in what proportion the two metals are combined: but, for greater accuracy, those who are in the way of having frequent occasion for this trial are provided with a sufficient number of small bars or needles, of which one is pure Gold, another pure Silver, and all the rest consist of these two metals mixed together in different proportions, varied by carats, or even by fractions of carats, if greater exactness be required.

The fineness of each needle being marked on it, that needle, whose colour seems to come nearest the colour of the metalline streak on the Touch-stone, is rubbed on the stone by the side of that streak. This needle likewise leaves a mark; and if there appear to be no difference between the two metalline streaks, the metalline mass is judged to be of the same fineness as the needle thus compared with it. If the eye discovers a sensible difference, another needle is sought for whose colour may come nearer to that of the metal to be tried. But though a man be ever so well versed in judging thus of the fineness of Gold by the eye only, he can never be perfectly and accurately

curately sure of it by this means alone. If such certainty be required, recourse must be had to the parting assay; and yet when you have gone through it, there always remains a small quantity of the metal, which should have been dissolved, and yet escaped the action of the solvent. For example, if you make use of *Aqua Regis*, the Silver that remains after the operation still contains a little Gold; and, if you make use of *Aqua Fortis*, the Gold that remains after the operation still contains a little Silver. And therefore if you resolve to carry the separation of these two metals still further by solvents, it will be necessary, after you have gone through one parting process, to perform a second the contrary way. For example, if you begin with *Aqua Fortis*, then, after it has dissolved all the Silver in the metalline mass that it is capable of taking up, dissolve the remaining Gold in *Aqua Regis*; by which means you will separate the small portion of Silver left in it by the *Aqua Fortis*. The contrary is to be done if you made use of *Aqua Regis* first.

CHAP. III.
Of COPPER.

PROCESS I.
To separate Copper from its Ore.

BEAT your Copper ore to a fine powder, having first freed it as accurately as possible, by washing and roasting, from all stony, earthy, sulphureous, and arsenical parts. Mix your ore thus pulverized with thrice its weight of the black flux; put the mixture into a crucible; cover it with common salt to the thickness of half an inch, and press
the

the whole down with your finger. With all this the crucible muſt be but half full. Set it in a melting furnace; kindle the fire by degrees, and raiſe it inſenſibly till you hear the Sea-ſalt crackle. When the decrepitation is over, make the crucible moderately red-hot for half a quarter of an hour. Then give a conſiderable degree of heat, exciting the fire with a pair of good perpetual bellows, ſo that the crucible may become very red-hot, and be perfectly ignited. Keep the fire up to this degree for about a quarter of an hour; then take out the crucible, and with a hammer ſtrike a few blows on the floor whereon you ſet it. Break it when cold. If the operation hath been rightly and ſuccefsfully performed, you will find at the bottom of the veſſel a hard Regulus, of a bright yellow colour, and ſemi-malleable; and over it a ſcoria of a yellowiſh brown colour, hard and ſhining, from which you may ſeparate the Regulus with a hammer.

OBSERVATIONS.

COPPER in the ore is often blended with ſeveral other metallic ſubſtances, and with volatile minerals, ſuch as Sulphur and Arſenic. Copper ores alſo frequently participate of the nature of the pyrites, containing a martial and an unmetallic earth, both of which are entirely refractory, and hinder the ore from melting. In this caſe you muſt add equal parts of a very fuſile glaſs, a little borax, and four parts of the black flux, to facilitate the fuſion, The black flux is moreover neceſſary to furniſh the Copper with the Phlogiſton it wants, or reſtore ſo much thereof as it may loſe in melting. For the ſame reaſon, when any ore, but that of Gold or Silver, is to be ſmelted, it is a general rule to add ſome black flux, or other matter abounding with Phlogiſton,

The

The Regulus produced by this operation is not malleable, becaufe it is not pure Copper, but a mixture of Copper with all the other metallic fubftances that were in the ore; except fuch as were feparated from it by roafting, of which it contains but little.

According to the nature of the metallic matters that remain combined with the Copper after this fufion, the colour of the Regulus is either like that of pure Copper, or a little more whitifh: it is alfo frequently blackifh, which has procured it the name of *Black Copper*. In this ftate, and even in general, it is ufual enough to call this Regulus by the name of Black Copper, when alloyed with other metallic fubftances that render it unmalleable, whatever its colour be.

Hence it appears that there may be feveral different forts of Black Copper. Iron, Lead, Tin, Bifmuth, and the reguline part of Antimony, are almoft always combined with the ores of Copper, in a multitude of different proportions; and all thefe fubftances, being reduced by the black flux in the operation, mix and precipitate with the Copper. If the ore contain any Gold or Silver, as is pretty often the cafe, thefe two metals alfo are confounded with the reft in the precipitation, and become part of the Black Copper.

Pyritofe, fulphureous, and arfenical Copper ores may be fufed, in order to get rid of the groffer heterogeneous parts, without previoufly roafting them: but in this cafe no alkaline flux muft be mixed with the ore; becaufe the Alkali in combination with the Sulphur would produce a Liver of Sulphur, and fo diffolve the metalline part; by which means all would be confounded together, and no Regulus, or very little, be precipitated. On this occafion therefore nothing muft be added to promote the fufion, but fome tender fufile glafs, together with a fmall quantity of borax.

This

This firſt fuſion may alſo be performed amidſt the coals, by caſting the ore upon them in the fur-nace, without uſing a crucible; and then an earthen veſſel, thoroughly heated, or even made red-hot; muſt be placed under the grate of the fire-place, to receive the metal as it runs from the ore.

The Regulus obtained by this means is much more impure and brittle than Black Copper, becauſe it contains moreover a large quantity of Sulphur and Arſenic; as theſe volatile ſubſtances have not time to evaporate during the ſhort ſpace requiſite to melt the ore, and as they cannot be carried off by the action of the fire after the ore is once melt-ed, whatever time be allowed for that purpoſe. However, ſome part thereof is diſſipated; and the Iron which is in pyritoſe ores, having a much greater affinity than Copper, and indeed than any other metallic ſubſtance, with Sulphur and Arſenic, abſorbs another part thereof, and ſeparates it from the Regulus.

This Regulus, it is plain, ſtill contains all the ſame parts that were in the ore, but in different proportions; there being more Copper, combined with leſs Sulphur, Arſenic, and unmetallic earth, which have been either diſſipated or turned to ſlag. Therefore if you would make it like Black Copper, you muſt pound it, roaſt it over and over, to free it from its Sulphur and Arſenic, and then melt it with the black flux.

If this Regulus contain much Iron, it will be adviſeable to melt it once or twice more, before all the Sulphur and Arſenic are ſeparated from it by roaſting; for as the Iron, by uniting with theſe volatile ſubſtances, ſeparates them from the Copper, with which they have not ſo great an affinity; ſo alſo the Sulphur and Arſenic, by uniting with the Iron, help in their turn to ſeparate it from the Copper.

PRO-

PROCESS II.

To purify Black Copper, and render it malleable.

BREAK into small bits the Black Copper you intend to purify; mix therewith a third part in weight of granulated Lead, and put the whole into a cupel set under the muffle in a cupelling furnace, and previously heated quite red. As soon as the metals are in the cupel raise the fire considerably, making use, if it be needful, of a pair of perpetual bellows, to melt the Copper speedily. When it is thoroughly melted, lower the fire a little, and continue it just high enough to keep the metalline mass in perfect fusion. The melted matter will then boil, and throw up some *scoriæ*, which will be absorbed by the cupel.

When most of the Lead is consumed, raise the fire again till the face of the Copper become bright and shining, thereby shewing that all its alloy is separated. As soon as your Copper comes to this state, cover it with charcoal dust conveyed into the cupel with an iron ladle: then take the cupel out of the furnace and let it cool.

OBSERVATIONS.

OF all the metals, next to Gold and Silver, Copper bears fusion the longest without losing its phlogiston; and on this property is founded the process here delivered for purifying it.

It is necessary the Copper should melt as soon as it is in the cupel, because its nature is to calcine much more easily and much sooner, when it is only red-hot, than when it is in fusion. For this reason the fire is to be considerably raised, immediately on putting the Copper under the muffle, that it may melt

as

as foon as poffible. Yet too violent a degree of fire muft not be applied to it: for when it is expofed to fuch a degree of heat only as is but juft neceffary to keep it in fufion, it is then in the moft favourable condition for lofing as little as may be of its phlogifton; and if the heat be ftronger, a greater quantity thereof will be calcined. As foon therefore as it flows it is proper to weaken the fire, and reduce it to the degree juft requifite to keep up the fufion.

The Lead added on this occafion is intended to facilitate and expedite the fcorification of the metallic fubftances combined with the Copper. So that the event is here nearly the fame as when Gold or Silver is refined on the cupel. The only difference between this refining of Copper, and that of the perfect metals, is that the latter, as hath been fhewn, abfolutely refift the force of fire and the action of Lead, without fuffering the leaft alteration; whereas a good deal of Copper is calcined and deftroyed, when it is purified in this manner on the cupel. Indeed it would be wholly deftroyed, if a greater quantity of Lead were added, or if it were left too long in the furnace. It is with a view to fave as much of it as poffible that we order it to be covered with charcoal-duft as foon as the fcorification is finifhed.

The Lead ferves moreover to free the Copper expeditiously from the Iron with which it may be alloyed. Iron and Lead are incapable of contracting any union together: fo that as faft as the Lead unites with the Copper, it feparates the Iron, and excludes it out of the mixture. For the fame reafon if Iron were combined in a large proportion with Copper, it would prevent the Lead from entering into the compofition. Now, as it is neceffary to give the more heat, and to keep the Copper to be incorporated with Lead the longer in fufion, as that Copper is alloyed with a greater proportion of Iron,

some

some black flux muſt be added on this occaſion, to prevent the Copper and the Lead from being calcined before their aſſociation can be effected.

Copper purified in the manner here directed is beautiful and malleable. It is now alloyed with no other metalline ſubſtance but Gold or Silver, if there were any in the mixed maſs. If you deſire to extract this Gold or Silver, recourſe muſt be had to the operation of the cupel. The proceſs here given for purifying Copper is not uſed in large works, becauſe it would be much too chargeable. In order to purify their Black Copper, and render it malleable, the ſmelters content themſelves with roaſting it, and melting it repeatedly, that the metallic ſubſtances, which are not ſo fixed as copper, may be diſſipated by ſublimation, and the reſt ſcorified by fuſion.

PROCESS III.

To deprive Copper of its Phlogiſton by calcination.

PUT your Copper in filings into a teſt, and ſet it under the muffle of a cupelling furnace; light the fire, and keep up ſuch a degree of heat as may make the whole quite red, but not enough to melt the Copper. The ſurface of the Copper will gradually loſe its metalline ſplendour, and put on the appearance of a reddiſh earth. From time to time ſtir the filings with a little rod of copper or iron, and leave your metal expoſed to the ſame degree of fire till it be entirely calcined.

OBSERVATIONS.

IN our obſervations on the preceding proceſs we took notice that Copper, in fuſion, calcines more ſlowly, and leſs eaſily, than when it is expoſed to a degree of fire barely ſufficient to keep it red-hot,

without melting it; and therefore, the defign here
being to calcine it, we have directed that degree of
heat only to be applied.

The cupelling furnace is the fitteft for this operation, becaufe the muffle is capable of receiving
fuch a flat veffel as ought to be ufed on this occafion, and communicating to it a great deal of heat;
while, at the fame time, it prevents the falling in
of any coals, which, by furnifhing the Copper
with frefh phlogifton, would greatly prejudice and
protract the operation.

As Copper calcines with great difficulty, this
operation is extremely tedious: nay, though Copper hath ftood thus expofed to the fire for feveral
days and nights, and feems perfectly calcined, yet
it frequently happens that, when you try afterwards
to melt it, fome of it refumes the form of Copper:
a proof that all the Copper had not loft its phlogifton. Copper is much more expeditioufly deprived of its phlogifton by calcining it in a crucible
with Nitre.

The calx of Copper perfectly calcined is with
great difficulty brought to fufion: yet, in the focus
of a large burning-glafs, it melts and turns to a reddifh and almoft opaque glafs.

By the procefs here delivered, you may likewife
calcine all other metalline fubftances, which do not
melt till they be thoroughly red-hot. As to thofe
which melt before they grow red, they are eafily
enough calcined, even while they are in fufion.

PROCESS IV.

*To refufciate the Calx of Copper, and reduce it to
Copper, by reftoring its Phlogifton.*

MIX the Calx of Copper with thrice as much
of the black flux; put the mixture into a
good crucible, fo as to fill two thirds thereof, and
over it put a layer of Sea-falt a finger thick. Cover
the crucible, and fet it in a melting furnace; heat
it gradually, and keep it moderately red till the
decrepitation of the Sea-falt be over. Then raife
the fire confiderably by means of a good pair of perpetual bellows; fatisfy yourfelf that the matter is in
perfect fufion, by dipping into the crucible an iron
wire; continue the fire in this degree for half a quarter of an hour. When the crucible is cold, you will find
at its bottom a button of very fine Copper, which
will eafily feparate from the faline fcoria at top.

OBSERVATIONS.

WHAT hath been faid before on the fmelting of
Copper ores may be applied to this procefs, as being
the very fame. The obfervations there added fhould
therefore be confulted on this occafion.

PROCESS V.

To diffolve Copper in the Mineral Acids.

ON a fand-bath, in a very gentle heat, fet a
a matrafs containing fome Copper filings; pour
on them twice their weight of Oil of Vitriol. That
Acid will prefently attack the Copper. Vapours
will rife, and iffue out of the neck of the matrafs.
A vaft number of bubbles will afcend from the fur-

face

face of the metal to the top of the liquor, and the liquor will acquire a beautiful blue colour. When the Copper is diffolved, put in a little and a little more, till you perceive the Acid no longer acts upon it. Then decant the liquor, and let it ftand quiet in a cool place. In a fhort time great numbers of beautiful blue cryftals will fhoot in it. Thefe cryftals are called *Vitriol of Copper*, or *Blue Vitriol*. They diffolve eafily in Water.

OBSERVATIONS.

THE Vitriolic Acid perfectly diffolves Copper, which is alfo foluble in all the Acids, and even in many other menftruums.

This Acid may be feparated from the Copper which it hath diffolved by diftillation only : but the operation requires a fire of the utmoft violence. The Copper remaining after it muft be fufed with the black flux, to make it appear in its natural form ; not only becaufe it ftill retains a portion of the Acid, but alfo becaufe it hath loft part of its phlogifton by being diffolved therein. The black flux is very well adapted both to abforb the Acid that remains united with the Copper, and to reftore the phlogifton which the metal hath loft.

The moft ufual method of feparating Copper from the Vitriolic Acid is by prefenting to that Acid a metal with which it hath a greater affinity than with Copper. Iron being fo qualified is confequently very fit to bring about this feparation. When therefore plates of Iron well cleaned are laid in a folution of Blue Vitriol, the Acid foon begins to act upon them, and by degrees, as it diffolves them, depofites on their furfaces a quantity of Copper in proportion to the quantity of Iron it takes up. The Copper thus precipitated hath the appearance of fmall leaves or fcales, exceeding thin, and of a beautiful copper-colour. Care muft be taken

5   to

to fhake the Iron-plates now and then, to make the fcales of Copper fall off, which will otherwife cover them entirely, hinder the Vitriolic Acid from attacking the Iron, and fo put a ftop to the precipitation of the remaining Copper.

When thefe fcales of Copper ceafe to fettle on the clean Iron-plates, you may be fure all the Copper that was in the liquor is precipitated, and that this liquor, which was a folution of Copper before the precipitation, is a folution of Iron after it. So that here two operations are performed at one and the fame time; to wit, the precipitation of the Copper, and the diffolution of the Iron.

The Copper thus precipitated requires only to be feparated from the liquor by filtration, and melted with a little black flux, to become very fine malleable Copper.

The Copper may alfo be precipitated out of a folution of Blue Vitriol by the interpofition of a Fixed Alkali. This precipitate is of a greenifh blue colour, and requires a much greater quantity of the black flux to reduce it.

Copper diffolves in the Nitrous Acid, in the Marine Acid, and in *Aqua Regis*; from all of which it may be feparated by the fame methods as are here ordered with regard to the Vitriolic Acid.

CHAP. IV.
Of IRON.

PROCESS I.

To separate Iron from its Ore.

POUND into a coarse powder the martial
stones or earths out of which you design to extract the Iron: roast this powder in a test under the
muffle for some minutes, and let your fire be brisk.
Then let it cool, beat it very fine, and roast it a
second time, keeping it under the muffle till it emit
no more smell.

Then mix with this powder a flux composed
of three parts of Nitre fixed with Tartar, one part
of fusile glass, and half a part of Borax and charcoal-dust. The dose of this reducing flux must be
thrice the weight of the ore.

Put this mixture into a good crucible; cover it
with about half a finger thick of Sea-salt; over the
crucible put its cover, and lute it on with Windsor-
loam made into a paste with water. Having thus
prepared your crucible, set it in a melting furnace,
which you must fill up with charcoal. Light the
fire, and let it kindle by gentle degrees, till the
crucible become red-hot. When the decrepitation
of the Sea-salt is over, raise your fire to the highest
by the blast of a pair of perpetual bellows, or rather
several. Keep up this intense degree of heat for
three quarters of an hour, or an whole hour, taking care that during all this time the furnace be
kept

kept conftantly filling up with frefh coals as the former confume. Then take your crucible out of the furnace; ftrike the pavement on which you fet it feveral times with a hammer, and let it ftand to cool: break it, and you will find therein a Regulus of Iron covered with flag.

OBSERVATIONS.

IRON ore, like all others, requires roafting, to feparate from it, as much as poffible, the volatile minerals, Sulphur and Arfenic, which being mixed with the Iron would render it unmalleable. Indeed it is fo much the more neceffary to roaft thefe ores, as Iron is, of all metallic fubftances, that which has the greateft affinity with thofe volatile minerals; on which account no metallic fubftance whatever is capable of feparating it from them by fufion and precipitation.

Fixed Alkalis, it is true, have a greater affinity than Iron with Sulphur; but then the compofition which a Fixed Alkali forms with Sulphur is capable of diffolving all metals. Confequently, if you do not diffipate the Sulphur by roafting, but attempt to feparate it from the Iron by melting the ore with a Fixed Alkali, the Liver of Sulphur formed in the operation will diffolve the martial part; fo that after the fufion you will find little or no Regulus.

All Iron ores in general are refractory, and lefs fufible than any other; for which reafon a much greater proportion of flux, and a much more violent degree of fire, is required to fmelt them. One principal caufe why thefe ores are fo refractory is the property which Iron itfelf has of being extremely difficult to fufe, and of refifting the action of the fire fo much the more as it is purer, and further removed from its mineral ftate. Among all the metallic fubftances it is the only one that is lefs fufible when combined with that portion of

phlo-

phlogifton which gives it the metalline form, than when it is deprived thereof, and in the form of a calx.

In fmelting-houfes Iron ore is fufed amidft charcoal, the phlogifton of which combines with the martial earth, and gives it the metalline form. The Iron thus melted runs down to the bottom of the furnace, from whence it is let out into large moulds, in which it takes the fhape of oblong blocks, called *Pigs* of Iron. This Iron is ftill very impure, and quite unmalleable. Its want of ductility after the firft melting arifes partly from hence, that, notwithftanding the previous roafting which the ore underwent, there ftill remains, after this firft fufion, a confiderable quantity of Sulphur or Arfenic combined with the metal.

A certain quantity of quick-lime, or of ftones that will burn to lime, is frequently mixed with Iron ore on putting it into the fmelting furnace. The lime being an abforbent earth, very apt to unite with Sulphur and Arfenic, is of ufe to feparate thofe minerals from the Iron.

It is alfo of ufe to mix fome fuch matters with the ore, when the ftones or earths which naturally accompany it are very fufible; for, as the Iron is of difficult fufion, it may happen that the earthy matters mixed with the Iron fhall melt as eafily as the metal, or perhaps more eafily. In fuch a cafe there is no feparation of the earthy from the metalline part, both of which melt and precipitate together promifcuoufly: now quick-lime, being extremely refractory, ferves on this occafion to check the melting of thofe matters which are too fufible.

Yet quick-lime, notwithftanding its refractory quality, may fometimes be of ufe as a flux for Iron. This is the cafe when the ore happens to be combined with fubftances which, being united with lime, render it fufible: fuch are all arfenical matters, and even

fome

some earthy matters, which being combined with quick-lime make a fusible compound.

When the ore of an Iron Mine is found difficult to reduce, it is usually neglected even though it be rich; because Iron being very common, people chuse to work those mines only whose ores are smelted with the most ease, and require the least consumption of wood.

Yet refractory ores are not to be altogether rejected, when another Iron ore of a different quality is found near them. For it often happens that two several Iron ores, which being worked separately are very difficult to manage, and yield at last but bad metal, become very tractable, and yield excellent Iron, when smelted together : and accordingly such mixtures are often made at Iron-works.

The Iron obtained from ores by the first fusion may be divided into two sorts. The one, when cold, resists the hammer, doth not easily break, and is in some measure extensible on the anvil; but if struck with a hammer when red-hot flies into many pieces : this sort of Iron hath always a mixture of Sulphur in it. The other sort on the contrary, is brittle when cold, but somewhat ductile when red-hot. This iron is not sulphurated, is naturally of a good quality, and its brittleness arises from its metalline parts not being sufficiently compacted together.

Iron abounds so much, and is so universally diffused through the earth, that it is difficult to find a body in which there is none at all : and this hath led several Chymists, even men of great fame, into the error of thinking that they had transmitted into Iron several sorts of earths in which they suspected no Iron, by combining them with an inflammable matter; whereas, in fact, all they did was to give the metalline form to a true martial earth which happened to be mixed with other earths.

PRO-

PROCESS II.

To render Pig-iron and brittle Iron malleable.

INTO an earthen veſſel widening upwards put ſome charcoal-duſt, and thereon lay the Pig-iron which you propoſe to render duⱪile; cover it all over with a quantity of charcoal; excite the fire violently with a pair, or more, of perpetual bellows till the Iron melt. If it do not readily flow and form a great deal of ſlag on its ſurface, add ſome flux, ſuch as a very fuſible ſand.

When the matter is in fuſion keep ſtirring it from time to time, that all the parts thereof may be equally aⱪed on by the air and the fire. On the ſurface of the melted Iron *ſcoriæ* will be formed, which muſt be taken off as they appear. At the ſame time you will ſee a great many ſparkles darted up from the ſurface of the metal, which will form a ſort of fiery ſhower. By degrees, as the Iron grows purer, the number of theſe ſparkles dimi-niſhes, though they never vaniſh entirely. When but few ſparkles appear, remove the coals which cover the Iron, and let the ſlag run out of the veſ-ſel; whereupon the metal will grow ſolid in a mo-ment. Take it out while it is ſtill red-hot, and give it a few ſtrokes with a hammer, to try if it be duⱪile. If it be not yet malleable, repeat the ope-ration a ſecond time, in the ſame manner as before. Laſtly, when it is thus ſufficiently purified by the fire, work it for a long time on the anvil, extending it different ways, and making it red-hot as often as there is occaſion. Iron thus brought to the neceſ-ſary degree of duⱪility, ſo as to yield to the ham-mer, and ſuffer itſelf to be extended every way, either hot or cold, without breaking to bits, or even cracking in the leaſt, is very good and very

pure.

pure. If it cannot be brought to this degree by the method here prescribed, it is a proof that the ore from which this Iron was extracted ought to be mixed with other ores; but it frequently requires a great number of trials to obtain an exact knowledge of the quality and proportion of those other ores with which it is to be mixed.

OBSERVATIONS.

The brittleness and shortness of Pig-iron arises from the heterogeneous parts which it contains, and which could not be separated from it by the first fusion. These extraneous matters are usually Sulphur, Arsenic, and unmetallic earth, and also a ferruginous earth; but such as could not be combined with the phlogiston as it ought to be, in order to have the properties of a metal, and must therefore be considered as heterogeneous, with respect to the other well-conditioned martial particles.

The Pig-iron, by undergoing repeated fusions, is freed from these heterogeneous matters; those which are volatile, such as Sulphur and Arsenic, being dissipated, and the unmetallic matters being scorified. As to the ferruginous earth, which did not at first acquire the metalline form, it becomes true Iron at last; because, among the coals with which it is encompassed, it meets with a sufficient quantity of phlogiston to reduce it to metal. Charcoal is also necessary on this occasion, that it may continually furnish phlogiston to the Iron, which would otherways be converted into a calx.

Hammering the red-hot Iron, after each fusion, serves to force out from amongst the martial parts such earthy matters as may happen to remain there, and so to bring into closer contact the metalline parts which were separated before by the interposition of those heterogeneous matters.

P R O C E S S III.

To convert Iron into Steel.

TAKE fmall bars of the beft Iron; that is,
of fuch as is malleable both hot and cold;
fet them on their ends in a cylindrical earthen veffel, whofe depth is equal to the length of the bars,
and in fuch a manner that they may be an inch diftant from each other, and from the fides of the crucible. Fill the veffel with a cement compounded
of two parts of charcoal, one part of bones burnt
in a clofe veffel till they become very black, and
one half part of the afhes of green wood; having
firft pulverifed and thoroughly mixed the whole
together. Take care to lift up the Iron bars a little, to the end that the cement may cover the bottom of the veffel, and fo that there be about the
depth of half an inch thereof under every bar : cover the crucible and lute on the cover.

Set the crucible thus prepared in a furnace, fo
contrived that the crucible may be furrounded with
coals from top to bottom : for eight or ten hours
keep up fuch a degree of fire that the veffel may be
moderately red; after this take it out of the furnace; plunge your little Iron bars into cold water,
and you will find them converted into Steel.

O B S E R V A T I O N S.

The principal difference between Iron and Steel
confifts in this, that the latter is combined with a
greater quantity of phlogifton than the former.

It appears by this experiment that, to make
Iron unite with an inflammable matter, it is not neceffary

ceſſary it ſhould be in fuſion; it is ſufficient that
it be ſo red-hot as to be opened and ſoftened by
the fire.

Every kind of charcoal is fit to be an ingredient
in the compoſition of the cement employed to make
Steel, provided it contain no vitriolic Acid. However, it hath been obſerved that animal coals produce a ſpeedier effect than others: for which reaſon it is proper to mix ſomething of that kind with
charcoal-duſt, as above directed.

The following ſigns ſhew that the operation
hath ſucceeded, and that the Iron is changed into
good Steel.

This metal being quenched in cold water, as pro-
poſed above, acquires ſuch an extraordinary degree
of hardneſs, that it will by no means yield to any
impreſſion of the file or hammer, and will ſooner
break in pieces than ſtretch upon the anvil. And
here it is proper to obſerve, that the hardneſs of
Steel varies with the manner in which it is quenched.
The general rule is, that the hotter the Steel is
when quenched, and the colder the water is in
which you quench it, the harder it becomes. It
may be deprived of the temper thus acquired, by
making it red-hot, and letting it cool ſlowly; for
it is thereby ſoftened, rendered malleable, and the
file will bite upon it. For this reaſon the artiſans
who work in Steel begin with untempering it, that
they may with more eaſe ſhape it into the tool they
intend to make. They afterwards new-temper the
tool when finiſhed, and by this ſecond temper the
Steel recovers the ſame degree of hardneſs it had
acquired by the firſt temper.

The colour of Steel is not ſo white as that of
Iron, but darker, and the grains, facets, or fibres,
which appear on breaking it, are finer than thoſe
obſerved in Iron.

If

If the bars of Iron thus cemented in order to convert them into Steel be too thick, or not kept long enough in cementation, they will not be turned into Steel throughout their whole thickness: their surfaces only will be Steel to a certain depth, and the center will be mere iron; because the Phlogiston will not have thoroughly penetrated them. On breaking a bar of this sort, the difference in colour and grain between the Steel and the Iron is very visible.

It is easy to deprive Steel of the superabundant quantity of Phlogiston which constitutes it Steel, and thereby reduce it to Iron. For this purpose it need only be kept red-hot some time, observing that no matter approach it all the while that is capable of refunding to it the Phlogiston which the fire carries off. The same end is still sooner obtained by cementing it with meagre hungry matters, capable of absorbing the Phlogiston; such as bones calcined to whiteness, and cretaceous earths.

Steel may also be made by fusion; or Pig-iron may be converted into Steel. For this purpose the same method must be employed as was above directed for reducing Pig-iron into malleable Iron; with this difference, that, as Steel requires more Phlogiston than is necessary to Iron, all the means must be made use of that are capable of introducing into the Iron a great deal of Phlogiston; such as melting but a small quantity of Iron at a time, and keeping it constantly encompassed with abundance of charcoal; re-iterating the fusions; taking care that the blast of the bellows directed along the surface of the metal do not remove the coals that cover it, &c. And here it must be observed, that there are some sorts of Pig-iron which it is very difficult to convert into Steel by this method, and that there are others which succeed very readily, and with scarce any trouble at all. The ores which yield the last men-

5   tioned

tioned fort of Pig-iron are called *Steel Ores.* Steel made by this means muſt be tempered in the ſame manner as that made by cementation *.

PROCESS IV.

The Calcination of Iron.   Sundry Saffrons of Mars.

TAKE filings of Iron, in what quantity you pleaſe; put them into a broad unglazed earthen veſſel; ſet it under the muffle of a cupelling furnace; make it red-hot; ſtir the filings frequently; and keep up the ſame degree of fire till the Iron be wholly turned into a red powder.

OBSERVATIONS.

IRON eaſily loſes its phlogiſton by the action of fire. The calx that remains after its calcination is exceeding red; which makes this be thought the natural colour of the earth of that metal. It hath accordingly been obſerved that all the earths and ſtones, which either are naturally red, or acquire that colour by calcination, are ferruginous.

The yellowiſh red colour which every calx of Iron hath, in whatever manner it be prepared, hath procured the name of *Crocus* or *Saffron* to every preparation of this kind. That made in the manner above directed is called in Medicine *Crocus Martis aſtringens.*

The ruſt produced on the ſurface of Iron is a ſort of calx of Iron made by the way of diſſolution.

* Mr. Réaumur hath obliged the publick with a treatiſe on the means of converting Iron into Steel, in which he hath exhauſted the ſubject. Such as deſire the ampleſt and moſt uſeful inſtructions, on that part of metallurgy, would do well to conſult his Work.

The

The moifture of the air acts upon the metal, dif-folves it, and robs it of fome of its Phlogifton. This ruft is called in Medicine *Crocus Martis Aperiens*; becaufe it is thought that the faline parts, by means whereof the humidity diffolves the Iron, remain united with the metal after its diffolution, and give it an aperitive virtue. The Apothecaries prepare this fort of Saffron of Mars by expofing Iron filings to the dew, till they be turned entirely to ruft; which is then called *Saffron of Mars by dew.*

Another Saffron of Mars is alfo prepared in a much fhorter manner, by mixing filings of Iron with pulverized Sulphur, and moiftening the mix-ture, which after fome time ferments and grows hot. It is then fet on the fire; the Sulphur burns away, and the mafs is kept ftirring till it become a red matter. This Saffron is nothing but Iron dif-folved by the Acid of Sulphur, which is known to be of the fame nature with that of Vitriol; and confequently this Saffron of Mars is no way diffe-rent from Vitriol calcined to rednefs.

PROCESS V.

Iron diffolved by the mineral Acids.

PUT any mineral Acid whatever into a matrafs with fome water; fet the matrafs on a fand-bath gently heated; drop into the veffel fome filings of Iron: the phenomena which ufually accompany metalline diffolutions will immediately appear. Add more filings, till you obferve the Acid hath loft all fenfible action upon them: then remove your matrafs from the fand-bath; you will find in it a folution of Iron.

OBSER-

OBSERVATIONS.

IRON is very eafily diffolved by all the Acids. If you make ufe of the Vitriolic Acid, care muft be taken to weaken it with water, in cafe it be concentrated; becaufe the diffolution will fucceed the better. The vapours that rife on this occafion are inflammable; and if a lighted paper be held to the mouth of the matrafs, efpecially after keeping it ftopt for fome time and fhaking the whole gently, the fulphureous vapours take fire with fuch rapidity as to produce a confiderable explofion; which is fometimes ftrong enough to burft the veffel into a thoufand pieces. This folution hath a green colour; and is in fact a fluid Green Vitriol, which wants nothing but reft to make it fhoot into cryftals.

If you make ufe of the Nitrous Acid, you muft ceafe adding more filings when the liquor, after ftanding ftill fome moments, becomes turbid; for, when this Acid is impregnated with iron to a certain degree, it lets fall fome of that which it had diffolved, and becomes capable of taking up frefh filings. Thus by conftantly adding new fupplies of Iron, this Acid may be made to diffolve a much greater quantity thereof than is neceffary to faturate it entirely. This folution is of a ruffet colour, and doth not cryftallize.

If the weather be not extremely cold, and the Acids have a proper degree of ftrength, the fand-bath is unneceffary; as the diffolution will fucceed very well without it.

Iron diffolved by Acids may be feparated therefrom, like all other metallic fubftances in the fame circumftances, either by the action of fire, which carries off the Acid and leaves the Martial earth, or by the interpofition of fubftances which have a greater affinity than metallic fubftances have with Acids; that is, by Abforbent Earths and Alkaline Salts. By

what-

whatever means you feparate Iron from an Acid folvent, it conftantly appears, after the feparation, in
the form of a yellowifh red powder; becaufe it is
then deprived of moft of the phlogifton to which it
owed its metalline form : whence it is reafonable to
think that this is the proper colour of Martial Earth.

All thefe precipitates of Iron are true Saffrons of
Mars, which, as well as thofe prepared by calcination, are fo much the further removed from the
nature of a metal, the more they are deprived of
their phlogifton. Thence it comes that they are
more or lefs foluble by Acids, and more or lefs attracted by the magnet: as no ferruginous earth, perfectly deprived of all inflammable matter, is at all
attracted by the magnet, or foluble by Acids.

CHAP. V.

Of Tin.

PROCESS I.

To extract Tin from its Ore.

BREAK your Tin ore into a coarfe powder,
and by wafhing carefully feparate from it all
the heterogeneous matters, and ores of a different
kind, that may be mixed therewith. Then dry it,
and roaft it in a ftrong degree of fire, till no more
Arfenical vapour rife from it. When the ore is
roafted, reduce it to a fine powder, and mix it thoroughly with twice its weight of the black flux well
dried, a fourth part of its weight of clean iron filings,
together with as much borax and pitch : put the
mixture into a crucible; over all put Sea-falt to the

thick-

thickness of four fingers; and cover the crucible close.

Set the crucible thus prepared in a melting furnace: apply at firft a moderate and flow degree of fire, till the flame of the pitch, which will efcape through the joint of the cover, difappear entirely. Then fuddenly raife your fire, and urge it with rapidity to the degree neceffary for melting the whole mixture. As foon as the whole is in fufion take the crucible out of the furnace, and feparate the Regulus from the fcoria.

OBSERVATIONS.

ALL Tin ores contain a confiderable quantity of Arfenic, and no Sulphur at all, or at moft very little. Hence it comes that, though Tin be the lighteft of all metals, its ore is neverthelefs much heavier than any other; Arfenic being much heavier than Sulphur, of which the ores of every other kind always contain a pretty large proportion. This ore is moreover very hard, and is not brought to a fine powder with fo much eafe as the reft.

Thefe properties of Tin ore furnifh us with the means of feparating it eafily by lotion, not only from earthy and ftony parts, but even from the other ores which may be mixed with it. And this is of the greater advantage on two accounts, viz. becaufe Tin cannot endure, without the deftruction of a great part thereof, the degree of fire neceffary to fcorify the refractory matters which accompany its ore; and again becaufe this metal unites fo eafily with Iron and Copper, the ores of which are pretty commonly blended with Tin ore, that, after the reduction, it would be found adulterated with a mixture of thefe two metals, if they were not feparated from it before the fufion.

But fometimes the Iron ore confounded with that of Tin is very heavy, and is not eafily pulverized;

whence

whence it comes to pass that it cannot be separated therefrom by washing only. In that case the magnet must be employed to separate it, after the ore hath been roasted.

Roasting is moreover necessary for Tin ore, in order to dissipate the Arsenic which volatilizes, calcines, or destroys one part of the Tin, and reduces the rest to a short, brittle substance, like a Semimetal. The ore is known to be sufficiently roasted when no more fumes rise from it; when it has lost the smell of garlic; and when it does not whiten a clean plate of Iron held over it.

Tin being one of those metals which are most easily calcined, it is necessary in reducing its ore to employ such matters as may furnish it with phlogiston. In order to defend it from the contact of the air, which always accelerates the calcination of metallic substances, the mixture is to be covered with Sea-salt; and the addition of pitch helps to increase the quantity of phlogiston.

PROCESS II.

The Calcination of Tin.

INTO an unglazed earthen dish put the quantity of Tin you intend to calcine; melt it, and keep stirring it from time to time. Its surface will be covered with a greyish white powder: continue the calcination till all your Tin be converted into such a powder, which is the *Calx of Tin.*

OBSERVATIONS.

THOUGH the calcination of metalline substances is promoted by exposing them, in powder, or in fileings, to the action of fire, and by ordering it so that they may not melt, because they present a much smaller surface when melted than when unmelted;

I   yet

yet we have not directed this precaution to be ufed in calcining Tin. The reafon is, this metal is fo fufible that it cannot endure the degree of fire requifite to deftroy its phlogifton without melting, and of courfe, though Tin calcines eafily, the operation is neverthelefs tedious, becaufe the melted metal prefents but a fmall furface to be acted on by the fire and the air. This inconvenience may be partly remedied, and the operation greatly expedited, by dividing the quantity of Tin to be calcined into feveral fmall parcels, and expofing them to the fire in feparate veffels, fo that they may not re-unite when melted, and form one fingle mafs.

Leaf Tin caft on Nitre in actual fufion caufes it to deflagrate and fulminate; and from this mixture there rifes a white vapour, which is converted into flowers when it meets with any obftacle to impede its flying off entirely.

Mr. Geoffroy, who went through a courfe of experiments on Tin, an account whereof may be feen in the Memoirs of the Academy of Sciences, found that from the colour of the calx of that metal a judgment may be formed of its degree of purity, and nearly of the quantity and quality of the metallic fubftances with which it is alloyed. The experiments tried on this fubject by that eminent Chymift are very curious.

He performed the calcination in a crucible, which he heated to a cherry-red, and kept up the fame degree of fire from the beginning to the end of the operation. The calx which formed upon his metal, in that degree of heat, appeared like fmall white fcales, a little reddifh on the under fide. He pufhed it to one fide as it formed, to the end that it might not cover the furface of the metal, which, like all others, requires the contact of the air to turn it into a calx.

" While

" While he was making thefe calcinations, he had
" an opportunity of obferving a curious fact, of
" which nobody before him had ever taken notice;
" probably becaufe nobody had ever calcined Tin
" by the fame method. The fact is, that during
" the calcination of the Tin, whether you break the
" pellicle which forms on the furface of the metal
" while in red-hot fufion, or whether you let it re-
" main without touching it, you perceive in feveral
" places a fmall fwell of a certain matter, which burfts
" and makes its way through the pellicle. This
" matter puffs up, grows red, at the fame inftant
" takes fire, and darts out a fmall whitifh flame,
" as vivid and as brilliant as that of Zinc, when
" urged by a fire ftrong enough to fublime it into
" flowers. The vivednefs of this flame may be fur-
" ther compared to that of feveral fmall grains of
" phofphorus of urine fired and gently dropped on
" boiling water. From this bright flame a white
" vapour exhales; after which the fwelled mafs
" partly crumbles down, and turns to a light white
" powder, fometimes fpotted with red, according
" to the force of the fire. After this momentary
" ignition, there arife ftronger, more numerous, or
" more frequent heavings of matter, out of which
" iffues a good deal of white fume, that may be in-
" tercepted by a cover of tin-plate or copper fitted
" to the crucible, and appears to be the flowers of
" Tin, which in fome meafure corrode thefe metals.
" Hence Mr. Geoffroy conjectures, with a great deal
" of probability, that their fublimation is promoted
" by a portion of Arfenic. When the cruft formed
" by this calx comes to be too thick, or in too great
" a quantity, to be pufhed on one fide, fo as to leave
" part of the metal uncovered, Mr. Geoffroy puts
" out the fire, becaufe no more calx would be form-
" ed; the communication of the external air with
" the Tin in fufion being abfolutely neceffary there-
" to,

" to, as hath been already faid.   In this operation
" it is to be obferved that, if the fire be too flow,
" neither the inflammation of the fulphureous par-
" ticles, nor the white fumes that rife, will be fo
" diftinctly perceived, as when the fire is of the
" degree requifite to keep the crucible juft of a
" cherry-red heat.

" Mr. Geoffroy having taken off this firft calx
" began the calcination anew.   In this fecond heat
" the buddings or heavings were more confiderable,
" and fhot up in the form of cauli-flowers; but were
" ftill compofed of little fcales. The thoroughly cal-
" cined portion of this vegetation was likewife white
" and red; and the inferiour furfaces of fome little
" bits thereof were wholly red.   When thefe calci-
" nations are continued, fulphureous vapours rife
" feemingly of another kind than thofe which ap-
" peared in the beginning; for all the calx made
" by the firft heat was perfectly white; whereas in
" the fecond it begins to be fpotted here and there
" with a tinge of black.   Mr. Geoffroy was obliged
" to go through a courfe of twelve feveral calcina-
" tions before he could convert two ounces of Tin
" into a calx.   He had the opportunity, during
" thefe feveral calcinations, to obferve that after
" the fourth, and fometimes after the third, the
" red fpots of the calx decreafe, and the black in-
" creafe; that the germinations ceafe; that the cruft
" of the calx remains flat; that in the twelfth fire
" the Tin yields no more of this fcaly cruft; that
" towards the end the undulations of the fufed
" metal appear no longer; and that the fmall re-
" mainder of calx is mixed with feveral very minute
" grains of metal, which feem much harder than
" Tin.   Mr. Geoffroy could not collect a fufficient
" quantity thereof to cupel them, and fatisfy him-
" felf whether or no they were Silver."

Though

Though Tin, and all the imperfect metals in general, feem converted to a calx, and lofe the metalline form, by one fingle calcination, and that a flight one; yet they are not wholly deprived of their phlogifton : for if the calx of Tin, for inftance, prepared according to the procefs above delivered, be caft upon Nitre in fufion, it will make that falt deflragrate very perceptibly; a convincing proof that it ftill contains much inflammable matter. If therefore a calx be required abfolutely free from phlogifton, this firft calx muft be re-calcined by a more violent fire, and the calcination continued till all the phlogifton be diffipated.

" Mr. Geoffroy, being defirous of having his calx
" of Tin very pure and perfectly calcined, expofed
" once more to the action of fire the twelve portions
" of calx obtained by his former calcinations. But,
" as it would have been too tedious to re-calcine
" them all feparately, he made four parcels of the
" whole, each confifting of three taken according to
" the order in which they were firft calcined; and
" gave to each a fire fufficiently ftrong, and long
" enough continued, to calcine them as thoroughly
" as was poffible. After this fecond calcination he
" found them all of a moft beautiful white, except
" the firft parcel : as that confifted of the portions
" obtained by the three firft heats, in all of which
" there were fcales tinged with red, it ftill retained
" a ftain of carnation, though hardly perceptible.
" Agreeably to the general rule, the two ounces of
" Tin gained in weight by being thus calcined; and
" the increafe was two drams and fifty-feven grains,
" Mr. Geoffroy obferves that no Tin, but what
" is abfolutely pure, will yield a perfectly white
" calx. He calcined in this manner feveral other
" parcels of Tin that were impure and varioufly al-
" loyed; each of which produced a calx differently.
" coloured, according to the nature and quantity of
" its

" its alloy : whence he juftly concludes that calci-
" nation is a very good method of trying the fine-
" nefs of Tin, or its degree of purity." The particulars of Mr. Geoffroy's experiments on this fubject, which are very curious, may be feen in the
Memoirs of the Academy for 1738.

It is proper to take notice that a man fhould be
very cautious how he expofes himfelf to the vapours
of Tin, becaufe they are dangerous ; this metal being very juftly fufpected by Chymifts of containing
fomething Arfenical.

P R O C E S S   III.

*The diffolution of Tin by Acids. The Smoking liquor
of Libavius.*

PUT into a glafs veffel what quantity you pleafe
of fine Tin cut into little bits. Pour on it
thrice as much *Aqua Regis*, compounded of two
parts *Aqua Fortis* weakened with an equal quantity
of very pure water, and one part Spirit of Salt. An
ebullition will arife, and the Tin will be very rapidly
diffolved ; efpecially if the quantities of metal and
of *Aqua Regis* be confiderable.

OBSERVATIONS.

TIN is foluble by all the Acids; but *Aqua Regis* diffolves it beft of any. Yet in this diffolution it comes
to pafs that part of the diffolved Tin precipitates of
its own accord to the bottom of the veffel, in the form
of a white powder. This folution of Tin is very
fit for preparing the purple-coloured precipitate of
Gold. For this purpofe the folution of Tin muft
be let fall, drop by drop, into a folution of Gold.
Spirit of Nitre diffolves Tin nearly as *Aqua Regis*
does ; but it occafions a greater quantity of calx.

If

, . If two or three parts of Oil of Vitriol be poured on one part of Tin, and if the veſſel in which the mixture is made be expoſed to ſuch a degree of heat as to evaporate all the moiſture, there will remain a tenacious matter ſticking to the ſides of the veſſel. If water be poured on this matter, and it be then expoſed a ſecond time to the fire, it will diſſolve entirely, excepting a ſmall portion of a glutinous ſubſtance, which alſo may be diſſolved in freſh Oil of Vitriol.

The Acid of Sea-ſalt may be combined with Tin by the following proceſs. Mix perfectly, by trituration in a marble mortar, an amalgam of two ounces of fine Tin, and two ounces and a half of Quick-ſilver, with as much Corroſive Sublimate. As ſoon as the mixture is completed, put it into a glaſs retort, and diſtill with the ſame precautions as we directed to be uſed in preparing concentrated and ſmoking Acids. There will firſt come over into the receiver ſome drops of a limpid liquor, which will be ſoon followed by an elaſtic ſpirit that will iſſue out with impetuoſity. At laſt ſome flowers, and a ſaline tenacious matter, will riſe into the neck of the retort. Then ſtop your diſtillation, and pour into a glaſs bottle the liquor you will find in the receiver. This liquor continually exhales a conſiderable quantity of denſe, white fumes, as long as it is allowed to have a free communication with the air.

The product of this diſtillation is a combination of the Acid of Sea-ſalt with Tin. As the affinity of Tin with this Acid is greater than that of Mercury, the Acid contained in the Corroſive Sublimate quits the Mercury, wherewith it was united, to join the Tin; which it volatilizes ſo as to make it riſe with itſelf in a liquid form. We make uſe of the amalgam of Tin with Quick-ſilver, becauſe we are thereby enabled to mix the Corroſive Sublimate perfectly there-

therewith, as the fuccefs of the operation requires it fhould be.

In this experiment the Tin is volatilized, and the Acid of Sea-falt, which is exceedingly concentrated, flies off inceffantly in the form of white vapours. This compound is known in Chymiftry by the name of *the Smoking Liquor of Libavius*; a name derived from its quality, and from its Inventor. Tin diffolved by Acids is eafily feparated from them by Alkalis. It always precipitates in the form of a white calx.

CHAP. VI.

Of LEAD.

PROCESS I.

To extract Lead from its Ore.

HAVING roafted your Lead ore reduce it to a fine powder; mix it with twice its weight of the black flux, and one fourth of its weight of clean iron-filings and borax; put the whole into a crucible capable of containing at leaft thrice as much; over all put Sea-falt four fingers thick; cover the crucible; lute the juncture: dry the whole with a gentle heat, and fet it in a melting furnace.

Make the crucible moderately red: you will hear the Sea-falt decrepitate, and after the decrepitation a fmall hiffing in the crucible. Keep up the fame degree of fire till that be over.

Then throw in as many coals as are neceffary to complete the operation entirely, and raife the fire fuddenly, fo as to bring the whole mixture into perfect fufion. Keep up this degree of fire for a quar-

ter

ter of an hour, which is time fufficient for the
precipitation of the Regulus.

When the operation is finifhed, which may be
known by the quietnefs of the matter in the crucible, and by a bright vivid flame that will rife from
it, take the crucible out of the furnace, and feparate the Regulus from the fcoria.

O B S E R V A T I O N S.

ALL Lead ore contains a good deal of Sulphur,
which muft be firft feparated from it by roafting :
and as this kind of ore is apt to fly when firft expofed to the fire, it is proper to keep it covered till it be
thoroughly heated.   Another precaution to be
ufed, in roafting this ore, is not to give it too
great a heat, but to keep the veffel which contains
it juft moderately red ; becaufe it eafily turns
clammy, which occafions it to ftick to the veffel.

The Iron that is added, and mixed with the flux,
abforbs the Sulphur which may happen to remain,
even after roafting : it helps alfo to feparate from the
Lead fome portions of femi-metal, efpecially of Antimony, which are frequently mixed with this ore.

There is no fear left the Iron mix with the Lead
in fufion, and adulterate it : for thefe two metals are
incapable of contracting any union together, when
each has its metalline form.

Nor is there any reafon to apprehend left the
Iron fhould, by its refractory quality, obftruct the
fufion of the mixture; for though this metal be
not fufible when alone, yet, by the union it contracts with the matters it is defigned to abforb, it
becomes fo to fuch a degree as in fome meafure to
perform, on this occafion, the office of a flux.

The government of the fire is a point of great
confequence in this operation. It is neceffary to apply but a moderate degree of heat at firft : for, when
the metallic earth of the Lead, combining with the
phlogifton,

phlogifton, acquires the metalline form, it fwells up
in fuch an extraordinary manner, that there is great
danger left the matter fhould overflow, and run all
out of the containing veffel. With a view therefore to avoid this inconvenience, we direct a very
large crucible to be ufed. This heaving of the
Lead, at the inftant of its reduction, is attended
with a noife like the whiftling of wind.

Notwithftanding all the precautions that can be
ufed to prevent the reduction from taking place too
haftily, and fo occafioning the effufion of the matter, it often happens that, on raifing the fire in
order to bring the mixture into fufion, the hiffing
fuddenly begins again, and is very loud. In that
cafe all the apertures of the furnace muft immediately be fhut clofe, in order to choak and fuffocate
the fire : for, if this be neglected, the matter in
the crucible will fwell up, make its way through
the luting of the juncture, nay, pufh up the
cover, and run over. This accident is to be appre-
hended during the firft five or fix minutes after you
raife the fire in order to melt the mixture. This effufion of the matter is accompanied with a dull
flame, a thick, grey and yellow fmoke, and a noife
like that of fome boiling liquor. When you obferve
thefe feveral phenomena you may be fure the matter is run out of the crucible, either in the manner
above defcribed, or by making its way through
fome cracks in the veffel, and confequently that the
operation is fpoiled.

Moreover, this event infallibly follows whenever
a bit of coal happens to fall into the crucible ; and
this is one reafon why it is neceffary to cover it.

You may be certain that the operation hath fuc-
ceeded if the fcoria be fmooth when cold, and have
not in part efcaped through the lute ; if the Lead be
not difperfed in globules through the whole mafs of
the matter contained in the crucible, but is, on the
contrary,

contrary, collected at the bottom, in the form of a solid Regulus, not very shining, but of a bluish cast, and ductile. Moreover the scoria ought, in the present case, to be hard and black, and should not appear full of holes like a sieve, except only in that part which was contiguous to the Salt.

. Here it is proper to observe that the Sea-salt doth not mix with the scoria, but floats upon it. After the operation it is black; which colour it gets, no doubt, from the charred parts of the flux. The absence of these signs shews the operation to have miscarried.

When the ore to be smelted is pyritose and refractory, it may be roasted at first with a much stronger degree of fire than is used for ores that are fusible; because the martial earth, and the unmetallic earth, which are always mixed in pyritose matters, hinder it from growing readily soft in the fire. Besides, such an ore requires a greater quantity of the black flux and of borax to be mixed with it, and a higher degree of fire to fuse it.

It is generally needless to mix iron filings with this sort of ore; because the martial earth, with which pyritose matters are always accompanied, is reduced during the operation by the help of the black flux, which for that purpose is mixed with it in a large proportion, and furnishes a quantity of iron sufficient to absorb the heterogeneous minerals mixed with the Lead.

Yet, if it should be observed that the pyrites which accompany the Lead ore are arsenical, then, as such pyrites contain but a small quantity of ferruginous earth, iron filings must be added; which are, on this occasion, so much the more necessary for absorbing the Arsenic, as this mineral remains in part confounded with the ore, is reduced to a Regulus during the operation, unites with the Lead, and destroys

deftroys a great deal of it by procuring its vitrification.

, The Lead obtained from fuch pyritofe ores is
commonly not very pure; it is blackifh and fcarce
ductile; qualities communicated to it by a fmall
mixture of Copper in the pyrites, which always contain more or lefs thereof. We fhall prefently fhew
the method of feparating Lead from Copper.

Lead ore may alfo be reduced by melting it amidft
coals. For that purpofe firft kindle a fire in the
furnace in which you intend to melt your ore; then
put a layer of your ore immediately upon the lighted
coals, and cover it with another layer of coals.

Though the melting furnace ufed for this operation be capable of giving a confiderable heat, yet it
is neceffary further to increafe the force of the fire
by the means of a good pair of perpetual bellows,
which will produce an effect like that of a forge. The
ore melts, the earth of the Lead unites with the phlogifton of the coals, and fo is reduced to metal, which
runs through the coals, and falls into an earthen veffel placed at the bottom of the furnace to receive it.
Care muft be taken to keep this veffel well filled
with charcoal-duft, to the end that the Lead may be
in no danger of calcination while it continues there;
the charcoal-duft conftantly furnifhing it with phlogifton to preferve its metalline form.

The earthy and ftony matters that accompany the
ore are fcorified by this fufion, juft as they are by the
other which is performed in a clofe veffel. With regard to the Sulphur and Arfenic, they are fuppofed
to have been firft accurately feparated from the ore
by roafting. This is the method commonly employed for fmelting Lead ore at the works.

PROCESS II.

To separate Lead from Copper.

WITH luting earth and charcoal-dust make a flat vessel, widening upwards, and large enough to contain your metalline mass. Set it shelving downwards from the back towards the fore-part; and in the fore-part, at the bottom, make a little gutter communicating with another vessel of the same nature, placed near the former and a little lower. Let the mouth of the gutter within side the upper vessel be narrowed, by means of a small iron plate fixed across it, while the loam is yet soft; so as to leave a very small aperture, in the lower part of this canal, sufficient to discharge the Lead as it melts. Dry the whole by placing lighted coals around it.

When this apparatus is dry, put your mixed mass of Copper and Lead into the upper vessel: both in that, and in the other vessel, light a very gentle fire of wood or charcoal, so as not to exceed the degree of heat necessary to melt Lead. In such a degree of heat the lead contained in the mixed mass will melt, and you will see it run out of the upper vessel into the lower; at the bottom of which it will unite into a Regulus. When in this degree of heat no more Lead flows, increase the fire a little, so as to make the vessel moderately red.

When no more will run, collect the Lead contained in the lower vessel. Melt it over again in an iron ladle, with a degree of fire sufficient to make the ladle red; throw into it a little tallow or pitch, and while it burns keep stirring the metal; in order to reduce any part of it that may be calcined. Remove the pellicle or thin crust which will form on the surface; squeeze out all the Lead it contains, and then put it to the mass of Copper left in the upper
vessel.

veſſel. Check the fire, and in the ſame manner take off a ſecond ſkin that will form on the ſurface of the Lead. Laſtly, when the metal is ready to fix, take off the ſkin that will then appear on it. The Lead remaining after this will be very pure, and free from all alloy of Copper.

With regard to the Copper itſelf, you will find it in the upper veſſel covered with a thin coat of Lead : and if the Lead mixed with it was in the proportion of a fourth or a fifth part only, and the fire applied was gentle and ſlow, it will retain nearly the ſame form after the operation that the mixed maſs had before.

OBSERVATIONS.

L E A D frequently remains mixed with Copper after the reduction of its ore, eſpecially if the ore was pyritoſe. Tho' Copper be a much more beautiful and more ductile metal than Lead, yet the latter by being alloyed with the former is rendered eager and brittle. This bad quality is eaſily diſcovered by the eye on breaking it : for the ſurface of the broken part appears all granulated ; whereas when it is pure it is more evenly, and reſembles a congeries of ſolid angles. If the Lead be alloyed with a conſiderable quantity of Copper, its colour hath a yellowiſh caſt.

Conſidering the bad qualities which Copper communicates to Lead, it is neceſſary to ſeparate theſe two metals from each other. The method above laid down is the ſimpleſt and the beſt. It is founded on two properties belonging to Lead : the firſt is that of being much more fuſible than Copper ; ſo that it will melt and run in a degree of heat that is not capable of making the Copper even red-hot, which yet is very far from being able to melt it : the ſecond is, that Lead, though it hath an affinity with Copper, and unites very perfectly therewith, yet is not able to diſſolve it without a greater heat than the degree barely neceſſary to fuſe Lead. Hence it comes that Lead may be melted in a Copper veſ-ſel, provided no greater degree of heat be applied than that purpoſe requires. But when the Lead be-comes ſo hot as to be red, fume and boil, it in-ſtantly begins to diſſolve the Copper. For this rea-ſon, it is eſſential to the ſuccefs of our operation that a moderate degree of heat only be applied, and no greater than is requiſite to keep the Lead in fuſion.

Charcoal-duſt is made an ingredient in the com-poſition of the veſſels uſed on this occaſion, in order to prevent the calcination of the Lead. ·

The iron plate, with which the entrance of the gutter within the upper veſſel is narrowed, ſerves to prevent the larger pieces of Copper, which the Lead may carry along with it, from paſſing through : it ſtops them, and allows the Lead to run off alone.

But as theſe parcels of Copper may entirely choak the paſſage, care muſt be taken, when any happen to be ſtopt, to remove them from the entrance of the gutter, and puſh them back into the middle of the veſſel. It is alſo neceſſary to obſerve whether or no the Lead fixes any where in the paſſage; and, if it does, the heat of that part muſt be increaſed, in or-der to melt it and make it run off.

Notwithſtanding all the precautions that can be taken, to hinder the melted Lead from carrying off any Copper with it, it is impoſſible to prevent this inconvenience entirely: and therefore the Lead is melted over again, in order to ſeparate the ſmall por-tion of Copper with which it is ſtill adulterated.

As Copper is much lighter than Lead, if theſe two metals happen to be ſo blended together that the Copper, without being in fuſion and diſſolved by the Lead, is only interpoſed between the parts of the melted Lead, ſo as to ſwim therein, it is then pre-ciſely in the caſe of a ſolid body plunged into a fluid heavier than itſelf, and muſt riſe to the ſurface,

like

like wood thrown into water. It is proper to burn some inflammable matter on this melted Lead, in order to reduce such parts thereof as are conftantly calcining on its furface while it is in fufion; for without this precaution they would be taken off together with the Copper.

The Copper remaining after this feparation is, as we took notice before, ftill mixed with a little Lead. If you defire to feparate it entirely therefrom, you muft put it into a cupel, and expofe it under the muffle to fuch a degree of fire as may convert all the Lead into litharge. This cannot be fo done but that fome of the Copper alfo will be fcorified by the heat of the fire, and by the action of the Lead: but as there is a very great difference between the facility and readinefs with which thefe two metals calcine, the portion of Copper that is calcined, while the whole Lead is turning into litharge, is fcarce worth confidering.

The Lead, though carefully feparated from the Copper by the procefs here delivered, is not yet abfolutely pure: fometimes it is alloyed with Gold, and almoft always contains fome Silver. If you would free the Lead as much as poffible from any mixture of thefe two metals, you muft convert it into glafs, feparate the remaining bead, and afterwards reduce this glafs of Lead. But, as thefe two perfect metals are of no prejudice to the Lead, it is not ufual to feparate them from it, unlefs they be in a fufficient proportion to defray the charge, and produce fome profit befides.

When we examine by the cupel the juft proportion of Gold and Silver that an ore or a mixed metalline mafs will yield, we make a previous affay of the Lead to be employed in the operation, and afterwards, in our eftimate, deduct a proper allowance for the quantity of fine metal due to the Lead made ufe of.

PRO-

P R O C E S S III.

The Calcination of Lead.

TAKE what quantity of Lead you pleafe; melt it in one or more unglazed earthen pans: a dark grey powder will be found on its furface. Keep ftirring the metal inceffantly till it be wholly converted into fuch a powder, which is the *Calx of Lead*.

O B S E R V A T I O N S.

As Lead is a very fufible metal, and in that refpect greatly refembles Tin, moft of the obfervations we made on the calcination of Tin may be applied here.

In the calcination of all metals, and particularly in this of Lead, there appears a fingular phenomenon which is not eafily accounted for. It is this: though thefe matters lofe a great deal of their fubftance, either by the diffipation of their phlogifton, or becaufe fome of the metal, perhaps, exhales in vapours, yet when the calcination is over their calxes are found to be increafed in weight, and this increafe is very confiderable. An hundred pounds of Lead, for example, converted into Minium, which is nothing but a calx of Lead brought to a red colour by continuing the calcination, are found to gain ten pounds weight; fo that for an hundred pounds of Lead we have one hundred and ten pounds of Minium: a prodigious and almoft incredible augmentation, if it be confidered that, far from adding any thing to the Lead, we have on the contrary diffipated part of it.

To account for this phenomenon Natural Philofophers and Chymifts have invented feveral ingenious hypothefes, but none of them entirely fatisfactory.

As

As we have no eftablifhed theory to proceed upon, we fhall not undertake to explain this extraordinary fact.

PROCESS IV.

To prepare Glaſs of Lead.

TAKE two parts of Litharge, and one part of pure cryftalline Sand; mingle them together as exactly as poffible, adding a little Nitre and Seafalt: put this mixture into a crucible of the moft folid and moft compact earth. Shut the crucible with a cover that may perfectly clofe it.

Set the crucible thus prepared in a melting furnace; fill the furnace with coals; light the fire gradually, fo that the whole may be flowly heated: then raife the fire fo as to make the crucible very red, and bring the matter it contains into fufion; keep it thus melted for a quarter of an hour.

Then take the crucible out of the furnace, and break it: in the bottom thereof you will moft commonly find a fmall button of Lead, and over it a tranfparent Glafs, of a yellow colour nearly refembling that of amber. Separate this Glafs from the little button of metal, and from the faline matters which you will find above it.

OBSERVATIONS.

PURE Lead, being expofed to a ftrong fire without any additament, turns to Litharge; which is a fcaly fort of fubftance, more or lefs yellowifh, fhining, and foft to the touch. This is the firft advance to the Vitrification of Lead. The large refineries of Gold and Silver by the means of Lead furnifh a great quantity of this material. It is fometimes whitifh, and is then called *Litharge of Silver*; fometimes

times

times yellow, and then bears the name of *Litharge of Gold.* The difference of its colour depends on the degree of fire it hath undergone, and on the metalline substances vitrified with it.

- Litharge alone is very fusible, and being exposed to the fire is easily converted into Glass: but this Glass of Lead, made without additament, is so active, so penetrating, and so apt to swell, that it can scarcely be made use of when pure. We are obliged in some sort to clog it, by uniting it with some vitrifiable matter that is not so subtile, such as Sand; and it is for this reason, not to render the mixture more fusible, that we have directed the addition of one third part of Sand to two thirds of Litharge.

The Nitre and Sea-salt, prescribed as ingredients in the mixture, are designed to procure an equal fusion of the whole. For, as the Sand is lighter and less fusible than the Litharge, it will partly rise towards the upper part of the crucible when that matter first begins to flow; in consequence whereof the contents of the upper part will be much more difficult to melt, and form a Glass much more compact than that below: but the Nitre and Sea-salt possessing the upper part of the crucible, because they are still lighter than the Sand, and being in their own nature very efficacious fluxes, on account of their great fusibility, they quickly bring about the fusion of those particles of Sand, which, having escaped the action of the Litharge, may have risen unvitrified to its surface.

The most difficult thing to procure, and yet the most necessary to the success of this operation, is a crucible of earth so firm and compact as not to be penetrated by the Glass of Lead, which corrodes and makes its way through every thing.

The precaution of chusing a crucible, that shall contain a good deal more than the matter to be vitrified,

trified, is a neceffary one, becaufe Litharge and Glafs of Lead are very apt to fwell.

The rule to keep the crucible clofe fhut is alfo indifpenfably neceffary, to prevent any bit of charcoal, or other inflammable matter, from falling into it: for when this happens it occafions a reduction of the Lead, which is always attended with a fort of effervefcence, and fuch a confiderable heaving, that commonly moft of the mixture runs over the crucible. For the fame reafon it is very proper, before you expofe the mixture to the fire, to examine whether or no it contains any matter capable of furnifhing a phlogifton during the operation; and if it does, to remove that matter with great care.

The little button of Lead, found at the bottom of the crucible after the operation, comes from a fmall portion of Lead that is commonly left in Litharge, unlefs you prepare it carefully yourfelf, and do not take it from the fire till you are fure of having deftroyed all the Lead. Befides, this fmall portion of Lead can be of no prejudice to the operation, becaufe it cannot communicate its phlogifton to the reft of the matter.

The revivifying of Litharge, of the Calx, and of the Glafs of Lead, may be obtained by the fame proceffes as the reduction of its ore.

PROCESS V.

Lead diffolved by the Nitrous Acid.

PUT into a matrafs fome *Aqua Fortis* precipitated like that ufed to diffolve Silver; weaken it by mixing therewith an equal quantity of common water; fet the matrafs in a hot fand-bath; throw into it, little by little, fmall bits of Lead, till you fee that no more will diffolve. *Aqua Fortis* thus lowered will diffolve about a fourth of its weight of Lead.

There

There is gradually formed upon the Lead, as it, diffolves, firſt a grey powder, and afterwards a white cruſt, which at laſt hinder the folvent from acting on the remaining part of the metal; and therefore the liquor ſhould be made to boil, and, the veſſel ſhould be ſhaken to remove thoſe impediments, by which means all the Lead will be diſſolved.

OBSERVATIONS.

LEAD very much refembles Silver, with reſpect to the phenomena which attend its diſſolution in Acids. For example, the Nitrous Acid muſt be very pure and uncontaminated with the Vitriolic or, Marine Acid, to qualify it for keeping the Lead in ſolution : for, if it be mixed with either the one or the other of theſe Acids, the Lead will precipitate in the form of a white powder as faſt as it diſſolves; which is juſt the cafe with Silver.

If the Vitriolic Acid be mixed with the Nitrous, the precipitate will be a combination of the Vitriolic Acid with Lead; that is, a Neutral Metallic Salt, or Vitriol of Lead. If the Acid of Sea-falt be mixed therewith, the precipitate will be a *Plumbum corneum*; that is, a Metallic Salt refembling the *Luna cornea*.

When all the Lead is diſſolved as above deſcribed, the liquor appears milky. If it be kept warm over the fire till little cryſtals begin to appear on its ſurface, and afterwards left to ſtand quiet, in a certain time there will be found at the bottom a greyiſh powder, which being tried on Gold is Mercurial enough to whiten it. Little globules of Quick-filver are even diſcernibie in it.

We owe this obſervation, together with this manner of proving the exiſtence of Mercury in Lead, and of procuring it from thence, to M. Groſſe, who hath given an account of his proceſs in the Memoirs

of

of the Academy of Sciences, from whence we have copied the defcription of the operation in hand.

The folution being quickly poured off by inclination from the grey mercurial precipitate is ftill milky, and depofites another white fediment. When this fecond precipitate falls the liquor becomes clear and limped, and is then of a fine yellow colour, like a folution of Gold. On this gold-coloured folution, and on the two precipitates above-mentioned, M. Groffe made feveral obfervations, the chief of which we fhall here infert.

The yellow liquor affects the tongue at firft with a tafte of fweetnefs; but afterwards vellicates it very fmartly, and leaves on it a ftrong fenfation of acrimony, which continues for a long time.

Alkalis precipitate the Lead fufpended in this liquor, juft as they do all other metals diffolved by Acids; and this precipitate of Lead is white.

Sea-falt, or Spirit of Salt, feparates the Lead from its folvent, and precipitates it, as we obferved before, into a *Plumbum corneum:* but this precipitate differs from the *Luna cornea,* as being very foluble in water; whereas the *Luna cornea* will not diffolve in it at all; or at leaft diffolves therein with great difficulty, and in a very fmall quantity. This *Plumbum corneum* diffolved in water is again precipitated by the Vitriolic Acid. M. Groffe obferves that this forms an exception to the eighth column of Mr. Geoffroy's Table of affinities; in which the Acid of Sea-falt is marked as having a greater affinity than any other Acid with Metallic fubftances.

Our folution of Lead is alfo precipitated in a white powder by feveral Neutral Salts; fuch as Vitriolated Tartar, Alum, and common Vitriol. It is by the means of double affinities that thefe Neutral Salts effect this precipitation.

Even pure water alone is capable of precipitating the Lead of our folution, by weakening the Acid, and

and thereby difabling it from keeping the metal fufpended. ·

· Laftly, as all the folutions of metals in Acids are nothing but Neutral metallic Salts in a fluid' form, fo if the folution of Lead be evaporated over the fire, it will fhoot into very beautiful cryftals, about the bignefs of hemp-feed, fhaped like regular pyra-mids having fquare bafes. Thefe cryftals are yel-lowifh, and have a fweet faccharine tafte : but what is.moit.fingular in them is, that, as they confift of the Nitrous Acid combined with Lead, which ma-nifeftly contains a great deal of phlogifton, they conftitute a Nitrous Metallic Salt, which ·has the property of deflagrating in a crucible, without the addition of. any other inflammable matter. It is extremely hard to diffolve this Salt in water.

·.The grey mercurial precipitate which whitens Gold, and in which little globules of running Mer-cury are perceivable, is far from being pure Mer-cury. This metallic fubftance makes but a fmall part thereof : for it is an affemblage 1. of little cryf-tals of the fame nature with thofe afforded by the evaporated folution ; 2. of a portion of the white matter, or powder, which renders the folution milky ; 3. of a.grey powder which M. Groffe confiders as the only mercurial part ; 4. and laftly, of little particles of Lead that have efcaped the action of the folvent ; efpecially if a little more Lead than the Acid is capable of diffolving were added with a· view to faturate it entirely, as in the prefent procefs.

.By means of motion and heat the fmall parcels of Mercury may be amalgamated with the Lead.

. That Mercury fhould be found intire and in glo-bules in.the Spirit of Nitre, which very eafily dif-folves that metallic fubftance, will not be furprizing to thofe who reflect that, in the prefent cafe, the Acid is faturated with Lead, ·with which it has a greater affinity than with Mercury ; as appears by M.

M. Geoffroy's Table of Affinities, where, in the column that hath the Nitrous Acid at top, Lead is placed above Mercury. Agreeably to this, if Lead be prefented to a folution of Mercury in Spirit of Nitre, the Lead will be diffolved, and as the diffolution thereof advances the Mercury will precipitate.

Hence it appears that, in order to find any Mercury in the fpontaneous precipitate of Lead diffolved by the Nitrous Acid, it is neceffary that the Acid be entirely faturated with Lead; or elfe that portion of the Acid which remains unfaturated will diffolve the Mercury.

With regard to the white powder that renders the folution milky, and afterwards precipitates, it is nothing but a portion of the Lead, which, not being intimately united with the Acid, falls in part of its own accord. It is a fort of calx of Lead, which being expofed to the fire becomes partly glafs, and partly Lead, becaufe it ftill retains fome of its phlogifton.

C H A P. VII.

Of MERCURY.

PROCESS I.

To extract Mercury from its Ore, or to revivify it from Cinabar.

PULVERIZE the Cinabar from which you would extract the Mercury; with this powder mix an equal part of clean iron filings; put the mixture into a retort of glafs or iron, leaving at leaft one third part thereof empty. Set the retort thus prepared in a fand-bath, fo that its body may.be quite buried in the fand, and its neck decline confiderably downwards: fit on a receiver half filled with water, and let the nofe of the retort enter about half an inch into the water.

Heat the veffels fo as to make the retort moderately red. The Mercury will rife in vapours, which will condenfe into little drops, and fall into the water in the receiver. When you fee that nothing more comes over with this degree of heat, increafe it, in order to raife what Mercury may ftill be left. When all the Mercury is thus brought over, take off the receiver, pour out the water contained in it, and collect the Mercury.

OBSERVATIONS.

Mercury is never mineralized in the bowels of the earth by any thing but Sulphur; with which it forms a compound of a brownifh red colour, known by the name of *Cinabar*.

Sometimes it is only mixed with earthy and ftony matters that contain no Sulphur; but as this metallic fubftance is never deftitute of its phlogifton, it then has its metalline form and properties. When it is found in this condition, nothing is more eafy than to feparate it from thofe heterogeneous matters. For that purpofe no more is requifite than to diftill the whole with a fire ftrong enough to raife the Mercury in vapours. This mineral is volatile; the earthy and ftony matters are fixed; and a certain degree of heat will effect a complete feparation of what is volatile from what is fixed.

This is not the cafe when Mercury is combined with Sulphur: for this latter mineral is volatile as well as Mercury; and the compound refulting from the union of them both is alfo volatile: fo that if Cinabar were expofed to the fire in clofe veffels,

as.

as it muft be to fave the Mercury, it would be fublimed in fubftance, without being decompofed at all.

In order therefore to feparate thefe two fubftances from each other, we muft have recourfe to the interpofition of fome third, which hath a greater affinity with one of them than the other hath, and no affinity with that other.

Iron hath all the conditions requifite for this purpofe; feeing it hath, as may be feen in the Table, a much greater affinity with Sulphur than Mercury hath, and is incapable of contracting any union with Mercury.

Iron, however, is not the only fubftance that may be employed on this occafion : Fixed Alkalis, Abforbent earths, Copper, Lead, Silver, Regulus of Antimony, have all, as well as Iron, a greater affinity than Mercury with Sulphur. Nay, feveral of thefe fubftances, namely, the faline and earthy Alkalis, as well as Regulus of Antimony, cannot contract any union with Mercury : the reft, to wit, Copper, Lead, and Silver, are indeed capable of amalgamating with Mercury; but then the union which thefe metals contract with the Sulphur prevents it; and even tho' they fhould unite with this metallic fubftance, the degree of heat to which the whole mixture is expofed would foon carry up the Mercury, and feparate it with eafe from thofe fixed fubftances.

In this diftillation the fame cautions muft be obferved as in all others : that is, the veffels muft be flowly heated, efpecially if a glafs retort be ufed; the fire muft be raifed by degrees, and a much ftronger one applied at laft than at firft. This operation particularly requires a very ftrong degree of fire, when there is but a fmall quantity of Mercury left.

After the operation there remains in the retort a compound of Iron and Sulphur, which may eafily be be converted into a *crocus*, by calcining it and burning away the Sulphur. ...

If a Fixed Alkali be employed, a Liver of Sulphur will be found in the retort after the diftillation.

If the Cinabar from which you extract the Mercury be good, you will generally obtain feven eighths of its weight in Quick-filver.

In the prefent operation it is not neceffary to lute on the receiver, becaufe the water, in which the nofe of the retort is plunged, is fufficient to fix the Mercurial vapours. In cafe the Cinabar, from which you intend to feparate the Mercury, be mixed with a great quantity of heterogeneous, but fixed, matters, fuch as earths, ftones, &c. it may be feparated from them by fubliming it with a proper degree of heat, becaufe it is volatile.

The vapours of Mercury are prejudicial, and may excite a falivation, tremors, and palfies; they fhould therefore be always avoided by fuch as work on this mineral.

The oldeft and richeft mine of Mercury is that of Almaden in Spain. It is a fingular property of that mine that, though the Mercury found in it is combined with Sulphur, and in the form of Cinabar, yet no additament is required to procure the feparation of thefe two; the earthy and ftony matter, with which the particles of the ore are incorporated, being itfelf an excellent abforbent of Sulphur.

In the Quick-filver works carried on at this mine they make no ufe of retorts. They place lumps of the ore on an iron grate, which ftands immediately over the furnace. The furnaces which ferve for this operation are clofed at the top by a fort of dome, behind which ftands the fhaft of a chimney that communicates with the fire-place, and gives vent to the fmoke. Thefe furnaces have in their fore-fide fixteen apertures, to each of which is luted an aludel in a horizontal pofition, communicating with a

long

long row of other aludéls placed likewife in an horizontal direction; which aludels fo connected together form one long pipe or canal, the further end whereof opens into a chamber deftined to receive and condenfe all the mercurial vapours. Thefe rows of aludels are fupported from end to end by a terrafs, which runs from the body of the building, wherein the furnaces are erected, to that where the chambers are built that perform the office of receivers.

This is a very ingenious contrivance, and faves much labour, expence, and trouble, that would be unavoidable if retorts were employed.

. That part of the furnace which contains the lumps of ore, ferves for the body of the retort; the row of aludels for its neck; and the little chambers in which thefe canals terminate are actual receivers. The terrafs of communication, which reaches from the one building to the other, is formed of two inclined planes, the lower edges of which, meeting in the middle of the terrafs, rife from thence infenfibly; the one quite to the building where the furnaces are, and the other to that which forms the recipient chambers. By this means, when any Mercury efcapes through the joints of the aludels, it naturally runs down along thefe inclined planes, and fo is collected in the middle of the terrafs, where the inferiour fides of the planes meeting together form a fort of canal, out of which it is eafily taken up.

The celebrated M. de Juffieu having viewed the whole himfelf, in a journey he made to this mine, furnifhed us with this defcription of the work.

P R O C E S S · II.

*To give Mercury, by the aBion of Fire, the. appear-
ance of a Metalline Calx.*

PUT Mercury into feveral little glafs matraffes
· with long and narrow necks.   Stop the matraffes with a little paper, to prevent any dirt from
falling into them.   Set them all in one fand-bath,
fo that they may be furrounded with fand as high as
two thirds of their length.   Apply the ftrongeft degree of heat that Mercury can bear without fubliming : continue this heat without interruption, till all
the Mercury be turned to a red powder.   The operation lafts about three months.

O B S E R V A T I O N S.

' MERCURY treated according to the procefs here
delivered hath all the appearance of a metalline calx,
but it hath no more : for, if it be expofed to a pretty
ftrong degree of fire, it fublimes, and is wholly reduced to running Mercury, without the addition of
any other inflammable matter ; which proves that
during this long calcination it loft none of its phlogifton.

The volatile nature of Mercury, which permits it
not to bear a heat of any ftrength without fubliming, prevents our examining all the effects that fire
is capable of producing on it.   Yet there is reafon
to believe that, as this metallic fubftance refembles
the perfect metals in its weight, its fplendour, and
a brilliancy whch refifts all the impreffions of the
air without alteration, it would like them be un-
changeable by the greateft force of fire, if it were
fixed enough to bear it.

In order to give Mercury the form of a metalline
calx, it muft neceffarily be expofed for about three

months

months together, to the utmoſt heat it can bear without ſubliming, as is above directed. Boerhaave kept it digeſting in a leſs heat for fifteen years ſucceſſively, both in open and in cloſe veſſels, without obſerving it to ſuffer the leaſt change; except that there was formed upon its ſurface a ſmall quantity of a black powder, which was reduced to running Mercury by trituration alone.

Mercury thus converted to a red powder is known in chymiſtry and medicine by the name of *Mercury precipitated per ſe :* a title proper enough, as it is actually reduced to the form of a precipitate, and that without any additament ; but very improper on the other hand, conſidering that in reality this Mercury is not a precipitate, as not having been ſeparated from any menſtruum in which it was diſſolved.

PROCESS III.

To diſſolve Mercury in the Vitriolic Acid. Turbith Mineral.

PUT Mercury into a glaſs retort, and pour on it thrice its weight of good Oil of Vitriol. Set the retort in a ſand-bath ; fit on a recipient ; warm the bath by degrees till the liquor juſt ſimmer. With this heat the Mercury will begin to diſſolve. Continue the fire in this degree till all the Mercury be diſſolved.

OBSERVATIONS.

THE Vitriolic Acid diſſolves Mercury pretty well : but for this purpoſe the Acid muſt be very hot, or even boil ; and then too it is a very long time before the diſſolution is completed. We have directed the operation to be performed in a retort ; becauſe this ſolution is uſually employed to make another preparation called *Turbith Mineral,* which requires that as much as poſſible of the Acid ſolvent be abſtracted by

diftillation. Having therefore diffolved your Mercury in the Vitriolic Acid, if you will now prepare
the Turbith, you muft, by continuing to heat the
retort, drive over all the liquor into the receiver,
and diftill till nothing remains but a white pow-
dery matter : then break the retort : pulverize its
contents in a glafs mortar, and thereon pour common water, which will immediately turn the white
matter of a lemon colour ; wafh this yellow matter
in five or fix warm waters, and it will be what is
called in medicine *Turbith Mineral*; that is, a combination of the Vitriolic Acid with Mercury, five
or fix grains whereof is a violent purgative, and
alfo an emetick ; qualities which it poffeffes in
common with the Vegetable Turbith, whofe name
it hath therefore taken.

There rifes out of the retort, both while the
Mercury is diffolving, and while the folvent is abftracting, a weak Spirit of Vitriol ; becaufe a great
part of the Acid remains united with the Quick-
filver, which at laft appears in the form of a white
powder : fo that, if you do not incline to fave the
Acid which rifes on this occafion, you may, inftead
of drawing off the liquor in a retort, evaporate it
in a glafs bafon fet on a fand-bath, which will be
much fooner done.

It is very remarkable that, on this occafion, the
Mercury may be expofed, without any danger of
fublining, to a much greater heat than it is capable
of bearing when not combined with the Vitriolic
Acid ; which fhews that this Acid hath the property of fixing Mercury to a certain degree.

The white matter, that remains after the evaporation of the fluid, is one of the moft violent corrofives, and would prove an actual poifon if taken
internally. By wafhing it feveral times in warm
water it is freed from a great deal of its Acid, and
fo confiderably fweetened. The proof is this ; if
the water ufed in wafhing the Turbith be evapo-

rated, there remains after the evaporation a matter in form of a Salt, that being set in a cellar runs into a liquor called *Oil of Mercury*, which is a powerful corrosive. Several authors further direct Spirit of Wine to be burnt on the Turbith, to sweeten it still more.

· If instead of washing the white matter that remains after the moisture is drawn off, fresh Oil of Vitriol be poured on it, and then abstracted as before ; this treatment being repeated two or three times, there will at last remain in the retort a matter having the appearance of an oil, which resists the action of the fire, and cannot be desiccated : qualities which are owing to the great quantity of Acid particles thus united with the Mercury. This Oil of Mercury is one of the most violent corrosives. The Mercury may be separated therefrom, by precipitating it with an Alkali, or a metallic substance that hath more affinity than Mercury with the Vitriolic Acid : Iron, for instance, may be employed in this precipitation. Mercury thus separated from the Vitriolic Acid need only be distilled to recover the form of Quick-silver.

PROCESS IV.

To combine Mercury with Sulphur. Æthiops Mineral.

MIX a dram of Sulphur with three drams of Quick-silver, by triturating the whole in a glass mortar with a glass pestle. By degrees, as you triturate, the Mercury will disappear, and the matter will acquire a black colour. Continue the triture till you cannot perceive the least particle of running Mercury. The black matter you will then have in the mortar is known in medicine by the name of *Ethiops Mineral*. An Æthiops may also be made by fire in the following manner.

In

· In a fhallow unglazed earthen pan melt one part of flowers of Sulphur : add three parts of running Mercury, making it fall into the pan in the form of fmall rain, by fqueezing it through chamoy leather. Keep flirring the mixture with the fhank of a to-bacco-pipe all the while the Mercury is falling ; you will fee the matter-grow thick and acquire a black colour. When the whole is thoroughly mixed, fet fire to it with a match, and let as much of the Sulphur burn away as will flame.

O B S E R V A T I O N S.

MERCURY and Sulphur unite together with great eafe ; cold triture alone is fufficient to join them. By this means the Mercury is reduced into exceed-ing fmall atoms, and combines fo perfectly with the Sulphur that the leaft veftige thereof is not to be feen.

Sulphur is not the only matter which being rub-bed with Mercury will deftroy its form and fluidity: all fat fubftances that have any dégree of confiftence, fuch as the fat of animals, balfams, and refins, are capable of producing the fame effect. This metal-lic fubftance, being triturated for fome time in a mortar with thefe matters, becomes at laft invifible, and communicates to them a black colour. When thus divided by the interpofition of heterogeneous particles, it is faid to be *Killed.* But Mercury doth not contract fuch an intimate union with thefe other matters as it doth with Sulphur.

. The Æthiops prepared by fufion is a more per-fect and accurate combination of Mercury and Sul-phur than the other : for, the quantity of Sulphur directed to be ufed in making it being much greater than is abfolutely neceffary to fix the Mercury, the redundant Sulphur is deftroyed by burning, and none left but what is moft intimately united with the Mercury, and hindered by the union it hath

I   con-

contracted with that metallic fubftance from being fo eafily confumed. The Æthiops therefore, which is prepared by fufion and burning the Sulphur, contains a much greater proportion of Mercury than that which is made by fimple triture; fo that in Medicine it ought to be prefcribed in different cafes, and in fmaller dofes.

· If no more Sulphur than is juft neceffary to kill the Mercury be added to it at firft, it will be difficult to obtain a perfect mixture; becaufe that quantity is very fmall : it is better therefore to employ at once the quantity above directed.

PROCESS V.

To fublime the combination of Mercury and Sulphur into Cinabar.

GRIND to powder Æthiops mineral prepared by fire. Put it into a cucurbit; fit thereto a head; place it in a fand-bath, and begin with applying fuch a degree of heat as is requifite to fublime Sulphur. A black matter will rife, and adhere to the fides of the veffel. When nothing more will rife with this degree of heat, raife the fire fo as to make the fand and the bottom of the cucurbit red; and then the remaining matter will fublime in the form of a brownifh red mafs, which is true *Cinabar,*

OBSERVATIONS.

ÆTHIOPS Mineral requires nothing but fublima·tion to become true Cinabar, like that found in Quick-filver mines : but our Æthiops contains ftill more Sulphur than ought to be in the compofition of Cinabar; for which reafon we have directed the degree of fire applied at firft to be no greater than that which is capable of fubliming Sulphur. As Cinabar, though confifting of Mercury and Sul-

phur,

phur; is yet much lefs volatile than either of thefe
fubftances alone ; which probably arifes from the
Vitriolic Acid contained in the Sulphur; therefore, if there be any redundant Sulphur in the
Æthiops, which hath not contracted an intimate
union with the Mercury, it will fublime by itfelf
in this firft degree of heat. Some mercurial particles alfo will rife with it, and give it a black colour.

Cinabar contains no more Sulphur than about a
fixth or feventh part of its weight : fo that, inftead
of employing the common Æthiops to make it, it
would be better to prepare one on purpofe that
fhould contain much lefs Sulphur ; becaufe too
much Sulphur prevents the fuccefs of the operation
by blackening the Sublimate. Indeed in whatever
manner you go about it, the Cinabar always appears black at firft : but when it is well prepared,
and contains no more than its due proportion of
Sulphur, the blacknefs is only external. This black
coat therefore may be taken off : and then the internal part will appear of a fine red, and, if fublimed a fecond time, will be very beautiful.

As artificial Cinabar hath the fame properties
with the native, it may be decompofed by the
fame means : fo that, if you want to extract the
Mercury out of it, recourfe muft be had to the procefs above delivered for working on Cinabar ores.

P R O C E S S VI.

To diffolve Mercury in the Nitrous Acid. Sundry
Mercurial Precipitates.

PUT into a matrafs the quantity of Mercury
you intend to diffolve : pour on it an equal
quantity of good Spirit of Nitre, and fet the matrafs
in a fand-bath moderately heated. The Mercury
will diffolve with the phenomena that ufually attend
the

the diffolutions of metals in this Acid. When the
diffolution is completed let the liquor cool. You
will know that the Acid is perfectly faturated, if
there remain at the bottom of the veffel, notwithftanding the heat, a little globule of Mercury that
will not diffolve.

OBSERVATIONS.

MERCURY diffolves in the Nitrous Acid with
much more facility, and in much greater quantity,
than in the Vitriolic ; fo that it is not neceffary, on
this occafion, to make the liquor boil. This folution when cold yields cryftals, which are a Nitrous
Mercurial Salt. If you defire to have a clear limpid folution of Mercury, you muft employ an *Aqua
Fortis* that is not tainted with the Vitriolic or Marine
Acid : for, the affinity of thefe two Acids with
Mercury being greater than that of the Nitrous
Acid, they precipitate it in the form of a white
powder, when they are mixed with the folvent.

Mercury thus precipitated in a white powder, out
of a folution thereof in the Spirit of Nitre, is ufed
in Medicine. To obtain this precipitate, which is
known by the name of the *White Precipitate*, Sea-
falt diffolved in water together with a little Sal Ammoniac is ufed ; and the precipitate is wafhed feveral
times in pure water, without which precaution it
would be corrofive, on account of the great quantity of the Marine Acid which it would contain.

The preparation known by the name of *Red Precipitate* is alfo obtained from our folution of Mercury in Spirit of Nitre. It is made by abftracting
all the moifture of the folution, either by diftillation
in a retort, or by evaporation in a glafs bafon fet on
a fand-bath. When it begins to grow dry it appears
like a white ponderous mafs. Then the fire is made
ftrong enough to drive off almoft all the Nitrous
Acid, which, being now concentrated, rifes in the

form

form of red vapours. If thefe vapours be catched
in a receiver, they condenfe into a liquor, which is
a very ftrong and vaftly fmoking Spirit of Nitre.

By degrees, as the Nitrous Acid is forced up by
the fire, the mercurial mafs lofes its white colour,
and becomes firft yellow, and at laft very red.
When it is become entirely of this laft colour the
operation is finifhed. The red mafs remaining is a
Mercury that contains but very little Acid, in comparifon of what it did while it was white: and indeed the firft white mafs is fuch a violent corrofive,
that it cannot be ufed in Medicine ; whereas, when it
is become red, it makes an excellent efcharotic, which
thofe who know how to ufe it properly apply with
very great fuccefs, particularly to venereal ulcers.

This preparation is very improperly called a *Precipitate :* for the Mercury is not feparated from the
Spirit of Nitre by the interpofition of any other
fubftance, but only by evaporating the Acid. It
is alfo called *Arcanum Corallinum.*

It muft be obferved that Mercury, by its union
with the Nitrous Acid, acquires a certain degree of
fixity : for the red precipitate is capable of fuftaining, without being volatilized, a ftronger degree
of heat than pure Mercury can ; which, as we obferved before, is the property of Turbith mineral
alfo.

PROCESS VII.

*To combine Mercury with the Acid of Sea-falt.
Corrofive Sublimate.*

EVAPORATE a folution of Mercury in the
Nitrous Acid till there remain only a white
powder, as mentioned in our obfervations on the
preceding procefs. With this powder mix as much
Green Vitriol calcined to whitenefs, and as much
decrepitated Sea-falt, as there was Mercury in the
folution.

, folution. Triturate the whole carefully in a glafs mortar. Put this mixture into a matrafs, fo that two thirds thereof may remain empty, having firft cut off the neck to half its length: or inftead thereof you may ufe an apothecary's phial. Set your vef-fel in a fand-bath, and put fand round it as high as the contents reach. Apply a moderate fire at firft, and raife it by flow degrees. Vapours will begin to afcend. Continue the fire in the fame degree till they ceafe. Then ftop the mouth of the veffel with paper, and increafe the fire till the bottom of the fand-bath be red-hot. With this degree of heat a Sublimate will rife, and adhere to the infide and upper part of the veffel, in the form of white, femi-tranfparent cryftals. Keep up the fire to the fame degree till nothing more will fublime. Then let the veffel cool; break it, and take out what is fublimed, which is *Corrofive Sublimate.*

OBSERVATIONS.

In this operation the mineral Acids act, and arc acted upon, in a remarkable manner. Every one of the three is at firft neutralized, or united with a dif-ferent bafis; the Vitriolic being combined with Iron; the Nitrous with Mercury, forming therewith a Ni-trous Mercurial Salt; and the Marine with its natural Alkaline bafis. The Vitriolic and Nitrous Acids, which are united with metalline fubftances, being both ftronger than the Acid of Sea-falt, ftrive to expel it from its bafis, in order to combine with it themfelves; but the Vitriolic Acid, being the ftrongeft of the two, would take fole poffeffion of this bafis exclufive of the Nitrous, which would continue united with the Mercury, if the Marine Acid had not a greater affinity than the Nitrous with this metallic fubftance. This Acid therefore being expelled from its bafis by the Vitriolic Acid, and fo fet at liberty, muft unite with the Mercury, and feparate the Nitrous Acid from it; which now hath no refource but to unite with the Iron deferted

by

by the Vitriolic Acid. But as all thefe changes are brought about by the means of a confiderable heat, and as the Nitrous Acid hath not a very firm connection with the Iron, it is driven off by the force of the fire; and this it is which we fee rife in vapours during the operation. It alfo carries off with it fome parts of the other two Acids, but in a very fmall quantity. After the operation therefore there remains, 1. A combination of the Vitriolic Acid with the bafis of Sea-falt; that is a Glauber's Salt: 2. A red martial earth, being that which was the bafis of the Vitriol: thefe two fubftances are blended together, and remain at the bottom of the veffel becaufe of their fixity: 3. A combination of the Marine Acid with Mercury; both of which being volatile, they afcend together into the upper part of the veffel, and there form a Corrofive Sublimate.

If we reflect on this procefs with attention, and recollect diftinctly the affinities of the different fub-ftances employed in it, we fhall perceive that it is not neceffary to make ufe of all thofe matters, and that the operation would fucceed though feveral of them were left out.

Firft, the Nitrous Acid may be omitted; fince, as hath been fhewn, it is not an ingredient in the Sublimate, but is diffipated in vapours during the operation. From an accurate mixture therefore of Vitriol, Sea-falt, and Mercury, a Corrofive Subli-mate muft be obtained: for as the Acid of the Vi-triol will difengage the Acid of Sea-falt, the latter will be at liberty to combine with the Mercury, and fo form the compound we are in queft of.

Secondly, if we make ufe of Mercury diffolved by the Nitrous Acid, we may omit the Vitriol; be-caufe the Nitrous Acid having a greater affinity than the Marine Acid itfelf with the bafis of Sea-falt, and the Acid of Sea-falt having a greater affinity than the Nitrous Acid with Mercury, thefe two Acids will naturally make an exchange of the bafes with

which

which they are united: the Nitrous will lay hold on the bafis of Sea-falt, and form a quadrangular Nitre, while the Marine Acid will join the Mercury, and with it form a Corrofive Sublimate.

Thirdly, inftead of Sea-falt its Acid only may be employed; which being mixed with the folution of Mercury in the Spirit of Nitre, will, by virtue of its greater affinity with that metallic fubftance, feparate it from the Nitrous Acid, unite with it, and form a white mercurial precipitate, which need only be fublimed to become the combination required.

Fourthly, inftead of Mercury diffolved in the Nitrous Acid, Mercury diffolved by the Vitriolic Acid, or Turbith, may be ufed; only mixing Sea-falt therewith: for thefe two faline fubftances will mutually decompound each other, by virtue of the affinities of their Acids, and for the fame reafons that Sea-falt and the Mercurial Nitrous Salt decompound each other. The Vitriolic Acid quits the Mercury with which it is combined, to unite with the bafis of the Sea-falt; and the Acid of this Salt being expelled by the Vitriolic, combines with the Mercury, and confequently forms our Corrofive Sublimate. In this cafe a Glauber's Salt remains after the fublimation.

Thefe feveral methods of preparing Corrofive Sublimate are never ufed, becaufe each of them is attended with fome inconvenience; fuch as requiring too long triture, yielding a Sublimate lefs corrofive than it fhould be, or a fmaller quantity of it. We muft, however, except the laft; which was invented by the late Mr. Boulduc, of the Academy of Sciences, who found none of thefe inconveniencies attending it *.

Corrofive Sublimate may alfo be made only by mixing Mercury with Sea-falt, without any additament. This may appear furprizing when we confider that, as Acids have a greater affinity with Alka-

* See the Memoirs of the Academy for 1730.

lis than with metallic fubftances, the Acid of Sea-falt ought not to quit its bafis, which is Alkaline, to unite with Mercury.

In order to explain this phenomenon it muft be remembered that Sea-falt, when expofed to the fire without additament, fuffers a little of its Acid to efcape. Now this portion of the Marine Acid unites with the Mercury, and forms a Corrofive Sublimate. Moreover, as there is a pretty ftrong affinity between the Marine Acid and Mercury, this may help to detach from the Sea-falt a greater quantity of Acid than it would otherwife part with. Neverthelefs the quantity of Sublimate obtained by this means is not confiderable, nor is it very corrofive.

On this occafion we muft alfo mention another combination of the Marine Acid with Mercury; which is made by mixing that metallic fubftance perfectly with Sal Ammoniac. by the means of triture. Mercury, like all other metals except Gold, poffeffes the property of decompounding Sal Ammoniac, feparating the volatile Alkali which ferves it for a bafis, and combining, by the help of a very gentle heat, with its Acid, which is well known to be the fame with that of Sea-falt. This decompofition of Sal Ammoniac, by the metalline fubftances, is a full exception to the firft column of Mr. Geoffroy's Table of Affinities, and is the bafis of feveral new medicines invented by the late Comte de la Garàye *.

Corrofive Sublimate is the moft violent and the moft active of all corrofive poifons. It is never ufed in medicine, but in external applications. It is a powerful efcharotic; it deftroys proud flefh, and cleans old ulcers: but it muft be ufed by thofe only who know how to apply it properly, and requires an able hand to manage it. It is not commonly applied by itfelf, but mixed in the proportion of half a dram

* See the Memoir given in by me on this fubject to the Açademy of Sciences in the *Memoires l'Acadamie* 1754.

to a pound of lime-water. This mixture is yellowifh, and bears the name of *Aqua Phagadenica*.

Water diffolves Corrofive Sublimate, but in a fmall quantity. If a Fixed Alkali be mixed with this folution, the Mercury precipitates in the form of a red powder. If the precipitate be procured by a Volatile Alkali, it is white; if by Lime-water, it is yellow. This Mercurial Salt diffolves pretty eafily in boiling Spirit of Wine.

PROCESS VIII.

Sweet Sublimate.

TAKE four parts of Corrofive Sublimate; pulverife it in a glafs or marble mortar; add by little and little three parts of Mercury revivified from Cinabar; triturate the whole carefully, till the Mercury be perfectly killed, fo that no globule thereof can be perceived. The matter will then be grey. Put this powder into an apothecary's phial, or into a matrafs, whofe neck is not above four or five inches long, leaving two-thirds thereof empty. Set the veffel in a fand-bath, and put fand round it to one third of its heighth. Apply a moderate fire at firft; and afterwards raife it gradually till you perceive that the mixture fublimes. Keep it up to this degree till nothing more will rife, and then break the veffel. Reject, as ufelefs, a fmall quantity of earth which you will find at the bottom; feparate alfo what adheres to the neck of the veffel, and carefully collect the matter in the middle, which will be white. Pulverize it; fublime it a fecond time, in the fame manner as before; and in the fame manner feparate the earthy matter left at the bottom of the veffel, and what you find fublimed into the neck. Pulverize, and fublime a third time, the white matter you laft found in the middle.

dle. The white matter of this third fublimation is the *Sweet Sublimate,* called alfo *Aquila Alba.*

OBSERVATIONS.

THE Acid of Sea-falt in the Corrofive Sublimate is very far from being perfectly faturated with Mercury; and thence comes the corrofive quality of this faline compound. But though Mercury, as appears by this combination, is capable of imbibing a much greater quantity of Acid than is neceffary to diffolve it; nay, though it naturally takes up this fuperabundant quantity of Acid, yet it doth not follow from thence that this redundant Acid may not combine with Mercury to the point of perfect faturation, fo as to lofe its corrofive acidity.

This is the cafe in the operation here defcribed. A frefh quantity of running Mercury is mixed with Corrofive Sublimate; and the frefh Mercury, combining with the fuper-abounding Acid, deprives the Sublimate of its acrimony, and forms a compound which comes much nearer the nature of a Neutral Metallic Salt.

Trituration alone is not fufficient to produce an union between the newly added Mercury and the Acid of the Corrofive Sublimate; becaufe, generally fpeaking, the Acid of Sea-falt cannot diffolve Mercury without the help of a certain degree of heat, and unlefs it be reduced into vapours.

Thus, though the newly added Mercury becomes invifible by trituration, and feems actually combined with the Corrofive Sublimate, yet the union is not intimate. There is only an interpofition of parts, but no true diffolution of the newly added Mercury by the fuper-abundant Acid of the Corrofive Sublimate. For this reafon the mixture muft be fublimed; and by this fublimation only is the true union effected. Nor is one fingle fublimation fufficient: no lefs than three are neceffary to deprive the Sublimate of the corrofive quality which ren-

ders

ders it poifonous. After the third fublimation, the Sublimate being put upon the tongue gives no confiderable fenfation of acrimony; nor doth it retain any more of its former activity than is requifite to make it a gentle purgative, when adminiftered from fix to thirty grains for a dofe.

If a lefs quantity of Mercury than that above directed be mixed with the Corrofive Sublimate, the fuper-abundant Acid will not be fufficiently faturated; and the lefs Mercury is added, the more of its corrofive virtue will the Sublimate retain.

If, on the contrary, a greater quantity of Mercury be added, there will be more than the Acid can poffibly diffolve, and the fuperfluous quantity will remain in its natural form of Quick-filver. It is better therefore to err in the excefs than in the defect of the proportion of Mercury to be added; becaufe the Corrofive Sublimate will take up no more than is neceffary to dulcify it.

Part of the Acid of the Corrofive Sublimate is alfo diffipated in vapours during the operation; and it is neceffary to allow room for thefe vapours to circulate, and a vent to give them paffage, or elfe they will burft the veffels. Thefe are our reafons for leaving an empty fpace in the fubliming veffels, and for having their necks no more than five or fix inches long.

The matter which fublimes into the neck of the veffel is always very acrid; for which reafon it muft be feparated from the Sweet Sublimate. There remains alfo at the bottom of the matrafs an earthy, reddifh matter; which probably comes from the Vitriol employed in making the Corrofive Sublimate. This matter muft likewife be rejected as ufelefs after every fublimation.

P R O C E S S IX.

The Panacea of Mercury.

PULVERISE fome Sweet Sublimate, and fublime it in the fame manner as you did thrice before. Repeat this nine times. After thefe fublimations it will make no impreffion on the tongue. Then pour on it aromatic Spirit of Wine, and fet the whole in digeftion for eight days. After that decant the Spirit of Wine, and dry what remains, which is the *Panacea of Mercury.*

O B S E R V A T I O N S.

THE great number of fublimations, which the Sweet Sublimate is made to undergo, fweeten it ftill more, and to fuch a degree that it leaves no fenfation on the tongue, nor hath any purgative virtue.

The Spirit of Wine, in which it is digefted after all the fublimations, is defigned to blunt ftill more the fharpnefs of any acid particles that may not have been fufficiently dulcified by the preceding fublimations.

As Mercury is the fpecific remedy for venereal diforders, fundry preparations thereof have been attempted with a view to produce different effects. Sweet Sublimate is purgative; and for that reafon is not quite proper for procuring a falivation, becaufe it carries off the humours by ftool. The Panacea of Mercury, which, on the contrary, is not purgative, may raife a falivation when taken inwardly.

End of the Firft Volume.

Pl. I.

Fig. 1.

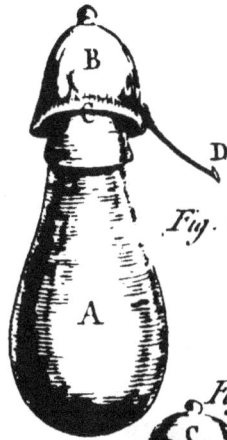

Fig. 2.

Fig. 3.

Pl.II:

D

C

Fig. 1.

B

E

A

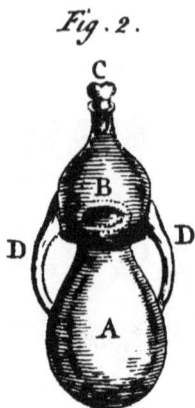

Fig. 2.

C

B

D   D

A

Fig. 4.

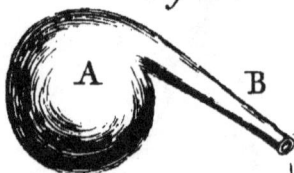

A   B

Fig. 3.

Fig. 5.

J. Mynde sculp.

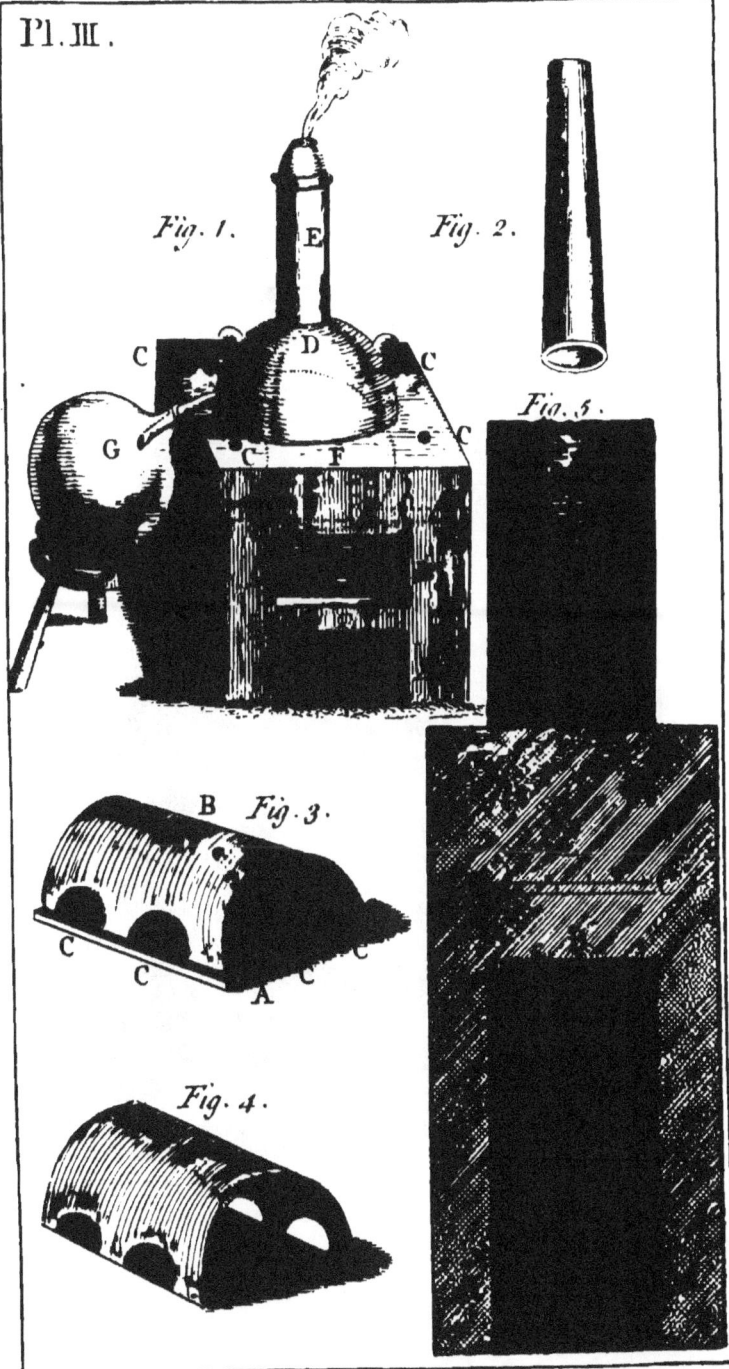

Pl. III.

Fig. 1.

E

D

C   C

G   C   F   C

Fig. 2.

Fig. 5.

B   Fig. 3.

C

C   A   C

Fig. 4.

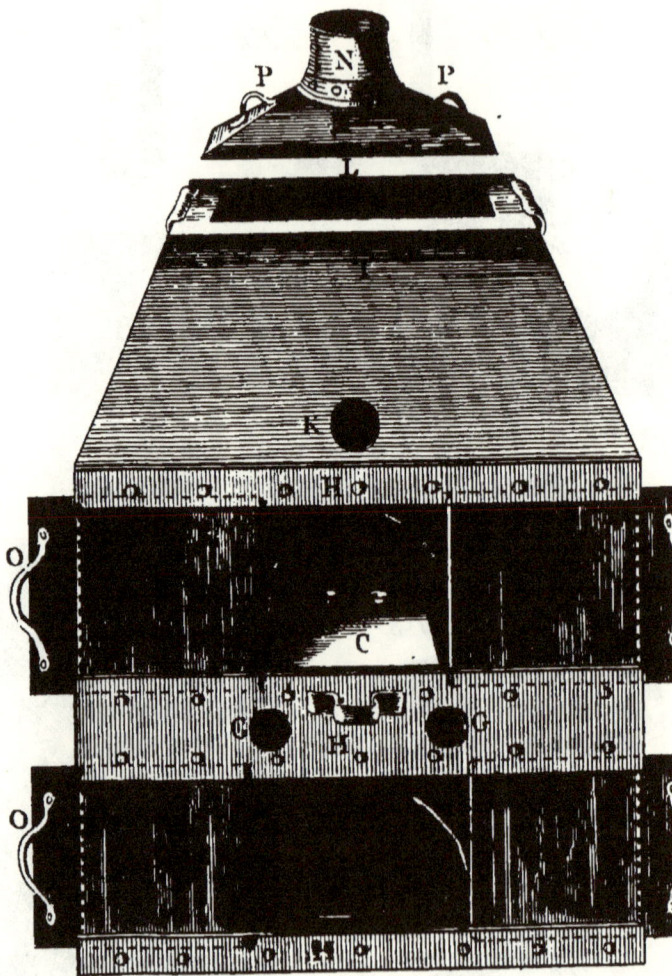

GEOFFROY'S TABLE of the
obſerved between

I.	II.	III.	IV.	V.	VI.	VII.	VIII.
⌢	⊖	⏀	⊕	▽	⊖v	⊖^	MS
⊖v	♃	♂	�	⊕	⊕	⊕	⊖
⊖^	♛	♀	⊖v	⏀	⏀	⏀	⊕
▽	♀	♄	⊖^	⊖	⊖	⊖	⏀
MS	☽	☿	▽		✚		✚
	☿	☽	♂		♁		
		♀					
		☽					
	☉						

Explanation of

⌢ *Acid Spirits*   ▽ *Abſorbent Earths*

⊖ *Marine Acid*   MS *Metallic Subſtances*

⏀ *Nitrous Acid*   ☿ *Mercury*

⊕ *Vitriolic Acid*   ♛ *Regulus of Antimony*

⊖v *Fixed Alkali*   ☉ *Gold*

⊖^ *Volatile Alkali*   ☽ *Silver*

Pl. VI.

COMPARATIVE AFFINITIES
sundry Substances.

IX.	X.	XI.	XII.	XIII.	XIV.	XV.	XVI.

the Characters.

♀ *Copper*   ♁ *Sulphur*
♂ *Iron*   △ *Phlogiston*
♄ *Lead*   ✠ *Spirit of Vinegar*
♃ *Tin*   ▽ *Water*
Ƶ *Zinc*   ⊖ *Neutral Salts*
LC *Calamine*   V *Ardent Spirits*.

EXPLANATION

OF THE

P L A T E S.

PLATE FIRST.
F I G. I. *A Copper Alembic.*

A. The Cucurbite or Body.
B. The Neck.
C. The Head
D. The Beak, Nofe, or Spout.
E. The Refrigeratory, or Cooler.
F. Its Cock.
G. The Receiver.

F I G. II. *A Glafs Alembic.*

A. The Cucurbite.
B. The Head.
C. The Gutter within the Head.
D. The Beak.

F I G. III. *A long-necked Glafs Alembic.*

A. The Body of the Matrafs.
B. The Neck.
C. The Head.

PLATE SECOND.

F I G. I. *A Glafs Alembic of one Piece.*

A. The Cucurbite
B. The Head.
C. The Aperture in the Head.
D. Its Stopple.
E. The Mouth of the Cucurbite.

FIG. II. *A Pelican.*

A. The Cucurbite.
B. The Head.
C. The Aperture in the Head, with its Stopple.
D. D. The two curved Spouts.

FIG. III. *A Row of Aludels.*

FIG. IV. *A Retort.*

A. Its Bowl.
B. Its Neck.

FIG. V. *An English Retort.*

PLATE THIRD.

FIG. I. *A Reverberating Furnace.*

A. The Ash-hole Door.
B. The Fire-place Door.
C. C. C. C. Regifters.
D. The Dome, or Reverberatory.
E. The Conical Funnel.
F. The Retort in the Furnace.
G. The Receiver.
H. H. Iron Bars to fuftain the Retort.

FIG. II. *The Conical Funnel by itfelf.*

FIG. III. *Back View of a Muffle.*

A. The bottom of the Muffle.
B. Its Arch.
C. C. C. Lateral apertures.

FIG. IV. *Fore View of a Muffle.*

FIG. V. *A Melting Furnace.*

A. A. The Bafe of the Furnace.
B. The Ash-hole.
C. D. The Grate for the Fire.
E. The Fire-place.

F.

F. G. H. Curvature of the infide of the upper part of the Fire-place.

I. The Shaft or Chimney.

PLATE FOURTH.

A Cupelling Furnace.

A. The Afh-hole.

B. B. Its fliding-Doors.

C. The Fire-place.

D. D. Its fliding Doors.

E. F. Small apertures in the Sliders.

G. G. Holes for Bars to bear the Muffles.

H. H. H. Iron braces in the forepart of the Furnace, which form grooves for the Doors of the Fire-place and Afh-hole to flide in.

I. The upper pyramidal part of the Furnace.

K. An aperture therein for managing the Coals.

L. The opening at top.

M. The Pyramidal Cover.

N. The Chimney or End of the Shaft, on which the conical Funnel may be fitted.

O. O. O. O. Handles for moving the fliding Doors.

P. P. Ears of the Pyramidal Cover.

N. B. The Furnaces, as reprefented in the two laft Plates, are not in due proportion to each other. The Cupelling Furnace is much larger than it fhould be, with refpeƈt to the Melting Furnace. Thefe dimenfions are here given it, only that all its parts might be more diftinƈtly exprefled, than could have been done if we had made it lefs.